Progre
Trypa
Sleepi

Springer

*Paris
Berlin
Heidelberg
New York
Barcelona
Hong Kong
London
Milan
Singapore
Tokyo*

Michel Dumas, Bernard Bouteille, Alain Buguet (Eds)

Progress in Human African Trypanosomiasis, Sleeping Sickness

 Springer

Michel Dumas
Université de Limoges, Faculté de Médecine
Institut d'Épidémiologie Neurologique et de
Neurologie Tropicale
2, rue du Dr. Marcland, 87025 Limoges, France

Bernard Bouteille
Université de Limoges, Faculté de Médecine
Institut d'Épidémiologie Neurologique et de
Neurologie Tropicale
2, rue du Dr. Marcland, 87025 Limoges, France

Alain Buguet
Centre de Recherches du Service de Santé des Armées Émile Pardé
Unité de Physiologie de la Vigilance
38702 La Tronche Cedex, France

Cover photo, © Stephan Schneider, Saint Goussaud, France

ISBN 2-287-59655-0 Springer-Verlag France, Berlin, Heidelberg, New York
© Springer-Verlag France 1999
Imprimé en France

Library of Congress Cataloging-in-Publication Data
Progress in human African trypanosomiasis, sleeping sickness / Michel
Dumas, Bernard Bouteille, Alain Buguet, eds.
P. cm.
Includes bibliographical references and index.
ISBN 2-287-59655-0 (alk. paper)
1. African trypanosomiasis. I. Dumas, Michel, 1934-. II. Bouteille, Bernard, 1952-. III. Buguet, Alain.
[DNLM: 1. Trypanosomiasis, African. 2. Trypanosoma-pathogenicity. 3. Trypanocidal Agents. 4. Host-Parasite Relations. WC 705
P964 1999]
RC 186. T82P76 1999
616. 9'363--dc21
DNLM/DLC
for Library of Congress   98-31567
                                                                                  CIP

Apart from any fair dealing for the purposes of the research or private study, or criticism or review, as permitted under the Copyright, Designs and Patents Act 1988, this publication may only be reproduced, stored or transmitted, in any form or by any means, with the prior permission in writting of the publishers, or in the case of reprographic reproduction in accordance with the terms of licences issued by the copyright licences issued by the copyright Licensing Agency. Enquiry concerning reproduction outside those terms should be sent to the publishers.

The use of registered names, trademarks, etc. in this publication does not imply, even in the absence of a specific statement, that such names are exempt from the relevant laws and regulations and therefore free for general use.

Product liability: the publisher can give no guarantee for information about drug dosage and application there of contained in this book. In every individual case the respective user must check its accuracy by consulting other pharmaceutical literature.

Spin: 10560434

Foreword

A new manual on human African trypanosomiasis (HAT) may seem inappropriate at the end of the twentieth century. Unfortunately, the life-threatening HAT remains an obscure disease which is, once again, spreading across intertropical Africa. Although there is no revolutionary breakthrough, some significant scientific advances are arising from field and laboratory research in the description of new clinical symptoms and diagnosis techniques as well as the search for an efficient and safe drug treatment. For example, modern pathogenic immunology has enhanced the understanding of the processes developed by the trypanosome to penetrate into the central nervous system. These breakthrough events are exposed in the following chapters written by several scientists and physicians who spend a great part of their research activities studying this grave disease.

Although fatal, HAT is largely neglected and tucked away from the major scientific interests, probably because it affects populations and countries economically weak. Because of its capacity to escape the host defense mechanisms, causing a disease with very diverse clinical potentialities, the trypanosome constitutes an excellent model of the host-parasite relationship. However, HAT still represents an enigma for the scientific community.

May this manual answer the reader's many questions, stimulate his curiosity and help his understanding of that disease.

Michel Dumas
Bernard Bouteille
Alain Buguet

Contents

Chapter 1 "Trypanosomiasis exists when it is searched for ..."
JL Frézil 1

Chapter 2 Identification of trypanosomes: from morphology to molecular biology
W Gibson, J Stevens, P Truc 7

Chapter 3 Antigenic variation in African trypanosomes
E Pays 31

Chapter 4 Carbohydrate metabolism
FR Opperdoes 53

Chapter 5 Polyamine metabolism
JC Breton, B Bouteille 81

Chapter 6 Cytokines and the blood-brain barrier in human and experimental African trypanosomiasis
VW Pentreath 105

Chapter 7 Cytokines in the pathogenesis of human African trypanosomiasis: antagonistic roles of TNF-α and IL-10
SG Rhind, PN Shek 119

Chapter 8 Immunology of African trypanosomiasis
P Vincendeau, MO Jauberteau-Marchan, S Daulouède, Z Ayed 137

Chapter 9 Pathology of African trypanosomiasis
K Kristensson, M Bentivoglio 157

Chapter 10 Hormones in human African trypanosomiasis
MW Radomski, G Brandenberger 183

Chapter 11 Sleeping sickness: a disease of the clock with nitric oxide involvement
A Buguet, R Cespuglio 191

Chapter 12 Electroencephalographic features and evoked potentials in human African trypanosomiasis
F Tabaraud, P Tapie 203

CHAPTER 13   Clinical aspects of human African trypanosomiasis
M Dumas, S Bisser 215

CHAPTER 14   Biological diagnosis of human African trypanosomiasis
N Van Meirvenne 235

CHAPTER 15   Present strategies in the treatment of human African trypanosomiasis
S Van Nieuwenhove 253

CHAPTER 16   The nitroimidazoles and human African trypanosomiasis
G Chauvière, J Périé 281

CHAPTER 17   Experimental models for new chemotherapeutic approaches to human African trypanosomiasis
B Bouteille, M Keita, B Enanga, J Mezui Me Ndong 289

CHAPTER 18   Prophylactic strategies in human African trypanosomiasis
A Stanghellini 301

CHAPTER 19   International co-operation: past and present
P de Raadt, J Jannin 315

Contributors

AYED Z.
Institut d'Epidémiologie Neurologique et de Neurologie Tropicale
Faculté de Médecine, 2, rue du Docteur Marcland
F-87025 Limoges Cedex - France

BENTIVOGLIO M.
Institute of Anatomy and Histology
University of Verona, Strada Le Grazie
I-37134 Verona - Italy

BISSER S.
Institut d'Epidémiologie Neurologique et de Neurologie Tropicale
Faculté de Médecine, 2, rue du Docteur Marcland
F-87025 Limoges Cedex - France

BOUTEILLE B.
Institut d'Epidémiologie Neurologique et de Neurologie Tropicale
Faculté de Médecine, 2, rue du Docteur Marcland
F-87025 Limoges Cedex - France

BRANDENBERGER G.
Laboratoire de Physiologie et Psychologie Environnementales
21, rue Becquerel
F-67087 Strasbourg Cedex - France

BRETON J.C.
Institut d'Epidémiologie Neurologique et de Neurologie Tropicale
Faculté de Médecine, 2, rue du Docteur Marcland
F-87025 Limoges Cedex - France

BUGUET A.
Centre de Recherches du Service de Santé des Armées Emile Pardé
Unité de Physiologie de la Vigilance
BP 87 F-38702 La Tronche Cedex - France

CESPUGLIO R.
Université Claude-Bernard Lyon I,
Département de Médecine Expérimentale, 8, avenue Rockefeller
F-69373 Lyon Cedex 08 - France

CHAUVIÈRE G.
Groupe de Chimie Organique Biologique - UMR 5623
Université Paul Sabatier Bât 2R1, 118, route de Narbonne
F-31032 Toulouse Cedex - France

DAULOUÈDE S.
Laboratoire de Parasitologie
Université de Bordeaux II, 146, rue Léo Saignat
F-33076 Bordeaux Cedex - France

DE RAADT P.
World Health Organization
Division of Control of Tropical Diseases, 20, avenue Appia
CH-1211 Geneva - Switzerland

DUMAS M.
Institut d'Epidémiologie Neurologique et de Neurologie Tropicale
Faculté de Médecine, 2, rue du Docteur Marcland
F-87025 Limoges Cedex - France

ENANGA B.
Institut d'Epidémiologie Neurologique et de Neurologie Tropicale
Faculté de Médecine, 2, rue du Docteur Marcland
F-87025 Limoges Cedex - France

FRÉZIL J.L.
Institut de Recherche pour le Développement
BP 5045, F-34032 Montpellier Cedex - France

GIBSON W.
Department of Biological Sciences
University of Bristol, Woodland Road
Bristol BS8 1UG - United Kingdom

JANNIN J.
World Health Organization
Division of Control of Tropical Diseases, 20, avenue Appia
CH-1211 Geneva - Switzerland

Jauberteau-Marchan M.O.
Institut d'Epidémiologie Neurologique et de Neurologie Tropicale
Faculté de Médecine, 2, rue du Docteur Marcland
F-87025 Limoges Cedex - France

Keita M.
Institut d'Epidémiologie Neurologique et de Neurologie Tropicale
Faculté de Médecine, 2, rue du Docteur Marcland
F-87025 Limoges Cedex - France

Kristensson K.
Department of Neuroscience
Karolinska Institute, Doktorsringen 17
S-171 77 Stockholm - Sweden

Mezui Me Ndong J.
Centre International de Recherche Médicale
BP 769, Franceville - Gabon

Opperdoes F.R.
ICP-TROP 74.39
Avenue Hippocrate 74
B-1200 Brussels - Belgium

Pays E.
Department of Molecular Biology
University of Brussels, 67, rue des Chevaux
B-1640 Rhode St Genèse - Belgium

Pentreath V.W.
Department of Biological Sciences
University of Salford
Salford M5 4WT - United Kingdom

Périé J.
Groupe de Chimie Organique Biologique - UMR 5623
Université Paul Sabatier Bât 2R1, 118, route de Narbonne
F-31032 Toulouse Cedex - France

Radomski M.W.
Defence and Civil Institute of Environmental Medicine
1133 Sheppard Avenue, West - P.O. Box 2000
Toronto, Ontario M3M 3B9 - Canada

Rhind S.G.
Defence and Civil Institute of Environmental Medicine
1133 Sheppard Avenue, West - P.O. Box 2000
Toronto, Ontario M3M 3B9 - Canada

Shek P.N.
Defence and Civil Institute of Environmental Medicine
1133 Sheppard Avenue, West - P.O. Box 2000
Toronto, Ontario M3M 3B9 - Canada

Stanghellini A.
Mission Française de Coopération
CP 707, Luanda - Angola

Stevens J.
Department of Biological Sciences
University of Bristol, Woodland Road
Bristol BS8 1UG - United Kingdom

Tabaraud F.
Service d'Exploration Fonctionnelle Neurologique
CHU Dupuytren, 2, avenue Martin Luther King
F-87042 Limoges Cedex - France

Tapie P.
Service d'Exploration Fonctionnelle Neurologique
CHU Dupuytren, 2, avenue Martin Luther King
F-87042 Limoges Cedex -France

Truc P.
Institut de Recherche pour le Développement
Conditions de vie et Développement
BP 5045
F-34032 Montpellier Cedex 01

Van Meirvenne N.
Institute of Tropical Medicine Prince Leopold
Nationalestraat 155
B 2000 Antwerpen - Belgium

Van Nieuwenhove S.
WHO Regional Office for Africa
Brazzaville - Republic of Congo

VINCENDEAU P.
Laboratoire de Parasitologie
Université de Bordeaux II, 146, rue Léo Saignat
F-33076 Bordeaux Cedex - France

CHAPTER 1

"Trypanosomiasis exists when it is searched for ..."

JL Frézil

"Trypanosomiasis exists when it is searched for ...": how many times in Africa have we heard this phrase sententiously stated during a lively discussion on the lack of resources or efficacity of the health services in charge of controlling this disease.

Indeed, during our work in the seventies and eighties investigating populations in collaboration with governmental health services in endemic areas, we have always been surprised to see the gap between the reality in the field and medical textbook descriptions of the disease and its epidemiology. When we arrived in a village whose inhabitants were known to be exposed, systematic examination of the villagers revealed only a small number of potential suspects: a few individuals with cervical lymph nodes due perhaps to human African trypanosomiasis (HAT), but which could also represent only secondarily infected tinea. In fact, we quickly noticed that clinical signs were of extremely limited value in the early stages of the disease, and that paradoxically they were generally searched for once the disease had been detected. Of course, we have seen patients in the advanced second phase in whom death was inevitable, but we have also seen a large number of people, detected by positive serologic tests and confirmed by visualization of the parasite, who were apparently doing so well that it was difficult to persuade them to undergo treatment. In the N'gabe region of the Congo river, we even saw the chief of a little village, in which 45% of the population was infected, oversee treatment of all of his subjects, and then refuse therapy himself ("because he was the chief"), even though his cerebrospinal fluid was highly abnormal. Several years later he was still doing well ...

Could sleeping sickness, like Chagas' disease, be relatively well tolerated by populations living in endemic areas?

The terrible epidemic at the beginning of this century which ravaged central Africa killing hundreds of thousands of people seems to invalidate this hypothesis [12]; however, according to the literature it also appears that there has been no record of an epidemic before the colonial period [1]. I, myself, have even heard an old Congolese inhabitant affirm: "Before the whites came, there wasn't any trypano!" Furthermore, the zones of predilection of sleeping sickness in central Africa appear to be singularly underpopulated ...

How can such apparently contradictory findings be interpreted?

We therefore attempted to observe what happens in areas of disease activity, not only in man, but also in domestic animals. The animal/parasite relationships enabled us by extrapolation to formulate several hypotheses to account for the natural course of the disease. During a campaign to put tsetse fly traps in the Niari region, we noticed, after several months, that swine protected by the traps spontaneously became free of their trypanosomes. The animal model demonstrates, in fact, that trypanosomal-tolerant livestock had to be continuously reinoculated to remain infected, and when the vector disappeared, the animal recovered by itself from the trypanosomiasis, perhaps when the trypanosome had exhausted its antigenic repertoire. There is no valid reason to believe that the selective pressure of trypanosomes in endemic zones is different in man than in animals. If man is thus subjected to the same phenomenon, the "doubtful" seropositivity which is well known in tsetse fly zones, most probably results from aborted infections of *Trypanozoon* and transient infections or repeated inoculations of *Duttonnella* and *Nannomonas* [2]. These repeated inoculations represent acquired immunity in addition to "innate immunity" (or natural resistance), which is unmistakably present in historically infested areas of HAT where the most susceptible individuals were progressively eliminated over the millennia.

After an extensive study of HAT in epidemic areas in Zaire, Wéry and Burke [3] were led to introduce in 1970 the concept of trypanosomal tolerance in man, which is probably more explicit than those commonly accepted such as an asymptomatic carrier state or a long latent period of the disease. This human trypanosomal tolerance was long considered to be exceptional. Only a few cases have been reported in the literature, for example, in African soldiers during the war in Indochina [4]. One of the great accomplishments of serological screening was to demonstrate that asymptomatic carriers were far from being a rare occurrence and indeed were the general rule in endemic areas of central Africa [5].

Bibliographical data on this problem are rare because, for obvious ethical reasons, no follow-up study of untreated patients with sleeping sickness exists and the few recent cases are from subjects who, for one reason or another, were not treated [6]. Nonetheless, we must pay tribute to the treatise by Janssens et al. [7], who devoted to this problem the attention it well deserved. This demonstrates, in fact, the variations in tolerance between different ethnic groups. The pygmies of central Africa have long had the reputation of being more or less refractory to infection. This information was confirmed by experiments by Lemesre [8] demonstrating that sera from populations of Bantus had varying degrees of trypanolytic activity according to the individual and that sera from pygmies were more trypanolytic than those from Bantu populations. Seed et al. [9] also noted that human sera, similar to those of animals, have varying degrees of trypanolytic activity (localized to the HDL fraction and a second undetermined factor) directly correlated with tsetse infestation, as a result of natural selection. Lastly, all the data from the colonial period confirm the high susceptibility to infection of European populations whose survival in certain zones of Africa did not exceed one or two years, all diseases included, at the beginning of this century.

Agreeing that there is an acquired immunity that keeps a low-grade infection at bay without eliminating it for a long period, one can extrapolate and attempt to apply to trypanosomiasis the concept of premunition or concomitant immunity as described in 1963 by Sergent for malaria [10]. With such an assumption, in patients in whom the disease is well tolerated, an equilibrium exists between the immune response and the parasites which can last for years, accounting for the long, clinically silent, latent period, or the "healthy carriers" so well-known in trypanosomiasis, especially in certain African populations at the beginning of this century, such as the inhabitants of Gorée Island. This equilibrium may sometimes be disrupted when the carrier is subjected to a stress (fatigue, famine, intercurrent illness) or a change in the strain of the parasite (a well-known phenomenon in N'dama cattle), which could account for the obvious relationship between HAT and socioeconomic disorders.

Analysis of past and present situations reveals that the great epidemics were intimately linked to spontaneous or forced population displacements during extensive work projects (construction of railways), the introduction of new strains of parasites, during wars or insurrection, or times of famine, etc.

We have repeatedly noted that subjects infected by *Trypanosoma gambiense*, during the clinically latent period when the infection is well tolerated, have high serum antibody levels (IgM, IgG) and very few parasites in the peripheral blood. These subjects had no idea that they were ill and thus went normally about their daily occupations and could thus infect tsetse flies. If this state can persist for several years, the question arises whether spontaneous recovery can occur as it has been observed in trypanosomal infections of animals. These spontaneous cures have been suspected for a long time ([4, 11]; Lapeyssonnie, personal communication) and without doubt they explain why positive serologic tests are found which subsequently become negative, a few months later.

With the onset of clinical signs, however, a fatal outcome appears irreversible. Trypanosomes invade the central nervous system, eliciting an intense immune response, while antibody titers in the blood decrease considerably. At this stage, the patient is undoubtedly aware of the illness, attributing it most often to a magical spell without recourse. Because of the traditional fatalistic perception of evil, individuals who are ill and conscious of their state paradoxically evade care during large-scale surveys, while healthy persons are present for testing, stressing the importance of lengthy discussions to convince the populations of the value of screening. Consequently, passive screening of patients often in the advanced phase is of course "better than nothing"; however, apart from using Jamot's method (mobile teams), it is illusory to believe that the disease can be controlled!

In general, areas of disease activity are brought to light through surveys undertaken after the detection of a patient somewhat haphazardly. When the socioeconomic conditions are relatively satisfactory, health workers can expect to find a few overt clinical cases, with a greater proportion of apparently healthy persons having adenopathy and a large number of people in apparent good health but with positive serologic tests and subsequent identification of the parasite. Prevalences of 10, 20, 50% or more can thus be attained, a perplexing finding since only a very small percentage of the flies are infected. The reason, in this case, probably stems from the fact that surveys may reveal people that have been infected for several years.

The following questions can logically be asked

How many cases would be detected without serologic screening? How should we consider individuals with positive serologic tests in whom parasites can not be detected in the blood? Does this truly represent an epidemic (a definite increase in the number of cases) or the logical outcome of an endemic situation without surveillance?

Outside the tsetse fly zones, sleeping sickness apparently does not exist. If the disease is to become endemic, the vector must necessarily be present, and because of its large range of hosts, it must be constrained to feed on man. Leaving aside the possible animal reservoir whose importance has never been truly demonstrated, the human reservoir constituted by patients who are asymptomatic (and in need of treatment!) undoubtedly plays an essential role. In all likelihood, the close contact between man and fly and the long clinical latent period account for the fact that trypanosomiasis always resurfaces in its old, historical sites. To this phenomenon, one must also add socioeconomic hardships and serious deficiencies in the health care systems.

Indeed, in endemic regions, one can easily understand why, "sleeping sickness really exists when you search for it!" And when the disease is discovered, measures must be taken against it. Of course at the present time there are excellent tests for screening patients and demonstrating the parasite, methods to control the vector, and relatively effective, albeit often dangerous, therapeutic agents. Nevertheless, no matter how simple and economical these techniques from the "north" appear, they remain out of reach for the majority of the countries in the south, a very unfortunate situation because pilot studies carried out in diverse geographical regions have shown that a combination of measures to control the tsetse fly and efficacious screening can sterilize an area with *T. gambiense* in less than two years. Of course, it is important that subsequent epidemiological surveillance be accomplished, and associated, if possible, with control of the tsetse fly in areas of contamination [13]. Nevertheless, the essential condition to be met, and perhaps the most difficult to accomplish, is to imperatively consider the socioeconomic impact of the spread of trypanosomiasis. It is important that the affected populations return to normal living conditions with proper hygiene and nutrition, i.e. a physical well-being in which the body's natural resistance can maximally inhibit the trypanosome's development. Undoubtedly, since these conditions have not been respected, the loss of life due to sleeping sickness is as great today as when it was discovered, almost a century ago.

References

1. Molyneux DH, Ashford RW (1983) The biology of *Trypanosoma* and *Leishmania*, parasites of man and domestic animals. Taylor and Francis, London
2. Blair DM, Smith EB, Gelfand M (1968) Human trypanosomiasis in Rhodesia. A new hypothesis suggested for the rarity of human trypanosomiasis. Centr Afr J Med 14 (7 Suppl): 1-12
3. Wéry M, Burke J (1970) Human "healthy carriers" of *Trypanosoma* (*brucei* type) discovered by immunofluorescence test in the République démocratique du Congo. Ann Soc Belge Méd Trop 50:613
4. Lapeyssonnie L (1960) Deuxième note concernant un cas exceptionnel de trypanosomiase. Parasitémie observée depuis 21 ans sans signes cliniques appréciables chez une malade traitée inefficacement pendant les 10 premières années. Bull Soc Path Ex 53:28-32
5. Frézil JL (1983) La trypanosomiase humaine en république populaire du Congo. Travaux et Documents de l'ORSTOM, n°155, p 165
6. Ginoux PY, Frézil JL (1980) Recherches sur la latence clinique et la trypanotolérance humaine dans le foyer du couloir du fleuve Congo. Cah. ORSTOM ser Ent Méd Parasitol XIX:33-40
7. Janssens PG, Kivits M, Vuylsteke J (1992) Médecine et hygiène en Afrique centrale de 1885 à nos jours (2 vol). Fondation Roi Baudouin, Bruxelles
8. Authié E, Cuisance D, Force-Barge P, Frézil JL, Gouteux JP, Jannin J, Lancien J, Laveissière C, Lemesre JL, Mathieu-Daudé F, Nitcheman S, Noireau F, Penchenier L, Tibayrenc M, Truc P (1991) Some new prospects in epidemiology and fight against human african trypanosomiasis. Res Rev Parasitol 51 (1-4):29-46
9. Seed JR, Sechelski JB, Ortiz JC, Chapman JF (1993) Relationship between human serum trypanocidal activity and host resistance to the African trypanosomes. J Parasitol 79:226-232
10. Sergent E (1963) Latent infection and premunition. Some definitions of microbiology and immunology. In: Wernsdorfer WH, Sir Mc Gregor (eds) (1988) Malaria. Churchill Livingstone
11. Gallais P (1953) In: Etudes sur la trypanosomiase humaine africaine. Méd Trop 13 (numéro spécial):800
12. Martin G, Leboeuf A, Roubaud E (1909) Rapport de la mission d'études de la maladie du sommeil au Congo français, 1906-1908. Masson & Cie, Paris
13. Challier A, Laveissière C (1973) Un nouveau piège pour la capture des glossines (*Glossina*, Diptera, Muscida): description et essais sur le terrain. Cah ORSTOM ser Ent Méd Parasitol XI:251-262

CHAPTER 2

Identification of trypanosomes: from morphology to molecular biology

W Gibson, J Stevens, P Truc

Introduction

One of the enduring problems in the epidemiology of sleeping sickness is that there are 3 morphologically indistinguishable subspecies of *Trypanosoma brucei* involved in a complex transmission cycle between humans, tsetse and reservoir hosts. Two subspecies, *T. b. gambiense* and *T. b. rhodesiense*, are infective to man and cause gambian and rhodesian sleeping sickness, respectively. The third subspecies *T. b. brucei* cannot by definition infect humans, but coexists with the other trypanosomes in reservoir hosts and vectors.

The advent of molecular methods for taxonomy brought the hope that unequivocal biochemical markers for the 3 subspecies would quickly be found. In particular, markers for the human infective trypanosomes would have enabled identification of reservoir hosts without recourse to experiments with human volunteers or serum resistance tests. However, the results of molecular characterization revealed a much more complex picture, with several subdivisions within *T. brucei*, rather than the 3 expected. Most isolates of *T. b. gambiense* fitted into one clearly demarcated group, but the other subdivisions did not correspond to the recognised subspecies. More significantly, analysis of the data using population genetics methods led to the discovery of genetic exchange in *T. brucei*. The importance of genetic exchange in generating diversity among *T. brucei* stocks in the field is still controversial, but the principle of gene flow destroys any remaining hope that stable markers for *T. brucei* subspecies exist.

In this chapter, we will first describe the methods for isolation and characterization of trypanosomes from the field and the mathematical methods of data analysis, before considering the implications of this work for epidemiology.

Isolation of trypanosomes stocks from the field

Before trypanosome characterization, stocks must be isolated from the field. The sampling process inevitably introduces bias: some regions may be geographically remote, some hosts may be easier to sample than others, some hosts may contain higher numbers of organisms. Most of the characterization methods described below require outgrowth from the original isolate and thus more slowly dividing trypanosomes may be lost from mixed isolates. Recent developments in isolation methods have tried to address these issues.

Rodent subinoculation

Since Bruce's original discovery, the standard method of isolation of bloodstream form *T. brucei* has been by inoculation of host blood, CSF or lymph into experimental rodents. The same method can be applied to infected saliva or macerated salivary glands from tsetse flies. Subsequent rodent subpassages will increase the virulence of the initial isolate until the rodent blood literally swarms with trypanosomes. This method works well for *T. b. rhodesiense* and *T. b. brucei*, even in subpatent parasitaemias, since in principle one organism is sufficient for infection. However, classical *T. b. gambiense* is of much lower virulence and therefore isolation by rodent subinoculation may fail unless susceptibility is increased by, for example, the use of neonatal rodents, immunosuppression or particular rodent species (e.g. *Mastomys natalensis*, [1]). These methods are not very practical for field use.

In vitro methods

Procyclic trypanosomes, as found in the tsetse midgut, grow readily in *in vitro* culture and a number of media have been described, of which the semi-defined liquid media, SDM-79 and Cunningham's medium, are probably most widely used [2, 3]. Bloodstream form trypanosomes (bsf's) are more difficult to establish *in vitro* and do not grow to such high densities as procyclics (typically 10^5 ml^{-1} for bsf's compared to 10^7 ml^{-1} for procyclics). However, bsf's readily transform to procyclics on transfer from 37°C to a lower temperature (~25°C), and this feature has been exploited to develop *in vitro* isolation methods. Dukes et al. [4] successfully isolated Group 1 *T. b. gambiense* from Cameroon by feeding cryopreserved blood from patients to experimental tsetse flies and then culturing the resultant midgut procyclics.

A more convenient and generally applicable method is the KIVI (Kit for In Vitro Isolation) developed by Aerts et al. [5]. The KIVI consists of a pre-prepared small bottle containing sterile growth medium plus antibiotics, into which 5-10 ml of sterile blood is introduced via an airtight rubber seal. After inoculation the bottles can be kept at ambient temperature, before transfer to the laboratory, where they can be opened and examined for trypanosome growth under sterile conditions. Since the trypanosomes take a few days to transform into procyclics and the volume of medium is large, 3 to 4 weeks may elapse between inoculation and necessity to subpassage. Thus, the KIVI is ideal for field isolation and has proved its worth in areas of gambian sleeping sickness [6-8, 127] and may also be used for isolation of *T. b. brucei* and *T. b. rhodesiense*.

Cryopreservation

Following collection and initial passage, trypanosome stocks can be cryopreserved, seemingly indefinately, in liquid nitrogen. Either glycerol (10% final volume) or DMSO (7.5% final volume) are used routinely as cryoprotectants. With care, there is little loss of viability; however, less virulent organisms may be lost from mixed isolates after freeze-thawing. Major collections of cryopreserved trypanosome isolates have been built up since the introduction of cryopreservation in the 1960's and these now form an important resource for longitudinal studies.

Determination of human infectivity

Unless a *T. brucei* stock has been isolated from a human patient, its status with regard to human infectivity is uncertain. This question was resolved in the past by the inoculation of human volunteers, and latterly by *in vitro* tests involving incubation of trypanosomes with human blood or serum. These tests rely on the differential killing of *T. b. brucei* by the trypanolytic factor (TLF) in human blood (reviewed by Hajduk, [9]). In the original Blood Incubation Infectivity Test (BIIT) developed by Rickman et al. [10], the viability of trypanosomes following incubation with human blood was tested by inoculation of rodents. However, TLF resistance was found to change with antigenic variation [11, 12], and a more reliable way of carrying out the test was developed using metacyclic trypanosomes direct from the fly, which have a smaller and more stable antigenic repertoire (Human Serum Resistance Test, HSRT; [13]). *In vitro* tests in their various guises have been used extensively to characterise field isolates from flies and non-human hosts for the trait of human infectivity [14]. These results will be linked with those from molecular characterization methods in the following section "Epidemiological implications".

Molecular characterization

Bloodstream form trypanosomes of the 3 principle tsetse-transmitted subgenera (*Trypanozoon*, *Nannomonas*, *Duttonella*) can be distinguished by characteristic morphological features visible by microscopy. Within the vector, however, these morphological differences disappear.

To address this problem, DNA probes based on repetitive DNA elements (e.g. satellite DNA) were developed and have proved very useful for the identification of trypanosomes in wild caught tsetse [15-18]. The repetitive DNA probes specific for *T. brucei* detect all 3 subspecies, however [19], indicating that they constitute a closely related group of organisms. Therefore, subspecific identification of *T. brucei* stocks relies on molecular "fingerprinting" techniques and comparison with reference isolates. The various methods that have been widely used are described below.

Isoenzyme electrophoresis

Isoenzyme analysis was the first technique to be widely applied to the characterization of isolates within the *T. brucei* species and well over 1000 stocks from all over the continent have now been characterized in this way [20-39]. The main conclusions from this data are discussed in the later section "Epidemiological implications". Developed in the 1950's, isoenzyme electrophoresis is comparatively cheap and robust and, given a good choice of enzymes, its usefulness for characterization purposes has not been significantly superseded by DNA-based techniques.

Isoenzyme analysis is performed on highly concentrated extracts of cytoplasmic proteins and thus requires large numbers of trypanosomes (minimum 100 million). The proteins are separated by gel electrophoresis and particular enzymes are visualised by specific staining reactions based on the catalytic properties of the enzyme. Multiple molecular forms of an enzyme (isoenzymes) give rise to multiple bands on the gel if the molecules differ in electrophoretic mobility. In practice, only about a quarter of amino acid substitutions give rise to changes in electrophoretic charge [40]. Therefore, a range of different enzymes need to be screened and the use of 10-20 enzymes, which give clear and reproducible results, is recommended [41, 42]. Note that the natural tendency to use enzymes which show up differences between isolates, rather than those which are conserved, accentuates the observed level of dissimilarity.

Since the metabolism of insect and bloodstream forms of *T. brucei* is different, the isoenzyme bands seen may differ in number, mobility or intensity depending on the lifecycle stage used; this is not a problem for DNA-based characterization. Both bloodstream forms or culture grown procyclics give satisfactory isoenzyme results [31, 43, 44]. Various media have been used for electrophoretic separation of *T. brucei* isoenzymes, with starch (both thick and thin layer) and cellulose acetate plates the most widely used for reasons of efficacy and economy [45].

Restriction fragment length polymorphisms

Like isoenzyme electrophoresis, RFLP analysis requires large numbers of trypanosomes, but is more costly in terms of materials and reagents. The starting material is DNA, which first has to be purified from protein and other cellular debris. Both nuclear and kinetoplast (= mitochondrial) DNA have been used. The purified DNA is digested with various restriction enzymes, each of which recognises a specific short sequence of bases and cuts the DNA at this point. The resulting DNA fragments are separated by gel electrophoresis, either in agarose or acrylamide, and are visualised by staining with ethidium bromide, which binds to the DNA. If 2 samples differ by a single base change in the recognition sequence of a particular restriction enzyme, the enzyme will no longer cut at this position and a change in fragment length will be observed.

Such RFLPs can be detected in purified kDNA by simple gel electrophoresis. The ~20 kb kDNA maxicircles, which correspond to the mitochondrial DNA of other eukaryotes, are homogeneous in sequence and therefore appear as one or more discrete fragments after electrophoresis depending on the restriction enzyme used [46, 47]. Analysis of maxicircles from 32 *T. brucei* ssp. stocks, including *T. b. gambiense* and *T. b. rhodesiense*, showed little variation, except for 2 subgroups of *T. b. brucei* (*kiboko* and *sindo*) with distinctive RFLPs [47]. The minicircles of *T. brucei* are heterogeneous in sequence. Linearisation by a single cut anywhere in the circle produces a 1 kb fragment, while enzymes which cut more than once yield a complex pattern of fragments smaller than 1 kb. The extreme heterogeneity of *T. brucei* minicircles makes them less useful for identification purposes than those of other trypanosomes (e.g. *T. cruzi*, [48]).

RFLPs are detected in nuclear DNA by Southern analysis. After electrophoresis the DNA is single-stranded *in situ* by alkali treatment of the gel and then transferred to a solid support (e.g. nitrocellulose or nylon membrane) by Southern blotting. The blot is incubated in a solution containing a labelled DNA probe, allowing the single-stranded DNA probe to hybridise with its complementary sequence on the blot. Unbound probe is then washed away and the position of hybridisation on the blot is visualised by autoradiography or other means, depending on the method used to label the probe. Various DNA probes have been used, but the largest data sets come from analysis of ribosomal DNAs [49-51], and variant surface glycoprotein (VSG) genes [52-55]. The repetitive nature of ribosomal and VSG genes means that multiple bands are produced for subsequent mathematical analysis. These results are discussed together with the isoenzyme data in the later section "Epidemiological implications".

Polymerase chain reaction-based methods

The chief advantage of polymerase chain reaction (PCR)-based methods is that far fewer trypanosomes are required than for isoenzyme or RFLP analysis. Two approaches have been tried: firstly the development of PCR identification methods for *T. b. gambiense* based on specific sequences, and secondly the random amplification of polymorphic DNA (RAPD) technique [56, 57].

Two PCR tests for Group 1 *T. b. gambiense* have been devised, based on a conserved VSG gene and kDNA minicircles, respectively [42, 58]. In this approach, the sequence of the target DNA must be known before suitable PCR primers can be chosen; these primers are then used to amplify the specific DNA fragment, which is visualised by gel electrophoresis and may be further characterized by hybridisation. For the first test, VSG gene AnTat 11.17 was the target, since this gene was found to be unique to Group 1 *T. b. gambiense* [54]. This PCR test was capable of distinguishing Group 1 *T. b. gambiense* from *T. b. brucei* in most foci of Gambian sleeping sickness, except north-west Uganda, and gave positive results only with 24 Group 1 stocks in a total sample of 39 *T. brucei* ssp. isolates of diverse origins [58]. In the second PCR test, the minicircle variable region was targeted for amplification by virtue of a conserved 122 bp sequence and the PCR product from 12 Group 1 *T. b. gambiense* stocks of 26 *T. brucei* ssp. stocks examined, was shown to be unique by hybridisation [42].

RAPD is a PCR-based technique, which uses arbitrary 10-mer primers to amplify random fragments from a genomic DNA template. A 10-mer primer has a theoretical chance of finding its complementary sequence roughly every million bases in random DNA. Thus, no sequence information about the target DNA is necessary and trial and error will show which 10-mer primers produce suitable amplification. Individual primers can yield a fingerprint consisting of 10 or so bands, and thus the use of several primers will rapidly generate large volumes of characterization data for strain comparison [59, 60]. The PCR reaction is carried out on purified template DNA in solution. Each reaction requires only about 10 - 20 ng of DNA, equivalent to tens of thousands rather than the billions of trypanosomes required for isoenzyme or RFLP analysis. Amplification products in the size range ~200 - 1000 bp are visualized by gel electrophoresis and staining with ethidium bromide. A typical result is shown in Figure 1.

RAPD is clearly the present method of choice for quick and easy characterization of trypanosome isolates. The results are reproducible and agree with those derived from isoenzyme or RFLP studies [59-62]. The main disadvantage is that the target sequences amplified are unknown and hence the data is not open to interpretation in terms of individual loci and alleles; RAPD data is therefore analysed by particular mathematical methods – see section "Mathematical analysis of characterization data". A further problem with *T. brucei* is that an unknown proportion of bands will derive from VSG genes, which comprise roughly 5% of the genome, have a non-diploid organisation and evolve rapidly [63]. In addition, contamination of trypanosome DNA with DNA from other sources must be avoided, since RAPD primers are not specific.

Molecular karyotype

Molecular karyotypes are produced by size fractionation of chromosomal DNAs by PFGE (pulsed field gel electrophoresis, [64]). Trypanosomes of the *T. brucei* species have over 100 chromosomes of sizes ranging from 50 kb to several Mb, which are subject to relatively frequent rearrangements and length alterations, thus giving rise to unique karyotypes [65-67]. In practice, karyotypes are highly variable and, as with kDNA minicircles, the results are more useful for identification of individual isolates by fingerprinting than characterization of populations. However, some karyotypic features have been found to be characteristic of Group 1 *T. b. gambiense*, e.g. size and number of minichromosomes [66, 68, 69].

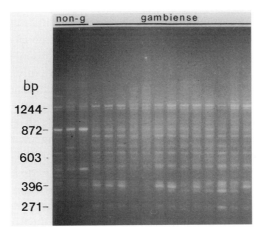

Fig. 1. Comparison of RAPD patterns obtained with a single primer from genomic DNAs of various *T. brucei* ssp. stocks. From left to right, lanes 1-3 non-gambiense stocks: *T. b. brucei*, Zaire; *T. b. brucei*, Côte d'Ivoire; *T. b. rhodesiense*, Zambia. Lanes 4-16 Group 1 *T. b. gambiense*: 3 stocks from Côte d'Ivoire, 2 stocks from Uganda, 6 stocks from Congo and Zaire, 2 stocks from Cameroon

Mathematical analysis of characterization data

Numerical methods for analysing molecular data derived from characterization studies of microorganisms fall into two basic categories: population genetics methods and phylogenetic methods. The former, as their name suggests, are used at the population level, while the latter are used across a broad evolutionary range from the population level upwards. In addition, a wide range of more general epidemiological and/or ecological methods (which lie outside the scope of this chapter) exist for mapping disease movements and population changes, certain of which have proved useful in the study of trypanosomiasis. The choice of method is generally related to the type of molecular data obtained.

The availability and use of numerical methods for studying trypanosomiasis has developed in relation to a range of factors, including the introduction of new characterization methods, the increased availability and processing power of computers, and the development of new theories in the field of trypanosome population genetics. In this section, we review those techniques commonly used in trypanosomiasis research, together with a number of newer, somewhat more sophisticated techniques which will undoubtedly be of use in the near future.

Population genetics methods

At the base of population genetics methods lies the concept of gene flow within and between populations. In particular, population genetics methods test for the presence of subdivisions within a given population, between which gene flow is either restricted or absent [70]. The null hypothesis (H_0) assumes that the population is randomly mixing i.e. panmictic; a significant variation from H_0 implies a non-panmictic population structure. Statistics used to evidence departures from panmixia consider either the lack of segregation or the lack of recombination of markers (see below), and thus equate to indirect measures of gene flow. The use of such measures to analyse field data can provide valuable information concerning the frequency and impact of sexual reproduction in natural populations of *T. brucei*. However, the possibility of population sub-structuring [71, 72], which can also affect gene flow, should also be considered. Furthermore, although all tests rely on the same basic principle (departure from panmixia), levels of resolution will differ between tests, possibly leading to divergent conclusions [73].

Segregation tests

Segregation tests are based on the concept of Hardy-Weinberg equilibrium, where there is random reassortment of different alleles at a given locus. Such tests require that alleles are identifiable and that the ploidy level of the organism being studied is known and greater than one. Although the ploidy of the smaller chromosomes of *T. brucei* is problematic, the larger chromosomes which contain housekeeping genes are diploid [28, 67, 126]. Therefore, diploidy can be used as a working hypothesis for population studies based on isoenzyme markers and segregation tests have been used in a number of important studies of *Trypanozoon* trypanosomes.

The seminal work of Tait [23] used classical Hardy-Weinberg statistics to reveal genetic exchange in African trypanosomes from the field, although a later study by Cibulskis [74] underlined some of the pitfalls associated with single-locus Hardy-Weinberg analysis when sample sizes are small. The randomization approach developed by Cibulskis [74] was extended by Stevens & Welburn [75] to study genetic exchange in epidemic populations, together with a multilocus approach. Multilocus analyses offer a robust extension to single locus methods [76] and form the basis of the next section covering recombination and linkage disequilibrium tests.

Recombination/linkage methods

Recombination tests offer a powerful alternative to segregation methods, with the advantage that mandatory requirements for segregation tests (e.g. knowledge of ploidy and allelic loci) can be avoided. Importantly, such tests can be performed not just on individual loci, but groups of loci, even when the alleles within such groups are not precisely defined. The only requirement is that the loci or groups of loci are independent from one another [70, 73].

Practically, the tests rely on evidencing departures from random assortment, where the expected frequency of a given genotype is simply the product of the observed frequencies of the individual genotypes which make it up. Data which are randomly assorted conform to a random distribution; this is the only state for which statistical criteria can be readily defined and is taken as the null hypothesis (H_0). Studies of organisms known to be undergoing regular genetic exchange also indicate that disequilibrium between loci is rare and that departures from equilibrium are not generally observed (reviewed by Cibulskis, [74]). Thus, a significant variation from H_0 implies a non-panmictic population structure. Such variation can be measured by any one of a number of statistics based on randomization methods (e.g. [77-79]), association indices [71, 80] or a combination of the two [81]. All explore different aspects of the same variation: departures from panmixia or linkage disequilibrium (non-random association among loci, where the predictions of expected probabilities for multilocus genotypes are no longer satisfied).

Of course, while such tests permit departures from panmixia to be demonstrated, the statistics say nothing about the underlying cause. Obstacles to gene flow can be classified under two main headings: physical (genetic isolation in either space or time) and biological (natural selection, physical linkage of different genes on the same chromosome, cryptic speciation, clonality). The relative importance of one or other will obviously vary, depending on the population being considered.

Many of these methods have been employed in population studies of African trypanosomes. Linkage analyses have been used for broad studies of the population structure of parasitic microorganisms including trypanosomes [60, 77, 78], while an extended Mantel test [82] has been used to study a range of *T. brucei* [61, 62, 79]. Association indices have also been employed for defining population structure in a range of *T. brucei* populations, e.g. Maynard Smith et al. [71], the Lambwe Valley, Kenya; Hide et al. [51], Uganda; Stevens & Tibayrenc [72], Côte d'Ivoire, Uganda, Zambia, and for investigating associations between parasite genotype and host/location [83].

Finally, it is important to realize the somewhat unconventional nature of tests for departures from equilibrium, as they are effectively tests of the null hypothesis [76] and are heavily dependent upon the richness of the data under study. Accordingly, as the richness of the data declines, H_0 is sometimes accepted when, in reality, all that has been shown is that there is not sufficient evidence for accepting the alternative hypothesis, H_1. This is not statistically valid and is known as a Type II error. For segregation methods, the probability (β) of making a Type II error can, in certain cases, be calculated from the number of arrangements of genotype frequencies which conform to H-W equilibrium by chance [74, 84]. Once β is known, steps can be taken to reduce it to less than 5%, usually by increasing the sample size. If β cannot be reduced, then at least a probability of having avoided a Type II error can be attached to results to provide some measure of confidence in the conclusions.

The very nature of randomization methods does not permit the calculation of such formal statistics and their use remains dependent on the richness of the data being analysed and on the discriminative power of the technique employed. For example, using RAPD analysis Stevens & Tibayrenc [79] identified thirteen genetically distinct populations, originating (cloned) from two primary isolates of *T. brucei* from tsetse; isoenzyme characterization of the same stocks revealed only eight zymodemes. Correspondingly, all linkage analyses of the RAPD data evidenced significant association, while only 70% of analyses of the isoenzyme data showed significant linkage; levels of significance obtained from analysis of the isoenzyme data were also much reduced, being at least one order of magnitude lower.

Phylogenetic methods

Phylogenetic techniques are complementary to population genetics methods and address higher levels of divergence between taxa, i.e. generally above the species level. Indeed, phylogenetic criteria are especially informative for defining taxa in microorganisms in general, for which the biological species concept [85] is often difficult or impossible to use [73].

At their most basic, phylogenetic methods can be thought of as trees describing evolutionary relationships between taxa. Due to the use of more general classes of marker for phylogenetic analysis, it has been possible to employ standard techniques for studies of African trypanosomes. Based on the level of phylogenetic divergence, to be explored, different classes of genetic marker (e.g. gene sequences, restriction sites or fragments), with different levels of resolution, have been used. It is, however, rare that one marker provides a level of resolution satisfactory across all levels of phylogenetic divergence, and many studies now combine data from a range of markers, with various 'clock' speeds. Details are available in standard texts, e.g. Avise [86], and a range of computer based methods are now available to elucidate relationships by means of phylogenetic trees [87, 88]. A measure of confidence can be attached to a given tree by additional bootstrap analysis [89]; the merits of such techniques for analysing molecular data remain much in debate.

Phylogenetic methods fall into two main categories: numerical methods and "true" phylogenetic methods.

Numerical methods

Numerical, or more appropriately phenetic methods, were developed primarily to explore bacterial taxonomy [90, 91]. Phenetic methods cluster taxa on the basis of overall similarity (or dissimilarity), e.g. the unweighted pair-group method using arithmetic averages (UPGMA). The measure of similarity is calculated according to the presence or absence of the characters chosen and an equal weight is given to all available characters, e.g. Jaccard's coefficient [92], simple matching coefficient [93]. Significantly, such methods assume an equal rate of evolution in all lineages. Relationships between the individuals in a population can be visualized by a phenogram (dendrogram). However, such dendrograms cannot be regarded as true phylogenetic trees unless all the individuals sampled represent discrete phylogenetic lineages between which gene flow is absent or severely reduced, e.g. natural clones or distinct biological species [73]. For *T. brucei* ssp., this is probably most often not the case, and such dendrograms provide only a pictorial representation of individual variability in the population being studied.

Nevertheless, due in part to their relatively early appearance in the field of biosystematics and to their ease of use, a range of phenetic measures continue to be widely used in trypanosome systematics. Since the first numerical taxonomic study of *Trypanozoon* [22], a host of other studies using phenetic classification systems have followed, e.g. Hide et al. [49, 50], Cibulskis [83], Stevens & Godfrey [41], Truc & Tibayrenc [94], Mathieu-Daudé & Tibayrenc [38], Enyaru et al. [95]. Surprisingly, despite improvements in both molecular and mathematical characterization methods, the broader relationships described by these phenetic based studies have remained largely unchanged.

True phylogenetic methods

Unlike phenetic methods, true phylogenetic methods do not assume a uniform evolutionary rate along all phylogenetic lines. Accordingly, many also offer the option to weight calculations according to the importance of different character state changes. There are two major classes of phylogenetic inference methods which can be applied to molecular data: distance methods and cladistic methods.

Phylogenetic distance methods, as for phenetic methods, use procedures which cluster intertaxon genetic distances derived from paired comparisons of the molecular data. Indeed, the two categories of method have much in common and differ primarily in the calculation of the underlying distance measure. Whereas phenetic distances rely simply on scoring the presence or absence of reaction products, e.g. all bands on an isoenzyme electrophoretic plate, phylogenetic distances are derived from a genetic (in the case of isoenzymes, allelic) interpretation of the patterns/fragments observed.

As noted, a range of methods exist including, the Fitch-Margoliash method [96], Nei's method [97], Neighbour joining [98] and the Wagner method [99] (see Nei [100] for a review of the relative efficiencies of different tree making methods). Of these, Nei's method and the Wagner method have been widely used for the study of African trypanosomes, with the Wagner method being used for studies of phylogenetic relationships in Kenyan and Ugandan trypanosomes [51, 74] and Nei's distance being employed for broader phylogenetic studies of *Trypanozoon* (e.g. Tait [27]; Stevens [35]; Hide [51]; Mathieu-Daudé [38]).

Cladistics (often referred to as the parsimony method [101]) and the closely allied technique of maximum likelihood, use discrete character data and work on changes in character states [102]; while cladistics evolved largely from studies of morphological data, maximum likelihood was developed specifically for molecular data [103]. The suitability of cladistics for analysing biochemical and molecular data is, therefore, much in debate as problems concerning the relative importance of varying levels of homoplasy, sequence gaps and alignments, variation in molecular clock speed between markers, deletions and insertions all remain to be resolved [104, 105]. Nevertheless, a number of apparently highly informative cladistic, parsimony based, studies of African trypanosomes have been undertaken. Parsimony analysis which, as the name suggests, seeks to define the phylogenetic tree requiring the least number of evolutionary changes (the most parsimonious tree) was first used to study evolutionary relationships within subgenus *Trypanozoon* using isoenzyme data [30]. Since then Mathieu-Daudé et al. [61] have extended this work with a RAPD-based parsimony analysis of *T. brucei*, while Fernandes et al. [106] and Maslov et al. [107] have used the technique with sequence data for broad studies of evolution in the Trypanosomatidae.

Maslov et al. [107] also used the maximum likelihood as an alternative to parsimony when reconstructing a phylogeny for a number of kinetoplastid species based on small and large subunit rDNA sequence data. To date, this method and cladistic methods in general, seem to have been little used in trypanosome taxonomy. In the case of maximum likelihood this is probably due to computing power requirements, while in the case of parsimony methods, the suitability of a procedure originally developed for morphological data remains unknown. Indeed, the phylogenetic value of electrophoretic data in general is still a subject of much debate (see Tibayrenc [70, 73]) and the high degree of homoplasy present in nearly all molecular data (e.g. DNA sequences, RAPD, RFLP, isoenzymes) must be considered.

Finally, it should be remembered that all mathematical methods for analysing biochemical and molecular data are totally reliant on the quality of the original data. Consequently, the inability of a particular mathematical approach to produce a meaningful result is probably due as much to the resolution of the chosen molecular marker as to the numerical method used - irrespective of the genetic variation being studied, a computer is always able to generate a dendrogram, even if the data have no phylogenetic value [73].

Epidemiological implications

The impact of molecular characterization techniques on our understanding of the epidemiology of trypanosomiasis has been considerable. In this section we will highlight the major points relevant to the clinical disease and control strategies: firstly, the question of the identity of *T. b. gambiense* and *T. b. rhodesiense*; secondly, the identification of animal reservoir hosts; and thirdly, the evolution of epidemics.

The identity of the trypanosomes

T. b. gambiense

A conclusion reiterated in turn by isoenzyme, RFLP and RAPD analyses is that the majority of *T. b. gambiense* isolates form a homogeneous group. This group (Group 1) conforms to the classical concept of *T. b. gambiense*, which runs a chronic course in the human patient and is of low virulence to experimental animals. Group 1 *T. b. gambiense* is characterized by particular isoenzyme patterns (Fig. 2; [21, 22, 24, 26, 29, 35], and by a set of RFLPs for VSG genes, notably AnTat 1.8 and 11.17 [52-55] and ribosomal genes [49]. Besides these molecular markers, Group 1 *T. b. gambiense* stocks have a restricted antigenic repertoire [108, 109], a small genome and fewer small chromosomes compared to other *T. brucei* ssp. [66, 68, 69]. This group of *T. b. gambiense* is widespread through tropical Africa, eastwards from Senegal to Zaire, and including adjoining foci in south-west Sudan and north-west Uganda [29, 36]. Considering biological characteristics, in the past *T. b. gambiense* was distinguished from *T. b. rhodesiense* by its susceptibility to tryparsamide; currently DFMO is effective for treatment of gambian but not rhodesian sleeping sickness, suggesting that Group 1 *T. b. gambiense* is also characterized by susceptibility to DFMO. *T. b. gambiense* is typically transmitted by tsetse flies of the *palpalis* rather than *morsitans* group and this association appears to be a further characteristic of Group 1 *T. b. gambiense* [110].

Group 2 *T. b. gambiense* is defined as a virulent form of *T. b. gambiense* from foci of gambian sleeping sickness [24, 66]. In contrast to Group 1, Group 2 *T. b. gambiense* grows well in experimental rodents and is easy to tsetse-transmit via *morsitans*-group flies in the laboratory [110]; whether it also differs in clinical features is presently under examination [128]. By molecular characterization this trypanosome falls outside homogeneous Group 1 and does not share its characteristic isoenzyme and RFLP markers (Fig. 2, [19, 24, 31, 49, 111]). Although often referred to as "*T. b. rhodesiense*-like", Group 2 *T. b. gambiense* is more akin to West African *T. b. brucei* stocks [49, 111] and indeed may represent a zoonotic form of sleeping sickness in West Africa [24, 128].

The homogeneity of Group 1 *T. b. gambiense* may reflect lack of sexuality and the wide distribution of this stable genotype has been interpreted as evidence of clonal expansion [19, 72, 77]. While genetic exchange has been demonstrated for Group 2 *T. b. gambiense* in the laboratory [112, 113], for Group 1 *T. b. gambiense* such experiments are rendered difficult by its poor tsetse transmissibility.

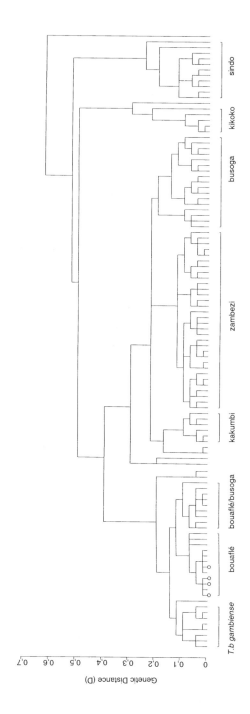

Fig. 2. Dendrogram summarizing the intraspecific taxonomy of *T. brucei*. Branch end points correspond to zymodemes defined by isoenzyme characterization studies. The major groups defined are: *T. b. gambiense*, corresponding to Group 1 *T. b. gambiense*; *bouaflé*, corresponding to *T. b. brucei* in West Africa, but including a number of stocks isolated from humans (o) which belong to Group 2 *T. b. gambiense*; five strain groups containing predominantly East African isolates, *busoga*, *kakumbi*, *kiboko*, *sindo* and *zambezi*, reflecting the complex relationships between *T. b. rhodesiense* and East African *T. b. brucei*. Of these, *busoga* and *zambezi* correspond to 'classical' *T. b. rhodesiense*, *busoga* being associated with northern sleeping sickness foci and *zambezi* with southern foci; the remaining three strain groups show differing degrees of man-infectivity and are strongly associated with wild animal reservoirs. See Gibson et al. [22, 47], Godfrey et al. [30], Stevens & Godfrey [41] for further information on strain groups and subspecific taxonomy. The dendrogram was constructed by UPGMA from a matrix of genetic distances (D). D values between zymodemes were calculated according to Nei's method [97] by allelic interpretation of ten enzyme systems (Stevens et al. [35]).

T. b. rhodesiense and T. b. brucei

Biochemical markers for *T. b. rhodesiense* and *T. b. brucei* have proved far more elusive than for Group 1 *T. b. gambiense*. By isoenzyme and RFLP analyses, *T. b. rhodesiense* and *T. b. brucei* stocks fall into several groups (Fig. 2), largely according to geographical origin [22, 25, 28, 30, 35, 49-51, 111]. Early on, isoenzyme analysis revealed that *T. b. brucei* stocks could be broadly divided into East and West African groups based on homozygosity at loci for phosphoglucomutase (PGM) and isocitrate dehydrogenase (ICD) [22], and this result was confirmed by RFLP analysis [28, 49, 111]. The human pathogens are more closely related to their sympatric *T. b. brucei* than each other. Thus, in East Africa *T. b. rhodesiense* remains difficult to distinguish from *T. b. brucei*, except by its ability to infect humans. In one way this is not surprising considering that representatives of these 2 subspecies can mate [115]; however, *T. b. brucei/T. b. rhodesiense* by no means constitutes a panmictic group in East Africa [72].

T. b. rhodesiense stocks in general are highly variable and each epidemic focus appears to have its own set of associated trypanosomes [22, 25, 27, 28, 30, 32, 33, 35, 37, 39, 51]. Ormerod [116] identified a trend in decreasing virulence of rhodesian sleeping sickness from north to south and associated this with northern and southern strains of *T. b. rhodesiense*. Characterization data supports this idea to some extent, since clear isoenzyme and RFLP differences have been shown between *T. b. rhodesiense* from the extremes of the range, Zambia and Uganda [22, 30, 35, 37, 50, 51]. However, results from endemic areas in between (Lambwe Valley, Kenya and Kigoma/Tabora, Tanzania) suggest that the observed dichotomy between northern and southern strains may be part of a larger mosaic of different genotypes in East Africa rather than a simple cline [28, 130].

Although the majority of East African *T. b. brucei* and *T. b. rhodesiense* isolates fall into one broadly similar group, there are in addition 2 or 3 highly distinctive groups of *T. b. brucei*, which have emerged from isoenzyme and kDNA characterization (*kiboko*, *sindo* and *kakumbi*, see Fig. 2; [22, 30, 35, 47]). These stocks appear to circulate in transmission cycles involving wild animals and tsetse, but not humans. Again, mating experiments indicate that these groups are probably not reproductively isolated [117].

Reservoir hosts

Isoenzyme and RFLP analyses have made possible the identification of animal reservoir hosts of both *T. b. gambiense* and *T. b. rhodesiense* without recourse to human infectivity experiments or serum resistance tests. The observation that sporadic infections with *T. b. rhodesiense* could be contracted by visitors to areas populated only by wild animals quickly established that rhodesian sleeping sickness was a zoonosis. However, the epidemiology of gambian sleeping sickness suggested that human infections alone maintained transmission. As described above, Group 1 *T. b. gambiense* has characteristic isoenzyme and RFLP markers and can thus be readily identified in hosts other than humans. In this way pigs, dogs, sheep, goats, cattle and also wild antelope have been incriminated as reservoir hosts of sleeping sickness in West Africa [24, 26, 34, 53, 118-121, 128]. There is now no doubt that these animals harbour *T. b. gambiense*, but their actual importance in maintaining transmission of sleeping sickness remains to be established. Figure 2 shows the similarity between Group 2 *T. b. gambiense* and *T. b. brucei* isolates of the *bouaflé* group, illustrating the truely zoonotic nature of this infection [24, 128, 129].

In East Africa, the existence of wild and domestic animal reservoir hosts of rhodesian sleeping sickness was proved by the experimental infection of human volunteers [122, 123]. The role of molecular fingerprinting has been to quantify the risk; thus cattle have been shown to constitute a significant reservoir of *T. b. rhodesiense* in epidemics in Kenya and Uganda [28, 51]. The presence of large numbers of infected animals in the peridomestic environment assumes particular importance if flies are also resident.

In the Kenyan study, trypanosomes tended to persist in the lymph glands and CNS of the cattle rather than the bloodstream, and may therefore not have been transmissible by tsetse after a few months; in addition, the cattle suffered heavy mortality from trypanosomiasis during this outbreak [28]. Reservoir hosts that remain contagious in the longterm and do not become sick themselves pose the greatest disease risk. Thus, secretive wild animal hosts, such as bushbuck or bushpig, may be as important as domestic reservoir hosts in sustaining endemic foci, especially considering that these animals are favoured hosts of tsetse.

Evolution of epidemics

Large numbers of trypanosome isolates from several sleeping sickness foci have now been characterized by isoenzyme or RFLP analysis. Some studies have included stocks isolated over extended time periods, up to 30 years in some cases. For gambian sleeping sickness, the picture is constant, with the Group 1 trypanosome predominant, but showing a degree of microheterogeneity within and between different foci [29, 35, 53, 55, 62, 94]. For example, stocks from Cameroon had a divergent gene for VSG LiTat 1.3, which is the antigen targeted in the CATT [124], while stocks from the Moyo focus in north-west Uganda lacked another *gambiense-*

specific VSG gene, AnTat 11.17 [36]. Thus, Group 1 *T. b. gambiense* appears to originate from a single strain, which spread widely and then diverged locally by mutations [19].

For rhodesian sleeping sickness, the story is more complex. Three areas have been intensively studied: Busoga in south-east Uganda [25, 27, 37, 50, 51, 75], the Lambwe valley in south-west Kenya [28, 32, 33], and the Luangwa valley in north-eastern Zambia [22, 35, 50]. Limited data is also available from foci in Tanzania and Ethiopia [22, 39, 130]. The general points that emerge are as follows. Firstly, each focus has its own associated *T. b. rhodesiense* strains. For example, although epidemics occurred simultaneously in Lambwe valley and Busoga in the 1980's, *T. b. rhodesiense* strains from these 2 geographically close foci were unrelated, indicating that each epidemic had a separate origin [25, 28].

Secondly, some strains of *T. b. rhodesiense* remain stable over many years in a given focus, while new strains also emerge. The stable strains are apparently well adapted to humans, but are nevertheless also found in reservoir hosts. An example is the *busoga* strain from the large focus spanning south-east Uganda and the neighbouring area of Kenya, which has been isolated repeatedly over the past 30 years or so and is characterized by both isoenzyme and RFLP data (Fig. 2; [25, 30, 37, 51]). This is an example of epidemic clonal propagation from an underlying sexual population [71, 72, 77].

On the other hand, new as well as old strains may be associated with an outbreak [25, 28, 37, 95]. The emergence of new strains may be interpreted as the adaptation to humans of strains previously circulating in reservoir hosts, or as the direct products of genetic exchange [33, 83]. These authors produced intriguing evidence of host selection for certain trypanosome genotypes, since cattle did not harbour all strains found in humans and had their own associated trypanosome genotypes. A further finding from epidemic foci in both Uganda and Kenya was that trypanosome isolates from tsetse were extremely variable [25, 28, 32, 33, 37, 95]. Since genetic exchange in *T. brucei* occurs in the fly [112], this was a possible explanation for the generation of new genotypes. Paradoxically, however, these new genotypes were rarely found among the isolates collected from vertebrate hosts, usually humans and domestic stock. This may reflect undersampling from wild animal hosts or, alternatively, may indicate a time lag between the generation of new genotypes in the fly and their appearance in an outbreak. The extent to which genetic exchange orchestrates strain variability in the field is still controversial [61, 72, 77, 83, 114]. Part of this controversy has arisen from attempting to combine data from unrelated epidemics, instead of considering each focus as a separate population [72].

Conclusion

From the panmictic view of *T. brucei* originally proposed by Tait [23], the consensus at present is that *T. b. brucei/T. b. rhodesiense* are basically sexual, but undergo periods of clonal expansion [71, 72, 125], while the majority of evidence suggests a lower degree of genetic exchange in classical *T. b. gambiense*.

Acknowledgments

We thank Michel Tibayrenc for valuable comments on an earlier draft of this manuscript and Georgette Kanmogne for kindly providing Fig. 1. WCG and JRS are funded by The Wellcome Trust.

References

1. Mehlitz D (1978) Untersuchungen zur Empfanglichkeit van Mastomys natalensis fur *Trypanosoma (Trypanozoon) brucei gambiense*. Tropenmed Parasit 29:101-107
2. Cunningham I (1977) New culture medium for maintenance of tsetse tissues and growth of trypanosomatids. J Protozool 24:325-329
3. Brun R, Schonenberger M (1979) Cultivation and in vitro cloning of procyclic culture forms of *Trypanosoma brucei* in a semi-defined medium. Acta Trop 36:289-292
4. Dukes P, Kaukus A, Hudson KM, Asonganyi T, Gashumba JK (1989) A new method for isolating *Trypanosoma brucei gambiense* from sleeping sickness patients. Trans R Soc Trop Med Hyg 83:636-639
5. Aerts D, Truc P, Penchenier L, Claes Y, Le Ray D (1992) A kit for in vitro isolation of trypanosomes in the field: first field trial with sleeping sickness patients in the Congo Republic. Trans R Soc Trop Med Hyg 86:394-395
6. Truc P, Aerts D, McNamara JJ, Claes Y, Allingham R, Le Ray D, Godfrey DG (1992) Direct isolation *in vitro* of *Trypanosoma brucei* from man and other animals, and its potential value for the diagnosis of gambian trypanosomiasis. Trans R Soc Trop Med Hyg 86:627-629
7. McNamara JJ, Bailey JW, Smith DH, Wakhooli S, Godfrey DG (1995) Isolation of *Trypanosoma brucei gambiense* from northern Uganda: evaluation of the kit for *in vitro* isolation (KIVI) in an epidemic focus. Trans R Soc Trop Med Hyg 89:388-389
8. Truc P (1996) A miniature kit for the *in vitro* isolation of *Trypanosoma brucei gambiense*: a preliminary field assessment on sleeping sickness patients in Côte d'Ivoire. Trans R Soc Trop Med Hyg 90:246-247
9. Hajduk SL, Hager K, Esko JD (1992) High density lipoprotein-mediated lysis of trypanosomes. Parasitol Today 8:95-98
10. Rickman LR, Robson J (1970) The testing of proven *Trypanosoma brucei* and *T. rhodesiense* strains by the blood incubation infectivity test. Bull WHO 42:911-916
11. Van Meirvenne N, Magnus E, Janssens PG (1976) The effect of normal human serum on trypanosomes of distinct antigenic type (ETat 1 - 12) isolated from a strain of *Trypanosoma brucei rhodesiense*. Ann Soc Belge Méd Trop 56:55-63
12. Rickman LR (1977) Variation in the test response of clone-derived *Trypanosoma (Trypanozoon) brucei brucei* and *T. (T.) b. rhodesiense* relapse antigenic variants, examined by a modified blood incubation infectivity test and its possible significance in rhodesian sleeping sickness transmission. Med J Zambia 11:31-41
13. Brun R, Jenni L (1987) Human serum resistance of metacyclic forms of *Trypanosoma brucei brucei*, *T. b. rhodesiense* and *T. b. gambiense*. Parasit Res 73:218-223
14. Rickman LR (1992) The significance of human serum sensitivity in the context of *T. b. rhodesiense* sleeping sickness epidemiology and control. East Af Med J 69:272-278
15. Kukla BA, Majiwa PAO, Young CJ, Moloo SK, Ole-Moiyoi OK (1987) Use of species-specific DNA probes for the detection and identification of trypanosome infections in tsetse flies. Parasitology 95:1-26

16. Gibson WC, Dukes P, Gashumba JK (1988) Species specific DNA probes for the identification of trypanosomes in tsetse. Parasitology 97:63-73
17. Majiwa PAO, Thatthi R, Moloo SK, Nyeko JHP, Otieno LH, Maloo S (1994) Detection of trypanosome infections in the saliva of tsetse flies and buffy coat samples from antigenaemic but aparasitaemic cattle. Parasitology 108:313-322
18. Masiga DK, McNamara JJ, Laveissière C, Truc P, Gibson WC (1995) A high prevalence of mixed trypanosome infections in tsetse flies in Sinfra, Côte d'Ivoire detected by DNA amplification. Parasitology 112:75-80
19. Gibson WC, Borst P (1986) Size-fractionation of the small chromosomes of *Trypanozoon* and *Nannomonas* trypanosomes by pulsed field gradient gel electrophoresis. Mol Biochem Parasitol 18:127-140
20. Bagster IA, Parr CW (1973) Trypanosome identification by electrophoresis of soluble enzymes. Nature 244:364-366
21. Godfrey DG, Kilgour V (1976) Enzyme electrophoresis in characterising the causative organism of Gambian trypanosomiasis. Trans R Soc Trop Med Hyg 70:219-224
22. Gibson WC, Marshall TF de C, Godfrey DG (1980) Numerical analysis of enzyme polymorphism: a new approach to the epidemiology and taxonomy of trypanosomes of the subgenus *Trypanozoon*. Adv Parasitol 18:175-246
23. Tait A (1980) Evidence for diploidy and mating in trypanosomes. Nature 287:536-538
24. Mehlitz D, Zillmann U, Scott CM, Godfrey DG (1982) Epidemiological studies on the animal reservoir of gambiense sleeping sickness. III. Characterization of *Trypanozoon* stocks by isoenzymes and sensitivity to human serum. Tropenmed Parasit 33:113-118
25. Gibson WC, Gashumba JK (1983) Isoenzyme characterization of some *Trypanozoon* stocks from a recent trypanosomiasis epidemic in Uganda. Trans R Soc Trop Med Hyg 77:114-118
26. Tait A, Babiker EA, Le Ray D (1984) Enzyme variation in *Trypanosoma brucei* spp. I. Evidence for the subspeciation of *T. b. gambiense*. Parasitology 89:311-326
27. Tait A, Barry JD, Wink R, Sanderson A, Crowe JS (1985) Enzyme variation in *Trypanosoma brucei* spp. II. Evidence for *T. b. rhodesiense* being a set of variants of *T. b. brucei*. Parasitol 90:89-100
28. Gibson WC, Wellde BT (1985) Characterization of *Trypanozoon* stocks from the South Nyanza sleeping sickness focus in Western Kenya. Trans R Soc Trop Med Hyg 79:671-676
29. Godfrey DG, Scott CM, Gibson WC, Mehlitz D, Zillmann U (1987) Enzyme polymorphism and the identity of *Trypanosoma brucei gambiense*. Parasitology 94:337-347
30. Godfrey DG, Baker RD, Rickman LR, Mehlitz D (1990) The distribution, relationships and identification of enzymic variants within the subgenus *Trypanozoon*. Adv Parasitol 29:1-74
31. Richner D, Schweizer J, Betschart B, Jenni L (1989) Characterization of West African *Trypanosoma (Trypanozoon) brucei* isolates from man and animals using isoenzyme analysis and DNA hybridization. Parasitol Res 76:80-85
32. Otieno LH, Darji N, Onyango P (1990) Electrophoretic analysis of *Trypanosoma brucei* subgroup stocks from cattle, tsetse and patients from Lambwe Valley, Western Kenya. Insect Sci App 11:281-287
33. Mihok S, Otieno LH, Darji N (1990) Population genetics of *Trypanosoma brucei* and the epidemiology of human sleeping sickness in the Lambwe Valley, Kenya. Parasitology 100:219-233
34. Truc P, Mathieu-Daudé F, Tibayrenc M (1991) Multilocus isozyme identification of *Trypanosoma brucei* stocks isolated in Central Africa: evidence for an animal reservoir of sleeping sickness in Congo. Acta Trop 49:127-135
35. Stevens JR, Lanham SM, Allingham R, Gashumba JK (1992) A simplified method for identifying subspecies and strain groups in *Trypanozoon* by isoenzymes. Ann Trop Med Parasit 86:9-28
36. Enyaru JC, Allingham R, Bromidge T, Kanmogne GD, Carasco JF (1993a) The isolation and genetic heterogeneity of *Trypanosoma brucei gambiense* from north-west Uganda. Acta Trop 54:31-39
37. Enyaru JC, Stevens JR, Odiit M, Okuna NM, Carasco JF (1993b) Isoenzyme comparison of *Trypanozoon* isolates from two sleeping sickness areas of south-eastern Uganda. Acta Trop 55:97-115
38. Mathieu-Daudé F, Tibayrenc M (1994) Isozyme variability of *Trypanosoma brucei* s.l.: genetic, taxonomic, and epidemiological significance. Exp Parasitol 78:1-19
39. Gashumba JK, Komba EK, Truc P, Allingham RM, Ferris V, Godfrey DG (1994) The persistence of genetic homogeneity among *Trypanosoma brucei rhodesiense* isolates from patients in north-west Tanzania. Acta Trop 56:341-348
40. Harris H, Hopkinson DA (1976) Handbook of enzyme electrophoresis in human genetics. North Holland Publishing Company, Amsterdam
41. Stevens JR, Godfrey DG (1992) Numerical taxonomy of *Trypanozoon* based on polymorphisms in a reduced range of enzymes. Parasitology 104:75-86
42. Mathieu-Daudé F, Bicart-See A, Bosseno MF, Brenière SF, Tibayrenc M (1994) Identification of *Trypanosoma brucei gambiense* Group 1 by a specific kinetoplast DNA probe. Am J Trop Med Hyg 50:13-19
43. Kilgour V (1980) The electrophoretic mobilities and activities of eleven enzymes of bloodstream and culture forms of *Trypanosoma brucei* compared. Mol Biochem Parasitol 2:51-62

44. Kaukus A, Gashumba JK, Lanham SM, Dukes P (1990) The substitution of procyclic for bloodstream form *Trypanosoma brucei gambiense* in isoenzyme studies. Trans R Soc Trop Med Hyg 84:242-245
45. Lanham SM, Grendon JM, Miles MA, Povoa MM, De Souza AAA (1981) A comparison of electrophoretic methods for isoenzyme characterization of trypanosomatids. 1: Standard stocks of *Trypanosoma cruzi* zymodemes from north-east Brazil. Trans R Soc Trop Med Hyg 75:742-750
46. Borst P, Fase-Fowler F, Gibson WC (1981) Quantitation of genetic differences between *Trypanosoma brucei gambiense* and *T. b. brucei* by restriction enzyme analysis of kinetoplast DNA. Mol Biochem Parasitol 3:117-131
47. Gibson WC, Fase-Fowler F, Borst P (1985a) Further analysis of intraspecific variation in *Trypanosoma brucei* using restriction site polymorphisms in the maxi-circle of kinetoplast DNA. Mol Biochem Parasitol 15:21-36
48. Morel C, Chiari E, Plessmann Camargo E, Mattei DM, Romanha AJ, Simpson L (1980) Strains and clones of *Trypanosoma cruzi* can be characterized by pattern of restriction endonuclease products of kinetoplast DNA minicircles. Proc Natl Acad Sci USA 77:6810-6814
49. Hide G, Cattand P, Le Ray D, Barry DJ, Tait A (1990) The identification of *Trypanosoma brucei* subspecies using repetitive DNA sequences. Mol Biochem Parasitol 39:213-226
50. Hide G, Buchanan N, Welburn S, Maudlin I, Barry JD, Tait A (1991) *Trypanosoma brucei* rhodesiense: characterization of stocks from Zambia, Kenya, and Uganda using repetitive DNA probes. Exp Parasitol 72:430-439
51. Hide G, Welburn SC, Tait A, Maudlin I (1994) Epidemiological relationships of *Trypanosoma brucei* stocks from South East Uganda: evidence for different population structures in human infective and non-infective isolates. Parasitology 109:95-111
52. Pays E, Dekerck P, Van Assel S, Babiker EA, Le Ray D, Van Meirvenne N, Steinert M (1983) Comparative analysis of a *Trypanosoma brucei gambiense* antigen gene family and its potential use in sleeping sickness epidemiology. Mol Biochem Parasitol 7:63-67
53. Paindavoine P, Pays E, Laurent M, Geltmeyer Y, Le Ray D, Mehlitz D, Steinert M (1986) The use of DNA hybridisation and numerical taxonomy in determining relationships between *Trypanosoma brucei* stocks and subspecies. Parasitology 92:31-50
54. Thi CDD, Aerts D, Steinert M, Pays E (1991) High homology between variant surface glycoprotein gene expression sites of *Trypanosoma brucei* and *T. gambiense*. Mol Biochem Parasitol 48:199-210
55. Kanmogne GD, Stevens JR, Asonganyi T, Gibson WC (1996a) Characterization of *Trypanosoma brucei gambiense* isolates using restriction fragment length polymorphisms in 5 variant surface glycoprotein genes. Acta Trop 61:239-254
56. Welsh J, McClelland M (1990) Fingerprinting genomes using PCR with arbitrary primers. Nucl Acids Res 18:7213-7218
57. Williams JGK, Kubelik AR, Livak KJ, Rafalski JA, Tingley SV (1990) DNA polymorphisms amplified by arbitrary primers are useful as genetic markers. Nucl Acids Res 18:6531-6535
58. Bromidge T, Gibson W, Hudson KM, Dukes P (1993) Identification of *Trypanosoma brucei gambiense* by PCR amplification of variant surface glycoprotein genes. Acta Trop 53:107-119
59. Waitumbi JN, Murphy NB (1993) Inter- and intra-species differentiation of trypanosomes by genomic fingerprinting with arbitrary primers. Mol Biochem Parasitol 58:181-186
60. Tibayrenc M, Neubauer K, Barnabe C, Guerrini F, Skarecky D, Ayala FJ (1993) Genetic characterization of six parasitic protozoa: parity between random-primer DNA typing and multilocus enzyme electrophoresis. Proc Natl Acad Sci USA 90:1335-1339
61. Mathieu-Daudé F, Stevens JR, Welsh J, Tibayrenc M, McClelland M (1995) Genetic diversity and population structure of *Trypanosoma brucei*: clonality versus sexuality. Mol Biochem Parasitol 72:89-101
62. Kanmogne GD, Stevens JR, Asonganyi T, Gibson WC (1996b) Genetic heterogeneity in the *Trypanosoma brucei gambiense* genome analysed by random amplification of polymorphic DNA. Parasitol Res 82:535-541
63. Van der Ploeg LHT, Valerio D, De Lange T, Bernards A, Borst P, Grosveld FG (1982) An analysis of cosmid clones of nuclear DNA from *Trypanosoma brucei* shows that the genes for variant surface glycoproteins are clustered in the genome. Nucl Acids Res 10: 5905-5923
64. Schwartz DC, Cantor CR (1984) Separation of yeast chromosome-sized DNAs by pulsed field gel electrophoresis. Cell 37:67-75
65. Van der Ploeg LHT, Schwartz DC, Cantor CR, Borst P (1984) Antigenic variation in *Trypanosoma brucei* analysed by electrophoretic separation of chromosome-sized DNA molecules. Cell 37:77-84
66. Gibson WC (1986) Will the real *Trypanosoma brucei gambiense* please stand up? Parasitol Today 2:255-257
67. Gottesdiener K, Garcia-Anoveros J, Lee G-SM, Van der Ploeg LHT (1990) Chromosome organization of the protozoan *Trypanosoma brucei*. Mol Cell Biol 10:6079-6083
68. Dero B, Zampetti-Bosseler F, Pays E, Steinert M (1987) The genome and the antigen gene repertoire of *Trypanosoma brucei gambiense* are smaller than those of *T. b. brucei*. Mol Biochem Parasitol 26:247-256
69. Kanmogne GD, Bailey M, Gibson W (1997) Wide variation in DNA content among *Trypanosoma brucei* ssp. isolates. Acta Trop 63:75-87

70. Tibayrenc M (1995) Population genetics of parasitic protozoa and other microorganisms. Adv Parasitol 36:47-115
71. Maynard Smith J, Smith NH, O'Rourke M, Spratt BG (1993) How clonal are bacteria? Proc Natl Acad Sci USA 90:4384-4388
72. Stevens JR, Tibayrenc M (1996) *Trypanosoma brucei* s.l.: evolution, linkage and the clonality debate. Parasitol 112:481-488
73. Tibayrenc M (1996) Towards a unified evolutionary genetics of microorganisms. Annu Rev Microbiol 50:401-429
74. Cibulskis RE (1988) Origins and organization of genetic diversity in natural populations of *Trypanosoma brucei*. Parasitology 96:303-322
75. Stevens JR, Welburn SC (1993) Genetic processes within an epidemic of sleeping sickness in Uganda. Parasitol Res 79:421-427
76. Workman PL (1969) The analysis of simple genetic polymorphisms. Human Biol 41:97-114
77. Tibayrenc M, Kjellberg F, Ayala FJ (1990) A clonal theory of parasitic protozoa: The population structures of *Entamoeba, Giardia, Leishmania, Naegleria, Plasmodium, Trichomonas* and *Trypanosoma* and their medical and taxonomical consequences. Proc Natl Acad Sci USA 87:2414-2418
78. Tibayrenc M, Kjellberg F, Arnaud J, Oury B, Brenière SF, Darda ML, Ayala FJ (1991) Are eucaryotic microorganisms clonal or sexual? A population genetics vantage. Proc Natl Acad Sci USA 88:5129-5133
79. Stevens JR, Tibayrenc M (1995) Detection of linkage disequilibrium in *Trypanosoma brucei* isolated from tsetse flies and characterized by RAPD analysis and isoenzymes. Parasitology 110:181-186
80. Brown AHD, Feldman MW (1981) Population structure of multilocus associations. Proc Natl Acad Sci USA 78:5913-5916
81. Souza V, Nguyen TT, Hudson RR, Piuero D, Lenski RE (1992) Hierarchical analysis of linkage disequilibrium in *Rhizobium* populations: evidence for sex? Proc Natl Acad Sci USA 89:8389-8393
82. Mantel N (1967) The detection of disease clustering and a generalized regression approach. Cancer Res 27:209-220
83. Cibulskis RE (1992) Genetic variation in *Trypanosoma brucei* and the epidemiology of sleeping sickness in the Lambwe Valley, Kenya. Parasitology:99-109
84. Fairburn DJ, Roff DA (1980) Testing genetic models of isoenzyme variability without breeding data: Can we depend on the X2? Can J Fish Aqu Sci 37:1149-1159
85. Dobzhansky T (1937) Genetics and the origin of species. Columbia University Press, New York
86. Avise JC (1994) Molecular Markers, natural history and evolution. Chapman & Hall, New York
87. Felsenstein J (1993) PHYLIP - Phylogeny Inference Package, version 3.5. University of Washington, Washington
88. Swofford DL (1993) PAUP - Phylogenetic Analysis Using Parsimony, version 3.1.1. Illinois Natural History Survey, Champaign, Illinois
89. Felsenstein J (1985) Confidence limits on phylogenies: an approach using the bootstrap. Evolution 39:783-791
90. Sneath PHA, Sokal RR (1973) Numerical Taxonomy. The principles and practice of numerical classification. WH Freeman & Co, San Francisco
91. Dunn G, Everitt BS (1982) An introduction to mathematical taxonomy. Cambridge University Press, Cambridge
92. Jaccard P (1908) Nouvelles recherches sur la distribution florale. Bull Soc Vaudoise Sci Nat 44:223-270
93. Sokal RR, Michener CD (1958) A statistical method for evaluating systematic relationships. Univ Kansas Sci Bull 38:1409-1438
94. Truc P, Tibayrenc M (1993) Population genetics of *Trypanosoma brucei* in Central Africa: taxonomic and epidemiological significance. Parasitology 106:137-149
95. Enyaru JC, Matovu E, Odiit M, Okedi LA, Rwendeire AJJ, Stevens JR (1997) Genetic diversity of *Trypanosoma brucei* isolates from mainland and Lake Victoria island populations in southeast Uganda: epidemiological and control implications. Ann Trop Med Parasit 91:107-113
96. Fitch W, Margoliash E (1967) Construction of phylogenetic trees. Science 155:279-284
97. Nei M (1972) Genetic distance between populations. Amer Nat 106:283-292
98. Saitou N, Nei M (1987) The neighbour-joining method: a new method for reconstructing phylogenetic trees. Mol Biol Evol 4:406-425
99. Farris JS (1970) Methods for computing Wagner trees. Syst Zool 19:83-92
100. Nei M (1991) Relative efficiencies of different tree-making methods for molecular data. In: Miyamoto MM, Cracraft J (eds) Phylogenetic analysis of DNA sequences. Oxford University Press, New York, pp 90-128
101. Hennig W (1966) Phylogenetic systematics. University of Illinois Press, Urbana
102. Scotland RW (1992) Cladistic theory. In: Forey PL, Humphries CJ, Kitching IL, Scotland RW, Siebert DJ, Williams DM (eds) Cladistics - A practical course in systematics. Oxford University Press, Oxford, pp 3-13
103. Cavalli-Sforza LL, Edwards AWF (1967) Phylogenetic analysis models and estimation procedures. Am J Hum Gen 19:233-257

104. Siebert DJ (1992) Tree statistics. In: Forey PL, Humphries CJ, Kitching IL, Scotland RW, Siebert DJ, Williams DM (eds) Cladistics - A practical course in systematics. Oxford University Press, Oxford, pp 72-78
105. Williams DM (1992) DNA analysis: theory. In: Forey PL, Humphries CJ, Kitching IL, Scotland RW, Siebert DJ, Williams DM (eds) Cladistics - A practical course in systematics. Oxford University Press, Oxford, pp 89-101
106. Fernandes AP, Nelson K, Beverley SM (1993) Evolution of nuclear ribosomal RNAs in Kinetoplastid protozoa: perspectives on the age and origins of parasitism. Proc Natl Acad Sci USA 90:11608-11612
107. Maslov DA, Lukes J, Jirku M, Simpson L (1996) Phylogeny of trypanosomes as inferred from the small and large subunit rRNAs: implications for the evolution of parasitism in the trypanosomatid protozoa. Mol Biochem Parasitol 75:197-205
108. Gray AR (1972) Variable and agglutinogenic antigens of *Trypanosoma brucei gambiense* and their distribution among isolates of the trypanosome collected in different places in Nigeria. Trans R Soc Trop Med Hyg 66:263-284
109. Jones TW, Cunningham I, Taylor AM, Gray AR (1981) The use of culture-derived metacyclic trypanosomes in studies on the serological relationships of stocks of *Trypanosoma brucei gambiense*. Trans R Soc Trop Med Hyg 75: 560-565
110. Richner D, Brun R, Jenni L (1988) Production of metacyclic forms by cyclical transmission of West African *Trypanosoma* (*Trypanozoon*) brucei isolates from man and animals. Acta Trop 45:309-319
111. Paindavoine P, Zampetti-Bosseler F, Coquelet H, Pays E, Steinert M (1989) Different allele frequencies in *Trypanosoma brucei brucei* and *T. b. gambiense* populations. Mol Biochem Parasitol 32:61-72
112. Jenni L, Marti S, Schweizer J, Betschart B, Lepage RWF, Wells JM, Tait A, Paindavoine P, Pays E, Steinert M (1986) Hybrid formation between African trypanosomes during cyclical transmission. Nature 322:173-175
113. Gibson W, Winters K, Mizen G, Kearns J, Bailey M (1997) Intraclonal mating in *Trypanosoma brucei* is associated with outcrossing. Microbiol 143: 909-920
114. Gibson WC (1990) Trypanosome diversity in Lambwe Valley, Kenya - Sex or selection? Parasitol Today 6:342-343
115. Gibson WC (1989) Analysis of a genetic cross between *Trypanosoma brucei rhodesiense* and *T. b. brucei*. Parasitology 99:391-402
116. Ormerod WE (1967) Taxonomy of the sleeping sickness trypanosomes. J Parasitol 53:824-830
117. Gibson WC, Garside LH (1991) Genetic exchange in *Trypanosoma brucei brucei*: variable location of housekeeping genes in different trypanosome stocks. Mol Biochem Parasitol 45:77-90
118. Gibson WC, Mehlitz D, Lanham SM, Godfrey DG (1978) The identification of *Trypanosoma brucei gambiense* in Liberian pigs and dogs by isoenzymes and by resistance to human plasma. Tropenmed Parasit 29:335-345
119. Scott CM, Frezil J-L, Toudic A, Godfrey DG (1983) The sheep as a potential reservoir of human trypanosomiasis in the Republic of the Congo. Trans R Soc Trop Med Hyg 77:397-401
120. Zillmann U, Mehlitz D, Sachs R (1984) Identity of *Trypanozoon* stocks isolated from man and domestic dog in Liberia. Tropenmed Parasit 35: 105-108
121. Noireau F, Paindavoine P, Lemesre JL, Toudic A, Pays E, Gouteux JP, Steinert M, Frezil J-L (1989) The epidemiological importance of the animal reservoir of *Trypanosoma brucei gambiense* in the Congo: Characterization of the *T. brucei* complex. Tropenmed Parasit 40:9-11
122. Heisch RB, McMahon JP, Manson-Bahr PEC (1958) The isolation of *Trypanosoma rhodesiense* from a bushbuck. Brit Med J 2:1203-1204
123. Onyango RJ, Van Hoeve K, De Raadt P (1966) The epidemiology of *Trypanosoma rhodesiense* sleeping sickness in Alego location, Central Nyanza, Kenya. I. Evidence that cattle may act as reservoir hosts of trypanosomes infective to man. Trans R Soc Trop Med Hyg 60:175-182
124. Dukes P, Gibson WC, Gashumba JK, Hudson KM, Bromidge TJ, Kaukus A, Assonganyi T, Magnus E (1992) Absence of the LiTat 1.3 (CATT antigen) gene in *Trypanosoma brucei gambiense* stocks from Cameroon. Acta Trop 51:123-134
125. Hide G, Tait A, Maudlin I, Welburn SC (1996) The origins, dynamics and generation of *Trypanosoma brucei* rhodesiense epidemics in East Africa. Parasitol Today 12:50-55
126. Gibson WC, Osinga KA, Michels PAM, Borst P (1985b) Trypanosomes of subgenus *Trypanozoon* are diploid for housekeeping genes. Mol Biochem Parasitol 16:231-242
127. Truc P, Bailey W, Dorra F, Laveissière C, Godfrey DG (1994) A comparison of parasitological methods for the diagnosis of Gambian trypanosomiasis in an aera of low endemicity in Côte d'Ivoire. Trans R Soc Trop Med Hyg 88:419-421
128. Truc P, Formenty P, Diallo PB, Komoin-Oka C, Lauginie F (1997a) Confirmation of two distinct classes of zymodemes of *Trypanosoma brucei* infecting patients and wild mammals in Côte d'Ivoire: suspected difference in pathologenicity. Ann Trop Med Parasit 91:951-956
129. Truc P, Formenty P, Duvallet G, Komoin-Oka C, Diallo PB, Lauginie F (1997b) Identification of trypanosomes isolated by KIVI from wild mammals in Côte d'Ivoire: diagnostic, taxonomic and epidemiological considerations. Acta Trop 67:187-196
130. Komba EK, Kibona SN, Ambwene AK, Stevens JR, Gibsons WC (1997) Genetic diversity among *Trypanosoma brucei rhodesiense* isolates from Tanzania. Parasitology 115:571-579

CHAPTER 3

Antigenic variation in African trypanosomes

E Pays

Changes of surface proteins during the life-cycle of the trypanosome

During their life-cycle African trypanosomes experience radical changes in the nature of their extracellular environment as they alternate between a variety of mammalian hosts and the *Glossina* fly vector. In order to survive and grow under these varying conditions, trypanosomes have developed strategies for altering the composition and organization of their plasma membrane. In particular, the bloodstream and procyclic (insect-specific) forms of *Trypanosoma brucei* express two different major surface glycoproteins, respectively the VSG (Variant Surface Glycoprotein) and procyclin (also termed PARP, for Procyclic Acidic Repetitive Protein) (Fig. 1). The VSG coat is acquired when trypanosomes undergo their final maturation in the fly salivary glands, during the metacyclic stage. This coat persists throughout the development in the bloodstream and is replaced by procyclin when the trypanosomes are ingested by the fly and differentiate into procyclic forms.

The VSG coat of the bloodstream form

The VSG consists of an N-terminal domain of 350 to 400 residues and a C-terminal domain of 50 to 100 residues, and is attached to the plasma membrane by a C-terminal glycosyl-phosphatidylinositol (GPI) anchor which contains two myristic acid residues. The VSG molecules associate as dimers, which adopt an extended configuration due to the folding of the N-terminal domain into two long antiparallel α-helices separated by a turn [1]. A comparison of several VSGs from different clones and isolates has shown that the entire N-terminal domain is extremely variable, while the C-terminal domain is more conserved, especially around cysteine residues which form disulphide bridges. Despite this sequence variability, the tertiary structure of different VSGs appears to be quite similar. All VSGs described so far are N-glycosylated near or within the C-terminal domain.

The main function of the VSG is to form a protective coat which is impermeable to macromolecules and covers the entire surface of the parasite. Each cell is surrounded by 10 million molecules of the same VSG. These molecules are so tightly packed that only a very limited stretch of their amino acid sequence is accessible to the extracellular environment

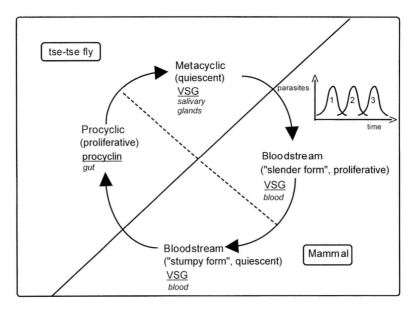

Fig. 1. The life-cycle of *Trypanosoma brucei*. The major developmental stages are named with indication of the major surface antigen. The insert *(top, right)* represents the parasitaemia waves developing in the bloodstream, which consist of successive trypanosome populations expressing different VSGs (1, 2, 3 etc)

and the C-terminal domain is completely buried. This organization not only prevents the immune recognition of conserved epitopes present in the C-terminal region of the VSG, but it also shields invariant proteins present on the surface of the cell [2]. In addition, it also protects the cell against lytic elements present in the serum, even from a non-immune host.

There is a specialized domain of the plasma membrane which is more accessible to extracellular macromolecules. This particular compartment is a membrane invagination at the base of the flagellum, termed the flagellar pocket. Such a small region appears to be the only site for endo- and exocytosis, presumably because it is not associated with the dense microtubule network underlying the rest of the surface membrane [3]. In the pursuance of this observation, the few surface receptors that have been characterized so far, such as the heterodimeric receptor for transferrin [4, 5], appear to be clustered in the flagellar pocket. Interestingly, at least some of these receptors resemble VSG molecules ([6]; Salmon and Pays, unpublished data). Moreover, recent results indicate that some VSG molecules that are released into the lumen of the flagellar pocket, and which are thus presumably more accessible to extracellular components, act as specific binding sites for cytokines of the host [7]. Therefore, the VSG plays several distinct roles on the parasite surface.

s a protective barrier and receptor for host com-
a highly immunogenic protein whose antigenic-
ed. Thus, the VSG is responsible for the process
ich enables the parasite to escape the immune
below).

Procyclin, the major surface protein of the procyclic form

When bloodstream forms are ingested by the fly, the VSG coat is rapidly lost and replaced by another major glycoprotein termed procyclin. Approximately 40% of the sequence of this protein consists of a Glu-Pro repeat which forms a rod-like structure, presumably allowing procyclin to transiently protrude through the layer of VSG during the process of differentiation [8]. As in the case of VSG, procyclin is also attached to the plasma membrane by a GPI anchor, but the structure of this glycolipid is different [9]. While the GPI of the VSG is cleavable by a GPI-specific phospholipase C (GPI-PLC), acylation of the inositol ring prevents this cleavage in the case of procyclin. Moreover, the GPI of procyclin is characterized by the presence of sialic acid.

Although the function of procyclin is unclear, it probably plays a role in the protection of the parasite against components from the fly such as proteases, present in the tsetse midgut. Trypanosomes depleted of procyclin develop poorly in the fly (Roditi, unpublished data).

Variation of the VSG: phenotypic aspects

The VSG is essential for the development of a long-lasting chronic infection by the parasite. The prolonged survival of the trypanosomes in their host is a consequence of both the high immunogenicity and the extraordinary variability of this antigen. The peptidic epitopes exposed at surface of the VSG coat are clearly recognized by the immune system, resulting in an efficient and rapid trypanolytic response. However, at any given time the trypanosome is able to change its VSG coat, which always allows some parasites to escape the antibody response directed against the previous VSG. Thus, the continuous interplay between the immune reaction of the host and the antigenic variation of the parasite shapes the pattern of parasitaemia in a succession of trypanosome populations which only differ in their VSG (see insert in Fig. 1). It seems as if the parasite makes use of the immune system to control its own growth, by exhibiting and then changing its visible surface antigen. This strategy leads to a persistent infection due to the presence of a relatively constant and tolerable number of parasites in the blood, rather than a rapid killing of the host which would occur in the case of an uncontrolled growth.

The antigenic variation potential of *T. brucei* seems to be unlimited. Results from experimental infections have shown that each parasite is able to sequentially express more than one hundred different VSGs. This variation does not depend on external stimuli such as antibodies, but occurs spontaneously, at a frequency which is strikingly different for laboratory-adapted and wild type strains, respectively about $10^{-7}/10^{-6}$ and 10^{-2} switches per cell and per generation [10, 11]. Interestingly, the VSGs are not produced totally at random, but occur with a loose hierarchy which prevents a rapid exhaustion of the total antigen repertoire. In a given trypanosome strain, some VSGs are frequently found early after the inoculation of parasites, whereas others appear somewhat later, probably the majority, and are only observed during the late development of chronic infection. This semi-programming is relatively conserved: in a naïve animal the early VSGs are generally the same, whatever the VSG expressed in the parasite inoculum.

Genetic mechanisms of antigenic variation

The genetic mechanisms involved in the switching of VSGs in *T. brucei* bloodstream forms have been largely characterized during the last 15 years. Several recent reviews should be consulted for detailed discussions on different aspects of this topic [12-16].

The VSG genes

To date several dozen VSG cDNAs have been largely characterized. While most of the sequence is extremely variable, the 3'-terminal region of the mRNAs is more conserved. This conservation is particularly important in the 3'-untranslated region (3'-UTR) and, to a lesser extent, in the sequence encoding the C-terminal domain of the protein. Such a conserved region is thus a signature for VSG genes and has been used as a tool to evaluate the number of these genes in the trypanosome genome. Based on the level of hybridization of probes from this region, it was estimated that the genomic DNA of *T. brucei* contains more than 1,000 VSG genes [17]. Therefore, a large fraction of the trypanosome genome (more than 2%) is devoted to coding for VSGs. While the majority of these genes are clustered in discrete locations in different chromosomes, many appear to be located in telomeres. This is particularly true for VSG genes located at the ends of minichromosomes of 50 to 150 kilobase pairs, which number in approximately 100 copies and appear to play the only role of reservoirs for VSG sequences [18]. The telomeric VSG genes are typically flanked by two extensive stretches of repeats, upstream imperfect 70 bp repeats and downstream telomeric repeats [19, 20]. These sequences are known as 'barren' regions, because they are devoid of most endonuclease restriction sites. Curiously, remnants of 70 bp repeats can be found upstream of some non-telomeric VSG genes, suggesting that this subset of VSG genes was originally translocated from a telomeric position. The variability in the number of these repeats between non-telomeric and telomeric VSG genes (typically 0 to a dozen and several dozen to hundreds, respectively) may influence the programming of VSG expression (see later).

The VSG gene expression sites

The non-telomeric VSG genes have never been found to be transcribed. In contrast, most, if not all, telomeric VSG genes of the large chromosomes, excepting minichromosomes, appear to be potentially transcribable *in situ*. These telomeric loci are usually termed as the VSG gene expression sites. A current estimate for the number of expression sites in trypanosome bloodstream forms is about 20 [21]. In addition, it is probable that a similar number of telomeric VSG genes, particularly from the largest chromosomes, is selectively transcribed in metacyclic forms [22, 23]. Up to now it is unclear to what extent, if any, the sets of bloodstream and metacyclic VSG gene expression sites overlap [22]. Therefore, it is possible that up to 40 telomeres from the large chromosome fraction, or all telomeres of this fraction, may contain VSG gene expression sites.

The bloodstream and metacyclic VSG gene expression sites differ in their length and general organization. While the former are polycistronic units of 45 to 60 kilobase pairs which contain about 10 genes [24-28], the latter are much shorter and appear to contain only the VSG gene [22, 23]. The genes contained in the polycistronic units of the bloodstream VSG gene expression sites were termed ESAGs, for Expression Site-Associated Genes [29]. Curiously, some of the metacyclic VSG expression sites are preceded by ESAGs, but these genes are transcribed independently [30, 31]. Similarly, in the bloodstream form genes related to some of the ESAGs are transcribed independently of the VSG expression sites [13]. The function of only some ESAGs is known (Fig. 2).

ESAG 10 is contained in a DNA fragment located between two identical copies of the transcription promoter, but this fragment is only present in half of the expression sites [32]. The protein encoded by this gene resembles a transmembrane transporter [33]. ESAG 7 and 6 encode the two subunits of a heterodimeric receptor for transferrin [4, 5]. ESAG 4 encodes a transmembrane adenylate cyclase with a receptor-like structure [27, 34, 35], while ESAG 3, 2 and 1 appear to encode minor surface proteins of unknown functions. Finally, the protein encoded by ESAG 8 is located in the nucleus and probably plays a regulatory role in gene expression [36-38]. As a rule, only a single telomere is transcribed at a time, leading to the synthesis of a single type of VSG, and consequently to a uniform surface coat. The significance of this observation is unclear, since it has been found experimentally that trypanosomes expressing two VSGs simultaneously are perfectly viable [39]. It is probable that the mechanisms ensuring the selective activity of a single expression site from the collection of 20-40 potentially transcribable telomeres are related to those allowing the alternative activation of different expression sites during chronic infection.

Genetic processes for VSG switching

Two basically different processes can lead to antigenic variation during the parasite development in the bloodstream. The VSG switches arise either by the alternative use of different VSG gene expression sites or by DNA recombination events which change the VSG gene present in the active expression site. These mechanisms are summarized in Figure 3.

The first process is often termed "*in situ* activation". It is manifested by the inactivation of the former expression site and the simultaneous activation of a new one from the repertoire. The only salient feature concerning this process is that it does not necessarily require DNA rearrangements in and around the promoter region. Occasionally, large deletions in this area may occur together with VSG switching, but these events are not always observed [32]. A current speculation relates the *in situ* inactivation of expression sites with the presence of a modified base, β-D-glucosyl-hydroxymethyluracil, in the non-transcribed telomeres of the bloodstream form only [40, 41]. Whether this modification is the

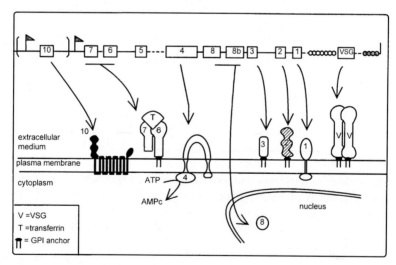

Fig. 2. Structure and function of an expression site for bloodstream VSG genes. On top, the boxes represent the different genes contained in this polycistronic transcription unit, with the expression site-associated genes (ESAGs) numbered from 1 to 10. The flags indicate the transcription promoters. The open and dark dots represent, respectively, the arrays of 70 bp repeats and telomeric repeats, and the terminal vertical bar represents the chromosome end. A schematic subcellular localization and function is proposed for the proteins encoded by these genes

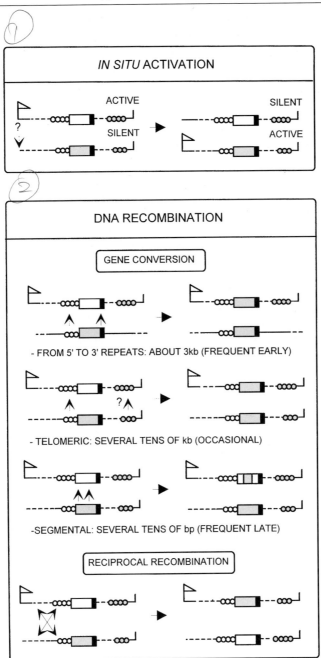

Fig. 3. The genetic mechanisms for antigenic variation in the bloodstream form. In each diagram, the symbols are the same as in Fig. 2, except that the boxes represent the VSG genes involved in antigenic switching, each one with a black bar depicting the conserved 3'-terminal sequence (see text for details)

cause or the consequence of the transcriptional silencing is up to now unclear. It has been proposed that when modified, the telomeres may be unable to interact with elements of the nuclear matrix, preventing them from having access to the transcription machinery. Specifically, the association of a telomeric expression site with the nucleolus could be necessary to trigger its transcription since the RNA polymerase active on the VSG expression site resembles the ribosomal pol I, which is present in the nucleolus [42]. More generally, the reversible (in)activation of the telomeric expression sites can be compared with the instable control of transcription operating in the direct vicinity of telomeres in yeast, the so-called position effect [12, 43]. This control reflects the variable spreading into non-telomeric DNA, of a locked chromatin structure mediated by the interaction of several proteins, called SIR for Silent Information Regulator, with the N-terminal domain of histones bound to telomeric DNA [44]. However, in the trypanosome system this type of control would have to account for the (in)activation of far distant promoters, since the VSG promoter is located at least 45 kilobases upstream from the telomeric repeats. Moreover, it seems difficult to reconcile this hypothesis with the high selectivity of the expression site control. Therefore, why a single expression site always escapes silencing and how this site is changed from time to time remains a mystery.

The case of the metacyclic VSGs deserves a comment. This subset of VSG genes, which is selectively transcribed during the final maturation of trypanosomes in the fly salivary glands, seems to be exclusively activated *in situ*. In metacyclic as in bloodstream forms, a single VSG expression site is active in any given cell. However, contrary to what occurs in bloodstream forms at this stage, all expression sites seem to be triggered simultaneously in the different cells so that the metacyclic population expresses the whole repertoire of metacyclic VSGs at the same time [45]. It is unknown if the mechanisms ensuring the selective activation of the metacyclic VSG gene expression sites are of the same type as those operating in the bloodstream. The simultaneous activation of the different metacyclic expression sites in the fly salivary glands, as well as their rapid inactivation in the bloodstream, strongly suggest the involvement of specific factors from the fly. However, a metacyclic expression site was found to be active in a bloodstream variant [22]. Whether this activation is accidental or forced by the selection is undetermined.

The DNA recombination events involved in VSG switching are of two types: gene conversion and reciprocal recombination (Fig. 3). Both processes are dependent on the recognition of sequence homology between two partners, the VSG gene expression site, on one hand, and another VSG gene-containing DNA fragment, on the other hand. They reflect the high potential of trypanosomatids for homologous recombination, which was revealed by the development of methods for transformation of the trypanosome genome with exogenous DNA [46-48]. However, the two recombinational processes differ fundamentally in their symmetry.

Gene conversion consists in the replacement of a DNA sequence by the copy of another. This process is asymmetric with a target -the VSG gene expression site- and a donor -one of the VSG genes contained in the silent repertoire. It is probable that the chromatin structure of the partner sequences dictates the orientation of the recombination, defining the respective roles of target and donor. The chromatin of the VSG expression site is the only one open for transcription, and which renders the DNA of this locus more susceptible to cleavage by nucleases [49], hence its designation as target. If this DNA is cut upstream from the VSG gene and downstream from the transcription promoter, the free double-stranded DNA end generated by the cleavage will tend to invade other sequences sharing homology, namely the upstream environment of silent VSG genes. Base pairing between the invading and the recipient strands triggers the synthesis of a copy of the invaded VSG gene. Therefore, a copy of another VSG gene is inserted downstream from the transcription promoter of the VSG expression site, while the former VSG gene present in this site is degraded and lost. The resolution of the conversion in the 3' region of the VSG genes involves recombination in a downstream region of homology (for details about this model, see [16, 50]).

Since this process is dependent on the recognition of sequence homology between the invading DNA and another DNA fragment, the location of sequences shared between the two partners will determine the end-limits of gene conversion, in other words, the length of the copy and replacement (often termed duplication-transposition). As illustrated in Figure 3, the location of these blocks of homology may vary considerably between different VSG genes, so that the extent of gene conversion is extremely variable. It is frequently about 3 kilobases long which separates the two blocks of homology commonly between VSG genes: the upstream 70 bp repeats and the downstream 3'-end region of the gene which includes the 3'-untranslated region (3'-UTR) of the mRNA. However, much longer and much smaller conversion events, from several tens of kilobase pairs to several tens of base pairs, can also be observed. In the latter case the process of copy/replacement occurs entirely within the gene coding region and hybrid VSG genes are constructed by the assembly of the residual fragments of the target and the copied sequences from the donor (see segmental gene conversion in Figure 3). Such type of chimaeric genes are selected, provided they encode VSGs with a new antigenic specificity. Therefore, gene conversion is a mechanism allowing for the generation of new VSG genes by sequence reassortment, which considerably extends the potential of the parasite for antigenic variability [51, 52].

Reciprocal recombination differs from gene conversion by its symmetry. As shown in Figure 3, this process allows telomeric VSG genes to be exchanged with the gene present in the active VSG expression site. This process does not involve either gene copy or gene replacement. It was believed to be a rare event because only a few cases were reported for a long time [53]. However, a detailed analysis of several VSG switches has recently led to the conclusion that this mechanism may contribute significantly to antigenic variation of *T. brucei* [54].

Programming of VSG expression

When analyzed in a well-defined trypanosome strain, the pattern of VSG appearance during a chronic infection is semi-predictable. Some VSGs are systematically expressed early, others can be observed later and many only appear very late in the infection. If a clone expressing a late variant is injected in a naïve animal, the early VSGs are again observed first; this clearly indicates that the different VSGs are not expressed randomly but according to a loose programming. Experimental evidence supports the view that this phenomenon does not primarily depend on the VSG sequence, but rather on its genomic environment. A VSG gene expressed late in the infection is then expressed early following a change of this environment [55, 56]. The characteristics of the genetic processes underlying VSG switching can explain these observations (Fig. 3).

At the beginning of the infection, the different VSG expression sites are alternatively activated *in situ*, until a given site is preferred. The switching of expression sites is obviously a simple way to trigger variation of the VSG, because the expression sites carry different VSG genes and are potentially functional as such, so that their alternative activation leads to the synthesis of different VSGs without requiring DNA rearrangements. It would thus appear that at least several expression sites have similar probabilities of being activated and are stochastically selected. This process may actually be driven by a selective pressure, since it could reflect a scanning for the best adapted battery of ESAGs, in particular those encoding the transferrin receptor (see section "Antigenic variation and growth control"). Therefore, the VSG genes present in potential expression sites frequently encode the earliest expressed VSGs.

The VSGs encoded by telomeric genes which are not contained in expression sites are also early sinthesized. These genes need to recombine with the active expression site to be transcribed, but the probability of this recombination is high because there is an extensive sequence homology between their genomic environment and the expression site. Both are telomeric sequences and share large arrays of upstream 70 bp repeats and downstream telomeric repeats. Presumably due to the presence of G-rich motifs, which form G-quartets, the latter may force chromosome ends to pair [57]. Indirect evidence suggests that these interactions favour the recombinations between telomeric VSG genes and the expression site [56].

Non-telomeric VSG genes are not flanked by extensive regions of homology with the expression site, so that their recombination with the expression site is less probable. Therefore, these genes are necessarily expressed later. Many of the non-telomeric VSG genes are preceded by some 70 bp repeats which allow them to recombine with the expression site. However, the number of these repeats is variable between genes, and directly influences the probability of gene activation. VSG genes preceded by only a few 70 bp repeats are expressed late in the infection [51, 55]. When the repeats are absent, the recombination of the VSG genes with the expression site cannot occur with the usual recognition sequences, and must rely on the fortuitous presence of unusual homologies in this site. These improbable events can only be selected late, when the expression of easily activable genes is prevented by the presence of antibodies against the early variants. A typical event of this kind is the recognition of sequence similarities existing within the coding region of the two partner VSG genes. This leads to intragenic recombination (segmental gene conversion) which creates new VSGs by sequence reassortment (see Fig. 3). Therefore, it is frequently observed that VSG genes expressed late during chronic infection are mosaïc sequences constructed from several donors, in particular VSG pseudogenes [58, 59]. The generation of new VSG sequences is a powerful way to expand the VSG gene repertoire, enabling the parasite to survive for long periods in the bloodstream [51, 52].

Mechanisms for VSG diversification

The VSG repertoires are subject to an extraordinarily rapid evolution. It is exceptional to find common VSGs (termed isotypes) between different trypanosome strains, even if these strains originate from the same region. Accordingly, the patterns of VSG-specific DNA sequences are highly variable between parasite isolates, with the notable exception of *T. b. gambiense* [60-62]. The existence of two basically different mechanisms for switching VSG genes may partly explain this observation. Indeed, the alternative use of these mechanisms allows the parasite to modify rapidly its VSG repertoire by a continuous creation, storage and deletion of different VSG genes.

In the active expression site, the VSG gene is prone to segmental gene conversion which may involve several VSG sequences as donors (see above). Consequently, the expression site may be considered as a workshop for the construction of novel VSG sequences by DNA rearrangement.

In addition, point mutations can be generated during the process of gene conversion [63]. Although this observation is still controversial [16, 64], it may be tentatively ascribed to the presence of the modified base β-D-glucosyl-hydroxymethyluracil in the DNA of silent telomeres [40]. It is tempting to speculate that when silent telomeric VSG genes, modified VSG genes, are used as templates for gene conversion, flaws are generated during the duplication of the donor. This suggestion is supported by the finding that, so far, point mutations have only been found in conversion events involving telomeric VSG genes as donors. Whatever the reason, the generation of point mutations also contributes to the hyperevolution of VSG sequences, especially since these mutations can alter the antigenic specificity of the VSGs [65].

There is a way for newly created VSG genes to be stored in the genomic repertoire. When gene conversion and *in situ* activation are successively used as mechanisms for antigenic variation during chronic infection, the change of the expression site allows the conservation, as a silent gene, of the VSG gene previously transcribed. Since this gene was the result of an earlier conversion event, its sequence may be original, due to either segmental conversion or point mutation, or both. In such a case the gene would encode a new VSG, expanding the antigen repertoire of the parasite. Transcriptional silencing would prevent this original gene from being rapidly lost by gene conversion since the structure of repressed chromatin inhibits the endonucleolytic cleavage required to trigger the recombination process[49,53].

While the succession of gene conversion followed by *in situ* activation allows for the acquisition of new VSG sequences, the reverse can lead to the loss of VSG genes. If a unique VSG gene present in a potential expression site is activated *in situ*, any gene conversion targeted to this site will inevitably lead to its replacement, and thus to the disappearance of this antigen from the repertoire. Therefore, the successive use of the different mechanisms for VSG gene activation continuously alters the antigen repertoire of the parasite by gain and loss of different genes.

It should be added that another phenomenon, involving genetic exchange between parasites, could further contribute to the variability of the VSG repertoires. It is clear that trypanosomes can sometimes exchange their genetic information during their development in the fly [66, 67]. This possibility allows for occasional transfer of VSG genes between trypanosome strains, permitting massive rearrangements of VSG repertoires.

Antigenic variation and growth control

Growth properties of trypanosomes differ considerably between variant clones of the same strain [68]. It is probable that this difference in growth reflects more the expression of different sets of ESAGs associated with the alternative activation of different VSG gene expression sites, rather than the nature of the VSG itself. Indeed, some ESAGs encode proteins which strongly influence the growth of trypanosome bloodstream forms. These include the transferrin receptor (ESAG 7 and 6) and a receptor-like adenylate cyclase (ESAG 4). Transferrin carries iron which is an essential requirement for the parasite [69] and adenylate cyclase synthesizes cAMP, which plays a crucial role in cell proliferation and/or differentiation [15]. The genes for these proteins are very similar but unidentical between different expression sites. These slight differences may be very important, as illustrated by the case of the transferrin receptor. The receptors encoded by different expression sites show large differences in affinity for transferrin from a given host [31]. Significantly, only a single amino acid difference in the most exposed surface loop of pESAG 6 can account for a more than 20-fold difference in the affinity of the receptor for the ligand (Salmon et al, unpublished data). Furthermore, the affinity of the same receptor may differ greatly for transferrins from different mammalian species [70]. Therefore, when antigenic variation occurs by *in situ* activation, not only the VSG is changed but also several other proteins which affect the growth parameters of the cell. The requirement for several host-specific ESAGs may actually be the basic reason for the existence of multiple expression sites. For example, depending on the species-specificity of transferrin, trypanosomes appear to switch expression sites until expression of the best adapted transferrin receptor is achieved [70]. In addition to being a mechanism involved in antigenic variation, *in situ* activation may also be a way for the parasite to select a set of receptors which is the most appropriate for the mammalian host.

Similarly, the relative resistance or susceptibility of trypanosomes to lytic factors present in human serum, at least in the *T. b. rhodesiense* subspecies, may also be related to *in situ* activation. Indeed, *T. b. rhodesiense* seems to acquire resistance to human serum only after switching to some VSGs [71]. It is tempting to speculate that this process is linked to the activation of VSG expression sites which contain a gene encoding resistance to human serum. According to this hypothesis, the latter gene would be an ESAG present only in some VSG expression sites. This conclusion appears to be supported by preliminary evidence (Chimfwembe and Hamers, unpublished observations).

Mechanisms of antigen switching during cellular differentiation

Surface antigens not only vary during the development of trypanosomes in the bloodstream, but a major rearrangement of the cell surface also occurs during the differentiation from the bloodstream to the insect-specific, procyclic form. Among other changes, procyclin quickly replaces the VSG as the predominant protein of the cellular surface. Since these two proteins and their mRNAs are by far the most abundant during their respective developmental stages, they constitute excellent markers for the study of the genetic mechanisms involved in antigen switching during cellular differentiation.

Transcriptional regulation

A remarkable characteristic of the genome organization of kinetoplastids is that genes are devoid of introns and are generally tightly packed into long polygenic transcription units [72]. According to the nature of the RNA polymerase used for their transcription, the protein-coding genes can be divided in two categories. While most of the genes are transcribed by the α-amanitin-sensitive pol II, the units containing the genes for the two major surface antigens, VSG and procyclin, are transcribed by the α-amanitin-resistant pol I [42]. It is possible that this property is related to the need for very high levels of transcripts in the latter cases, since pol I is more efficient than pol II.

So far, transcription promoters have only been characterized in the VSG and procyclin gene units. The available evidence points to the ribosomal nature of these promoters [73-78]. Promoters able to recruit pol II have yet to be described, although weakly active candidates have been proposed [79, 80].

While no evidence has been found for developmental control of any gene transcription by pol II, control of the pol I-mediated transcription of the VSG and procyclin units has been amply documented [81-83]. Strikingly opposite controls ensure the stage-specific accumulation of the major mRNAs of the parasite, those of the VSG in the bloodstream form and of procyclin in the procyclic form. In the case of the VSG, an additional regulation is also involved which limits transcription to a single expression site in the bloodstream form. Interestingly, it would appear that these controls do not primarily operate at the level of transcription initiation. Two types of observations support this conclusion:

a) detailed *in vivo* and *in vitro* studies on the RNA synthesis occurring immediately downstream from the promoters in the VSG and procyclin units have concluded that this region is transcribed even at the developmental stage where these units are not functionally expressed [84-86].

b) in both transient and permanent assays of a reporter function, the VSG and procyclin promoters can exhibit a comparable activity at both stages of the parasite life-cycle [74, 86-89].

These observations indicate that all the necessary components required to trigger recruitment of the RNA polymerase and initiate transcription on these promoters are present during both bloodstream and procyclic stages. While this is consistent with the ribosomal nature of these promoters, it should be pointed out that full promoter activity *in situ* is clearly dependent on the chromosomal context. 1-In the bloodstream form the activity of the ribosomal promoter is reduced when inserted in the procyclin locus, while the procyclin promoter is down-regulated when integrated at either the rRNA non-transcribed spacer or the procyclin locus [89]; 2-in both bloodstream and procyclic forms the VSG promoter is inactive when integrated within the tubulin transcription unit [87], while, in the same location, the procyclin promoter is inactive in the bloodstream form and active in the procyclic form [90]; 3-the VSG promoters are down-regulated in the VSG expression sites of the procyclic form, since their activity can be stimulated when they are inserted in the non-transcribed spacer of the rRNA locus [86]; 4-an upstream element of the procyclin promoter can stimulate the activity of the VSG promoter when placed in its vicinity, even in opposite orientation [91, 92]; 5-finally, the VSG promoters of the non-active expression sites of the bloodstream form seem to be down-regulated by telomeric silencing [43]. Therefore, the activity of the few promoters known to date in trypanosomes seems to be essentially constitutive, although a modulation is possible due to cis-acting and largely non-specific chromatin elements.

If regulation of promoter activity is not primarily responsible for the stage-specific transcription control of the VSG and procyclin units, where is this control effected? Analysis of transcription in isolated nuclei has shown that elongation of the RNA chains is subject to a stage-specific control operating in a strikingly inverse fashion in the VSG and procyclin units [93]. Abortive transcription thus occurs in the VSG unit in the procyclic form as well as in the procyclin units in the bloodstream form. The stimuli for cellular differentiation from the bloodstream to the procyclic form (in particular a cold shock and the presence of cis-aconitate in the medium) appear to change these RNA elongation controls very rapidly. Once again, such observations point to cis-acting controls of transcription, probably mediated by chromatin components. The opposite symmetry in these controls suggest a cross-talk between the two transcription units, but its nature remains obscure at present.

Post-transcriptional regulation

The maturation of the polygenic transcripts into individual mRNAs is achieved by cleavage in the intergenic region of the primary transcripts followed by processing of the RNA extremities: trans-splicing at the 5'-end and polyadenylation at the 3'-end. These processing events are coupled. Both seem to depend on the recognition of polypyrimidine tracts present in the intergenic regions [83, 94]. Accumulating evidence indicates that the nature of the intergenic RNA sequences (untranslated regions or UTRs), in particular the 3'-UTR, determines the final amount of the mature mRNAs in a stage-specific manner. This phenomenon seems to be primarily responsible for the differential expression of genes as a function of the developmental stage of the parasite [82, 83]. This particularly applies to the VSG and procyclin 3'-UTRs, which give rise to a high level of transcripts from reporter genes in the bloodstream and in the procyclic form, respectively [90, 95, 96]. The 3'-UTR of the VSG mRNA was found to cause two effects: it up-regulates the mRNA level in the bloodstream form by increasing the RNA stability and it down-regulates the mRNA level in the procyclic form by slowing down the rate of mRNA maturation [90]. The mechanisms by which the 3'-UTRs differentially influence the RNA amounts depending on the stage in the trypanosome development are presently unknown. Indirect evidence suggests that this effect is mediated by short-lived proteins [97-99].

In the cases of VSG and procyclin, where gene expression under steady-state conditions is already regulated at the primary transcription level, these post-transcriptional mechanisms probably play a role only during the transition between successive developmental stages. They presumably ensure a very fast change of mRNA for the major surface antigen when cellular differentiation is triggered. Indeed, when bloodstream forms are induced to transform into procyclics, the amount of VSG mRNA switches from 10% of the polyadenylated transcripts to almost nothing within 4 hours, whereas at the same time the exact opposite occurs for the procyclin mRNA [100]. Presumably, post-transcriptional controls acting on the stability and processing of the transcripts allow for accelerated disappearance of the VSG mRNA when the procyclic mRNA must rapidly accumulate.

Relevance for the adaptation of the parasite

The organization and expression of the trypanosome genome differs strikingly from that of higher eukaryotes. The genes are generally contained in long polycistronic units which appear to be continuously transcribed and cellular differentiation is dependent on stage-regulated post-transcriptional controls which determine the selective accumulation of some mRNAs, while the others are degraded. A priori, this system may seem less flexible and more wasteful than what generally occurs in other eukaryotes, namely controls of transcription initiation in monocistronic units. However, a remarkable feature of this system is the speed and the precision at which the concerted changes linked to cellular differentiation occur, as revealed in particular during the experimentally-induced switching from the bloodstream to the procyclic form *in vitro* (see [15] for a timing of these events). Therefore, it would appear that modulating gene expression at the level of RNA processing is at least as efficient and accurate as controlling transcription promoters. It may be hypothesized that the constitutive transcription of most genes is a system allowing for a continuous supply of all primary sequences, whatever the developmental stage of the parasite. This would provide an immediate source of different sets of transcripts in case of a sudden change of the environmental conditions, such as the abrupt transfer from the blood into the fly or vice versa. Of course, this system is not the most economical, but for parasites the supply of energy is probably not a major problem.

Conclusions

The cellular surface of African trypanosomes is continuously remodelled. During the bloodstream stage, the plasma membrane is completely surrounded by a thick barrier consisting of VSG. This antigen has two major functions: it elicits an efficient antibody response which downregulates the number of parasites to allow a prolonged survival of the host and, through a constant variation of the exposed epitopes, it also continuously allows some parasites to escape this antibody response which leads to new waves of infection. The antigenic variation is driven by two basically distinct processes, the alternative transcriptional activation of different VSG gene expression sites and frequent DNA rearrangements due to spontaneous homologous recombination in the active telomeric VSG expression site. The first mechanism probably also allows the parasite to express surface receptors that are the best adapted to the host, while the second also provides the trypanosome with a means to generate new VSG genes by sequence reassortment. The succession of these mechanisms during chronic infection can explain some of the puzzling features of this system, such as the relative programming of VSG expression and the hyperevolution of the VSG repertoires. During the transition between the bloodstream form and the insect-specific procyclic form, the VSG is very rapidly replaced by another major surface glycoprotein termed procyclin. This change is only one example of the profound physiological and morphological modifications linked to the cellular differentiation. Strikingly, the controls responsible for this major reprogramming of gene expression operate not primarily at the level of the transcription promoters, but rather during RNA elongation and processing of the transcripts. This originality is probably linked to the organization of the trypanosome genome in long polycistronic transcription units.

Many important questions remain unsolved in this field. What is the basis for the selective activation of a single VSG expression site at a given time? How is the switch from one expression site to another effected? What are the enzymes involved in the processes of homologous recombination and are these enzymes regulated during the parasite development? Do pol II transcription promoters exist in these organisms? How is differential RNA elongation controlled in the VSG and procyclin transcription units? What are the mechanisms and components involved in the stage-specific control of the mRNA abundance mediated by the 3'-UTRs of mRNAs? There is no doubt that in the future, as in the past, trypanosomes will continue to surprise us.

Acknowledgements

I thank D Nolan for useful comments on the manuscript and M Berberof for helping in the realisation of Figures. The work in my laboratory was supported by the Belgian FRSM and FRC-IM, a research contract with the Communauté Française de Belgique (ARC) and by the Agreement for Collaborative Research between ILRAD (Nairobi) and Belgian Research Centres.

References

1. Blum ML, Down JA, Gurnett AM, Carrington M, Turner MJ, Wiley D (1993) A structural motif in the variant surface glycoproteins of *Trypanosoma brucei*. Nature 362:603-609
2. Ziegelbauer K, Overath P (1993) Organization of two invariant surface glycoproteins in the surface coat of *Trypanosoma brucei*. Infect Immun 61:4540-4545
3. Webster P, Russell DG (1993) The flagellar pocket of trypanosomatids. Parasitol Today 9:201-206
4. Salmon D, Geuskens M, Hanocq F, Hanocq-Quertier J, Nolan D, Ruben L, Pays E (1994) A novel heterodimeric transferrin receptor encoded by a pair of VSG expression site-associated genes in *Trypanosoma brucei*. Cell 78:75-86
5. Steverding D, Stierhof YD, Fuchs H, Tauber R, Overath P (1995) Transferrin-binding complex is the receptor for transferrin uptake in *Trypanosoma brucei*. J Cell Biol 131:1173-1182
6. Carrington M, Boothroyd J (1996) Implications of conserved structural motifs in disparate trypanosome surface proteins. Mol Biochem Parasitol 81:119-126
7. Magez S, Geuskens M, Del Favero H, Beschin A, Radwanska M, Lucas R, Pays E, De Baetselier P (1997) Specific uptake of tumor necrosis factor is involved in growth control of *Trypanosoma brucei*. J Cell Biol 137:715-727
8. Roditi I, Pearson TW (1990) The procyclin coat of African trypanosomes. Parasitol Today 6:79-82
9. McConville MJ (1996) Glycosyl-phosphatidylinositols and the surface architecture of parasitic protozoa. In: Smith DF, Parsons M (eds) Molecular biology of parasitic protozoa. IRL Press, Oxford, pp 205-228
10. Turner CMR, Barry JD (1989) High frequency of antigenic variation in *Trypanosoma brucei rhodesiense* infections. Parasitology 99:67-75
11. Cross GAM (1990) Cellular and genetic aspects of antigenic variation in trypanosomes. Annu Rev Immunol 8:83-110
12. Borst P, Gommers-Ampt JH, Ligtenberg MJ, Rudenko G, Kieft R, Taylor M, Blundell P, Van Leeuwen F (1993) Control of antigenic variation in African trypanosomes. Cold Spring Harbor Symp Quant Biol 58:105-114
13. Pays E, Vanhamme L, Berberof M (1994) Genetic controls for the expression of surface proteins in African trypanosomes. Annu Rev Microbiol 48:25-52
14. Cross GAM (1996) Antigenic variation in trypanosomes: secrets surface slowly. BioEssays 18:283-291
15. Pays E, Rolin S, Magez S (1996) Cell signalling in trypanosomatids. In: Hide G, Mottram JC, Coombs GH, Holmes PH (eds) Trypanosomiasis and Leishmaniasis: biology and control. British Society for Parasitology/CAB International, Oxford, pp 199-225

16. Barry JD (1997) The relative significance of mechanisms of antigenic variation in African trypanosomes. Parasitol Today 13:212-218
17. Van der Ploeg LHT, Valerio D, De Lange T, Bernards A, Borst P, Grosveld FG (1982) An analysis of cosmid clones of nuclear DNA from *Trypanosoma brucei* shows that the genes for variant surface glycoproteins are clustered in the genome. Nucl Acids Res 10:5905-5923
18. Weiden M, Osheim YN, Beyer AL, Van der Ploeg LHT (1991) Chromosome structure: DNA nucleotide sequence elements of a subset of the minichromosomes of the protozoan *Trypanosoma brucei*. Mol Cell Biol 11:3823-3834
19. Liu AYC, Van der Ploeg LHT, Rijsewijk FAM, Borst P (1983) The transcription unit of variant surface glycoprotein gene 118 of *Trypanosoma brucei*. Presence of repeated elements at its border and absence of promoter associated sequences. J Mol Biol 167:57-75
20. Aline RFJ, MacDonald G, Brown E, Allison J, Myler P, Rothwell V, Stuart K (1985) (TAA) n within sequences flanking several intrachromosomal variant surface glycoprotein genes in *Trypanosoma brucei*. Nucl Acids Res 13:3161-3177
21. Navarro M, Cross GAM (1996) DNA rearrangements associated with multiple consecutive directed antigenic switches in *Trypanosoma brucei*. Mol Cell Biol 16:3615-3625
22. Alarcon CM, Son HJ, Hall T, Donelson JE (1994) A monocistronic transcript for a trypanosome variant surface glycoprotein. Mol Cell Biol 14:5579-5591
23. Graham SV, Barry JD (1995) Transcriptional regulation of metacyclic variant surface glycoprotein gene expression during the life cycle of *Trypanosoma brucei*. Mol Cell Biol 15:5945-5956
24. Kooter JM, Van der Spek HJ, Wagter R, d'Oliveira CE, Van der Hoeven F, Johnson P, Borst P (1987) The anatomy and transcription of a telomeric expression site for variant-specific surface antigens in *Trypanosoma brucei*. Cell 51:261-272
25. Johnson PJ, Kooter JM, Borst P (1987) Inactivation of transcription by UV irradiation of *T. brucei* provides evidence for a multicistronic transcription unit including a VSG gene. Cell 51:273-281
26. Alexandre S, Guyaux M, Murphy NB, Coquelet H, Steinert M, Pays E (1988) Putative genes of a variant-specific antigen gene transcription unit in *Trypanosoma brucei*. Mol Cell Biol 8:2367-2378
27. Pays E, Tebabi P, Pays A, Coquelet H, Revelard P, Salmon D, Steinert M (1989b) The genes and transcripts of an antigen gene expression site from *T. brucei*. Cell 57:835-845
28. Lips S, Revelard P, Pays E (1993) Identification of a new expression site-associated gene in the complete 30.5 kb sequence from the AnTat 1.3A variant surface protein gene expression site of *Trypanosoma brucei*. Mol Biochem Parasitol 62:135-138
29. Cully DF, Ip HS, Cross GAM (1985) Coordinate transcription of variant surface glycoprotein genes and an expression site associated gene family in *Trypanosoma brucei*. Cell 42:173-182
30. Graham SV, Matthews KR, Barry JD (1993) *Trypanosoma brucei*: unusual expression-site-associated gene homologies in a metacyclic VSG gene expression site. Exp Parasitol 76:96-99
31. Steverding D, Overath P (1996) *Trypanosoma brucei* with an active metacyclic variant surface gene expression site expresses a transferrin receptor derived from ESAG6 and ESAG7. Mol Biochem Parasitol 78:285-288
32. Gottesdiener KM, Goriparthi L, Masucci JP, Van der Ploeg LHT (1992) A proposed mechanism for promoter-associated DNA rearrangement events at a variant surface glycoprotein gene expression site. Mol Cell Biol 12:4784-4795
33. Gottesdiener KM (1994) A new VSG expression site-associated gene (ESAG) in the promoter region of *Trypanosoma brucei* encodes a protein with 10 potential transmembrane domains. Mol Biochem Parasitol 63:143-151
34. Ross DT, Raibaud A, Florent IC, Sather S, Gross MK, Storm DR, Eisen H (1991) The trypanosome VSG expression site encodes adenylate cyclase and a leucine-rich putative regulatory gene. EMBO J 10:2047-2053
35. Paindavoine P, Rolin S, Van Assel S, Geuskens M, Jauniaux JC, Dinsart C, Huet G, Pays E (1992) A gene from the VSG expression site encodes one of several transmembrane adenylate cyclases located on the flagellum of *Trypanosoma brucei*. Mol Cell Biol 12:1218-1225
36. Smiley BL, Stadnyk AW, Myler PJ, Stuart K (1990) The trypanosome leucine repeat gene in the variant surface glycoprotein expression site encodes a putative metal-binding domain and a region resembling protein-binding domains of yeast, *Drosophila*, and mammalian proteins. Mol Cell Biol 10:6436-6444
37. Revelard P, Lips S, Pays E (1990) A gene from the VSG gene expression site of *Trypanosoma brucei* encodes a protein with both leucine-rich repeats and a putative zinc finger. Nucl Acids Res 18:7299-7303
38. Lips S, Geuskens M, Paturiaux-Hanocq F, Hanocq-Quertier J, Pays E (1996) The ESAG 8 gene of *Trypanosoma brucei* encodes a nuclear protein. Mol Biochem Parasitol 79:113-117
39. Munoz-Jordan JL, Davies KP, Cross GAM (1996) Stable expression of mosaic coats of variant surface glycoproteins in *Trypanosoma brucei*. Science 272:1795-1797
40. Gommers-Ampt JH, Van Leeuwen F, de Beer ALJ, Vliegenthart JFG, Dizdaroglu M, Kowalak JA, Crain PF, Borst P (1993) b-D-glucosyl-hydroxymethyluracil: a novel modified base present in the DNA of the parasitic protozoan *Trypanosoma brucei*. Cell 75:1129-1136
41. Van Leeuwen F, Wijsman ER, Kuyl-Yeheskiely E, Van der Marel GA, Van Boom JH, Borst P (1996) The telomeric GGGTTA repeats of *Trypanosoma brucei* contain the hypermodified base J in both strands. Nucl Acids Res 24:2476-2482

42. Chung HM, Lee MGS, Van der Ploeg LHT (1992) RNA polymerase I-mediated protein-coding gene expression in *Trypanosoma brucei*. Parasitol Today 8:414-418
43. Horn D, Cross GAM (1995) A developmentally regulated position effect at a telomeric locus in *Trypanosoma brucei*. Cell 83:555-561
44. Gottschling DE, Aparicio OM, Billington BL, Zakian VA (1990) Position effect at *Saccharomyces cerevisiae* telomeres: reversible repression of pol II transcription. Cell 63:751-762
45. Le Ray D, Barry JD, Vickerman K (1978) Antigenic heterogeneity of metacyclic forms of *Trypanosoma brucei*. Nature 273:300-302
46. ten Asbroek ALMA, Ouellette M, Borst P (1990) Targeted insertion of the neomycin phosphotransferase gene into the tubulin gene cluster of *Trypanosoma brucei*. Nature 348:174-175
47. Lee MGS, Van der Ploeg LHT (1990) Homologous recombination and stable transfection in the parasitic protozoan *Trypanosoma brucei*. Science 250:1583-1587
48. Eid J, Sollner-Webb B (1991) Stable integrative transformation of *Trypanosoma brucei* that occurs exclusively by homologous recombination. Proc Natl Acad Sci USA 88:2118-2121
49. Pays E, Lheureux M, Steinert M (1981) The expression-linked copy of the surface antigen gene in *Trypanosoma* is probably the one transcribed. Nature 292:265-267
50. Pays E (1985) Gene conversion in trypanosome antigenic variation. Progr Nucleic Acid Res Mol Biol 32:1-26
51. Pays E (1989) Pseudogenes, chimaeric genes and the timing of antigen variation in African trypanosomes. Trends Genet 5:389-391
52. Barbet AF, Kamper SM (1993) The importance of mosaïc genes to trypanosome survival. Parasitol Today 9:63-66
53. Pays E, Guyaux M, Aerts D, Van Meirvenne N, Steinert M (1985) Telomeric reciprocal recombination as a possible mechanism for antigenic variation in trypanosomes. Nature 316:562-564
54. Rudenko G, McCulloch R, Dirks-Mulder A, Borst P (1996) Telomere exchange can be an important mechanism of variant surface glycoprotein gene switching in *Trypanosoma brucei*. Mol Biochem Parasitol 80:65-75
55. Laurent M, Pays E, Van der Werf A, Aerts D, Magnus E, Van Meirvenne N, Steinert M (1984) Translocation alters the activation rate of a trypanosome surface antigen gene. Nucl Acids Res 12:8319-8328
56. Van der Werf A, Van Assel S, Aerts D, Steinert M, Pays E (1990) Telomere interactions may condition the programming of antigen expression in *Trypanosoma brucei*. EMBO J 9:1035-1040
57. Fang G, Cech TR (1993) The b subunit of *Oxytricha* telomere-binding protein promotes G-quartet formation by telomeric DNA. Cell 44:875-885
58. Thon G, Baltz T, Eisen H (1989) Antigenic diversity by the recombination of pseudogenes. Genes Dev 3:1247-1254
59. Thon G, Baltz T, Giroud C, Eisen H (1990) Trypanosome variable surface glycoproteins: composite genes and order of expression. Genes Dev 9:1374-1383
60. Paindavoine P, Pays E, Laurent M, Geltmeyer Y, Le Ray D, Mehlitz D, Steinert M (1986a) The use of DNA hybridization and numerical taxonomy in determining relationships between *Trypanosoma brucei* stocks and subspecies. Parasitology 92:31-50
61. Dero B, Zampetti-Bosseler F, Pays E, Steinert M (1987) The genome and the antigen gene repertoire of *Trypanosoma brucei gambiense* are smaller than those of *T. b. brucei*. Mol Biochem Parasitol 26:247-256
62. Gibson W (1986) Will the real *Trypanosoma b. gambiense* please stand up. Parasitol Today 2:255-257
63. Lu Y, Hall T, Gay LS, Donelson JE (1993) Point mutations are associated with a gene duplication leading to the bloodstream reexpression of a trypanosome metacyclic VSG. Cell 72:397-406
64. Graham VS, Barry JD (1996) Is point mutation a mechanism for antigenic variation in *Trypanosoma brucei*? Mol Biochem Parasitol 79:35-45
65. Baltz T, Giroud C, Bringaud F, Eisen H, Jacquemot C, Roth CW (1991) Exposed epitopes on a *Trypanosoma equiperdum* variant surface glycoprotein altered by point mutations. EMBO J 10:1653-1659
66. Jenni L, Marti S, Schweizer J, Betschart B, Le Page RWF, Wells JM, Tait A, Paindavoine P, Pays E, Steinert M (1986) Hybrid formation between trypanosomes during cyclical transmission. Nature 322:173-175
67. Paindavoine P, Zampetti-Bosseler F, Pays E, Schweizer J, Guyaux M, Jenni L, Steinert M (1986b) Trypanosome hybrids generated in tsetse flies by nuclear fusion. EMBO J 5:3631-3636
68. Van Meirvenne N, Janssens PG, Magnus E (1975) Antigenic variation in syringe-passaged populations of *Trypanosoma brucei*. 1. Rationalization of the experimental approach. Ann Soc Belge Méd Trop 55:1-23
69. Schell D, Borowy NK, Overath P (1991) Transferrin is a growth factor of the bloodstream form of *Trypanosoma brucei*. Parasitol Res 77:558-560
70. Borst P, Bitter W, Blundell P, Cross M, McCulloch R, Rudenko G, Taylor MC, van Leeuwen F (1996) The expression sites for variant surface glycoproteins of *Trypanosoma brucei*. In: Hide

G, Mottram JC, Coombs GH, Holmes PH (eds) Trypanosomiasis and Leishmaniasis: biology and control. British Society for Parasitology/CAB International, Oxford, pp 109-131
71. Van Meirvenne N, Magnus E, Janssens PG (1976) The effect of normal human serum on trypanosomes of distinct antigenic types (ETat 1 to 12) isolated from a strain of *Trypanosoma brucei rhodesiense*. Ann Soc Belge Méd Trop 56:55-63
72. Pays E (1993) Genome organization and control of gene expression in trypanosomatids. In: Broda PM, Oliver SG, Sims P (eds) The Eukaryotic Microbial Genome. Cambridge University Press, pp 99-132
73. Zomerdijk JCBM, Kieft R, Duyndam M, Shiels PG, Borst P (1991a) Efficient production of functional mRNA mediated by RNA polymerase I in *Trypanosoma brucei*. Nature 353:772-775
74. Zomerdijk JCBM, Kieft R, Shiels PG, Borst P (1991b) Alpha-amanitin-resistant transcription units in trypanosomes: a comparison of promoter sequences for a VSG gene expression site and for the ribosomal RNA genes. Nucl Acids Res 19:5153-5158
75. Brown SD, Huang J, Van der Ploeg LHT (1992) The promoter for the procyclic acidic repetitive protein (PARP) genes of *Trypanosoma brucei* shares features with RNA polymerase I promoters. Mol Cell Biol 12:2644-2652
76. Janz L, Clayton C (1994) The PARP and rRNA promoters of *Trypanosoma brucei* are composed of dissimilar sequence elements that are functionally interchangeable. Mol Cell Biol 14:5804-5811
77. Vanhamme L, Pays A, Tebabi P, Alexandre S, Pays E (1995b) Specific binding of proteins to the noncoding strand of a crucial element of the variant surface glycoprotein, procyclin, and ribosomal promoters of *Trypanosoma brucei*. Mol Cell Biol 15:5598-5606
78. Rudenko G, Blundell PA, Dirks-Mulder A, Kieft R, Borst P (1995) A ribosomal DNA promoter replacing the promoter of a telomeric variant VSG gene expression site can be efficiently switched on and off in *T. brucei*. Cell 83:547-553
79. Ben Amar MF, Jefferies D, Pays A, Bakalara N, Kendall G, Pays E (1991) The actin gene promoter of *Trypanosoma brucei*. Nucl Acids Res 19:5857-5862
80. Lee MGS (1996) An RNA polymerase II promoter in the hsp70 locus of *Trypanosoma brucei*. Mol Cell Biol 16:1220-1230
81. Vanhamme L, Pays E (1995) Control of gene expression in trypanosomes. Microbiol Rev 59:223-240
82. Graham SV (1995) Mechanisms of stage-regulated gene expression in Kinetoplastida. Parasitol Today 11:217-223
83. Pays E, Vanhamme L (1996) Developmental regulation of gene expression in African trypanosomes. In: Smith DF, Parsons M (eds) Molecular biology of parasitic protozoa. IRL Press, Oxford, pp 88-114
84. Pays E, Coquelet H, Pays A, Tebabi P, Steinert M (1989a) *Trypanosoma brucei*: post-transcriptional control of the variable surface glycoprotein gene expression site. Mol Cell Biol 9:4018-4021
85. Pays E, Coquelet H, Tebabi P, Pays A, Jefferies D, Steinert M, Koenig E, Williams RO, Roditi I (1990) *T. brucei*: constitutive activity of the VSG and procyclin gene promoters. EMBO J 9:3145-3151
86. Rudenko G, Blundell PA, Taylor MC, Kieft R, Borst P (1994) VSG gene expression site control in insect form *Trypanosoma brucei*. EMBO J 13:5470-5482
87. Jefferies D, Tebabi P, Pays E (1991) Transient activity assays of the *Trypanosoma brucei* VSG gene promoter: control of gene expression at the post-transcriptional level. Mol Cell Biol 11:338-343
88. Zomerdijk JCBM, Ouellette M, ten Asbroek ALMA, Kieft R, Bommer AMM, Clayton CE, Borst P (1990) The promoter for a variant surface glycoprotein gene expression site in *Trypanosoma brucei*. EMBO J 9:2791-2801
89. Biebinger S, Rettenmaier S, Flaspohler J, Hartmann C, Pena-Diaz J, Wirtz LE, Hotz HR, Barry JD, Clayton C (1996) The PARP promoter of *Trypanosoma brucei* is developmentally regulated in a chromosomal context. Nucl Acids Res 24:1202-1211
90. Berberof M, Vanhamme L, Pays A, Tebabi P, Jefferies D, Welburn S, Pays E (1995) The 3'-terminal region of the mRNAs for VSG and procyclin can confer stage-specificity to gene expression in *Trypanosoma brucei*. EMBO J 14:2925-2934
91. Urményi TP, Van der Ploeg LHT (1995) PARP promoter-mediated activation of a VSG expression site promoter in insect form *Trypanosoma brucei*. Nucl Acids Res 23:1010-1018
92. Qi CC, Urményi T, Gottesdiener KM (1996) Analysis of a hybrid PARP/VSG ES promoter in procyclic trypanosomes. Mol Biochem Parasitol 77:147-159
93. Vanhamme L, Berberof M, Le Ray D, Pays E (1995a) Stimuli of differentiation regulate RNA elongation in the transcription units for the major stage-specific antigens of *Trypanosoma brucei*. Nucl Acids Res 23:1862-1869
94. Ullu E, Tschudi C, Günzl A (1996) Trans-splicing in trypanosomatid protozoa. In: Smith DF, Parsons M (eds) Molecular biology of parasitic protozoa. IRL Press, Oxford, pp 115-133
95. Hug M, Carruthers V, Sherman D, Hartmann C, Cross GAM, Clayton CE (1993) A possible role for the 3'-untranslated region in developmental regulation in *Trypanosoma brucei*. Mol Biochem Parasitol 61:87-96

96. Clayton C, Hotz HR (1996) Post-transcriptional control of PARP gene expression. Mol Biochem Parasitol 77:1-6
97. Ehlers B, Czichos J, Overath P (1987) RNA turnover in *Trypanosoma brucei*. Mol Cell Biol 7:1242-1249
98. Dorn PL, Aman RA, Boothroyd JC (1991) Inhibition of protein synthesis results in super-induction of procyclin (PARP) RNA levels. Mol Biochem Parasitol 44:133-140
99. Graham SV, Barry JD (1996) Polysomal, procyclin mRNAs accumulate in bloodstream forms of monomorphic and pleomorphic trypanosomes treated with protein synthesis inhibitors. Mol Biochem Parasitol 80:179-191
100. Pays E, Hanocq-Quertier J, Hanocq F, Van Assel S, Nolan D, Rolin S (1993) Abrupt RNA changes precede the first cell division during the differentiation of *Trypanosoma brucei* bloodstream forms into procyclic forms *in vitro*. Mol Biochem Parasitol 61:107-114

CHAPTER 4

Carbohydrate metabolism

FR Opperdoes

Introduction

The peculiarities of carbohydrate metabolism of the African trypanosome has received much attention and has been the subject of intense research for many years. The fact that bloodstream form trypanosomes can easily be grown in large numbers in the blood of infected rats has certainly contributed to this situation. With the advent of the tools of molecular biology even the number of cells available for research is often not a limiting factor anymore and this has led to an exponential growth of the literature of a more molecular nature in this area. A vast number of publications is available on the glycolytic pathway of the African trypanosome and its respective enzymes and it would be impossible to cover it all at this place. For a more detailed description of the earlier work the reader is referred to a number of previous reviews [1-5]. The more recent information about the bloodstream forms will be discussed here, together with what is known about the procyclic insect stage. The intermediate bloodstream stages, the short stumpy forms, as well as the epimastigotes and metacyclics present in the insects, can either not yet or not easily be obtained in sufficient quantities and, as a consequence, little to nothing is known about their metabolism.

The Embden-Meyerhof-Parnass (EMP) pathway of glycolysis

The long slender bloodstream form

The vertebrate stage of the African trypanosome dwells in the blood and tissue fluids of its mammalian host, where it has access to an unlimited source of glucose which is maintained at the relatively constant concentration of 5 mM. Because it is so abundant, glucose serves essentially as the sole source of carbon and energy for the trypanosome. It is metabolised at a rate exceeding that found in most other eukaryotic cells, but despite this fact the bloodstream form trypanosome utilises only a relatively small portion of its total protein for this activity. This is explained by the fact that while in most cells the glycolytic pathway takes place in the cytosol, in the trypanosome the first seven enzymes are localized within a membrane-bound organelle [1, 6]. Although this organelle resembles the microbody or peroxisome of other cell types, it lacks typical peroxisomal enzymes such as catalase and acylCoA oxidase, while 90% of its protein is glycolytic enzyme [7]. Therefore, these highly specialised microbodies have been called "glycosomes" [6]. Because neither a functional citric-acid cycle nor a classical mitochondrial respiratory chain are present, pyruvate, the end-product of the glycolytic pathway, is secreted as such into the host's bloodstream and is not metabolised to lactate or carbon dioxide *plus* water, as in most other organisms. Hence, little ATP is produced and this is done by substrate-level phosphorylation at the level of the phosphoglycerate kinase and pyruvate kinase steps.

Because bloodstream trypanosomes are completely dependent on glycolysis for their supply of ATP, and because the organisation of their glycolytic pathway is very different from that in the host, glycolysis has been identified as a promising target for the development of new drugs against African sleeping sickness [8]. For this reason the pathways of carbohydrate metabolism in the vertebrate stage of *Trypanosoma brucei* have been studied in great detail.

In the long slender bloodstream form the EMP is solely responsible for the oxidation of glucose to pyruvate, with the concomitant generation of 2 moles of ATP per mole of glucose consumed (Fig. 1). Glucose is the preferred energy source for the bloodstream form, but fructose and mannose can also be metabolised. Although probably not of physiological relevance, glycerol may also be used as an energy substrate, at a rate equal or even higher than that of glucose consumption. The ATP yield in this case is only half that of glucose, which explains why the rate of respiration with this substrate sometimes supersedes that observed with glucose. Due to the absence of a mitochondrial pyruvate dehydrogenase complex, the long slender form does not oxidise glucose beyond pyruvate and excretes it as such into the bloodstream. Since a functional citric-acid cycle and mitochondrial respiratory chain are absent, amino acids and fatty acids, cannot be utilized as energy substrates by these life-cycle stages. Moreover, bloodstream forms contain neither any carbohydrate stores, such as glycogen or other polysaccharides, nor any energy reserves of any significance, such as creatine phosphate or polyphosphates. Thus, depletion of trypanosomes of an energy substrate such as glucose will result in a rapid drop of cellular ATP levels and a total loss of motility.

The bloodstream form of *T. brucei* carries out an aerobic type of glycolysis. This means that the NADH that is produced in the glycosome by the glyceraldehyde-phosphate dehydrogenase reaction is reoxidized indirectly by molecular oxygen. For this oxidation the reducing equivalents have to cross the glycosomal membrane in order to reach the terminal oxidase located in the mitochondrial inner membrane. The African trypanosome transfers the reducing equivalents coming from NADH by a glycerol-3-phosphate (G3P): dihydroxyacetone-phosphate (DHAP) shut-

Carbohydrate metabolism

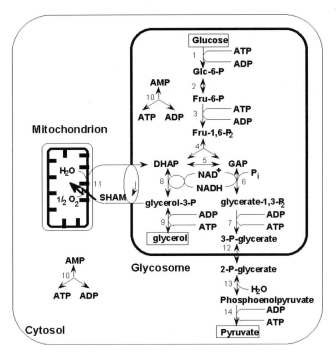

Fig. 1. Glycolysis in the bloodstream form trypanosome. Boxed metabolites are nutrients or end-products of metabolism. Enzymes: 1, hexokinase; 2, phosphoglucose isomerase; 3, phosphofructokinase; 4, aldolase; 5, triosephosphate isomerase; 6, glyceraldehyde-3-phosphate dehydrogenase; 7, phosphoglycerate kinase; 8, glycerol-3-phosphate dehydrogenase; 9, glycerol kinase; 10, adenylate kinase; 11, glycerol-3-phosphate oxidase; 12, phosphoglycerate mutase; 13, enolase; 14, pyruvate kinase

tle. This shuttle comprises two reactions, the reduction of DHAP to G3P by the glycosomal NAD-linked G3P dehydrogenase and the reversed reaction that is catalysed by a FAD-linked G3P dehydrogenase in the mitochondrial inner membrane. By analogy with the mammalian G3P dehydrogenases, this enzyme is most likely located at the outside of the mitochondrial inner membrane, so that the G3P does not have to cross this membrane in order to be oxidised. The latter dehydrogenase transfers the reducing equivalents to ubiquinone (UQ9) which is then reduced to ubiquinol. The above shuttle can only be operational when the glycosomal membrane is permeable to G3P and DHAP. However, latency measurements carried out with intact glycosomes have shown that the glycosomal membrane constitutes a diffusion barrier for the glycolytic intermediates. Therefore, Opperdoes and Borst [6] have proposed a specific DHAP: G3P antiport allowing for the entry of 1 molecule of DHAP for each molecule of G3P that leaves the glycosome. Although, so far, no direct experimental evidence has been brought forward for the presence of such an antiport, recent mathematical modelling of the glycolytic pathway also required a DHAP: G3P antiport in the glycosomal membrane in order to explain the observed sensitivity of anaerobic glycolysis to inhibition by glycerol through mass action [9].

Ubiquinol is oxidised by a ubiquinol: oxygen oxidoreductase present in the mitochondrial inner membrane. This enzyme, which is insensitive to cyanide, but is sensitive to salicyl hydroxamic acid (SHAM), resembles the so-called 'alternative oxidases' that have been described for plants [10] and certain fungi such as *Neurospora crassa* [11, 12]. In neither fungi nor trypanosomes, evidence has been found for an involvement of the enzyme in transmembrane proton translocation. The enzyme has also been detected in *Phytomonas* [13, 14], but it has been reported to be absent from *Leishmania*, where neither the activity, nor the gene encoding the enzyme, could be detected [14]. The mammalian host does not have such an alternative oxidase and transfers its electrons towards oxygen via the classical respiratory chain containing the cytochromes b, c, c_1 and aa_3. Thus, the *T. brucei* alternative oxidase has been suggested as an interesting target for chemotherapy.

Inhibition of the trypanosome alternative oxidase by 1 mM of SHAM mimicks the effect of a lack of oxygen on the carbohydrate metabolism of the bloodstream form. Under these conditions, long slender bloodstream forms continue to utilize glucose at about the same rate as under aerobic conditions, but because the glycerol 3-phosphate: dihydroxyacetone phosphate cycle is now inoperative, glycerol 3-phosphate accumulates inside the glycosome, while the glycosomal ATP concentration rapidly drops [6]. This situation leads to a condition where a reversal of an essentially irreversible glycerol kinase reaction by mass action becomes possible, leading to the production of glycerol as an end-product of anaerobic glycolysis with the synthesis of 1 mole of ATP for each glycerol produced [15, 16]. This compensates for the loss of one mole of ATP in the glycosomal phosphoglycerate kinase reaction, because now one mole of phosphoglycerate, rather than two, is produced per mole of glucose consumed. As a consequence, glucose is dismutated into equimolar amounts of pyruvate and glycerol, with net synthesis of 1 molecule of ATP [1, 6, 17]. Together with pyruvate and glycerol, trace amounts of alanine are excreted, probably as a result of the transamination of pyruvate in the cytosol.

Under anaerobic conditions (or with SHAM) bloodstream forms survive and remain motile, while cellular ATP levels drop to about 50% [18]. Because glycerol 3-phosphate cannot be oxidized to DHAP without molecular oxygen, glycerol cannot serve as a substrate in the absence of oxygen, while glucose does. Modeling of the aerobic/anaerobic transition of glycolysis suggests that it takes place at very low (micromolar) oxygen concentrations, close to that of the Km of trypanosomal glycerol-3-phosphate oxidase. The anaerobic pathway appears to be almost completely inoperative at oxygen tensions in the range of those found in venous and arterial blood [19].

Inhibition of respiration alone is not sufficient to kill the organism and the trypanosome is able to survive as long as glycerol does not accumulate in the medium [20], but once above several millimolar, glycerol becomes toxic. Due to mass action the reversal of the glycerol kinase reaction now becomes inhibited and glycolysis comes to a complete stop [20]. Since trypanosomes lack energy stores such as creatine phosphate or polyphosphates, cellular ATP levels rapidly drop to zero, leading to their complete immobilisation. This observation has been exploited for the development of a rational treatment of experimental animals infected with either *T. brucei*, *T. rhodesiense* or *T. vivax*. A simultaneous administration of SHAM and glycerol leads to an almost immediate lysis and disappearance of parasites from the circulation [21]. Permanent cures, however, were only obtained at concentrations of the drugs that were toxic to the animals [22, 23].

Many new and effective inhibitors of the trypanosome alternative oxidase have been described [24-26], of which ascofuranone was the most potent [27]. But so far, none of these have been developed into an effective anti-trypanosome drug.

The trypanosome alternative oxidase gene has recently been cloned and sequenced [28]. Its inferred amino acid sequence shows that the enzyme is homologous to the alternative oxidases of plants and fungi. Transformation of an *Escherichia coli* mutant, lacking a functional cytochrome c oxidase, with the alternative oxidase gene of the trypanosome was able to restore respiration in this bacterium [28].

Regulation of glycolysis

The glycolytic pathway of the bloodstream trypanosome is thought to be essentially unregulated. Hexokinase is not subject to any significant feedback inhibition by glucose 6-phosphate [29] ([29]; Opperdoes and Van Roy, unpublished), as are the hexokinases from most other eukaryotes. Phosphofructokinase (PFK) is also essentially unregulated. Although it is an allosteric enzyme with regard to the binding of its substrate fructose 6-phosphate (Fru6P), that is activated by AMP and ADP, it does not respond to fructose 1,6-bisphosphate or fructose 2,6-bisphosphate (Fru-2,6-P_2) or other compounds known to regulate PFKs [30, 31].

Van Schaftingen et al. [32] discovered that in trypanosomes Fru(2,6)P_2 modulates the activity of the cytosolic enzyme pyruvate kinase (PYK) rather than that of PFK. This peculiar way of regulating the glycolytic pathway is most likely related to the fact that PFK in the Kinetoplastida is present inside glycosomes, a sequestration that most likely renders it inaccessible to Fru(2,6)P_2, which is synthesized from Fru6P by the cytosolic enzyme phosphofructo-2-kinase (PFK2, [33]). The compartmentation of the first part of the glycolytic pathway has probably led to an adaptation of the original Fru(1,6)P_2-binding site of cytosolic PYK to accommodate Fru(2,6)P_2 as its most potent allosteric regulator [34, 35]. *T. brucei* PYK, together with the PYKs from the other glycosome-containing members of the Kinetoplastida, all differ from their homologous counterparts from other eukaryotes in that they constitute the only class of PYKs that are activated by Fru(2,6)P_2 [32, 33].

Van Schaftingen et al. [33] have shown that in the vertebrate stage of *T. brucei*, both with aerobic and anaerobic glycolysis, the cellular concentrations of Fru(2,6)P_2 and phosphoenolpyruvate (PEP) are inversely related. Since the conversion of PEP to pyruvate is the only known fate of this glycolytic metabolite in the vertebrate stage, it is not unexpected that a decrease in the concentration of Fru(2,6)P_2 will be compensated for by an increase in PEP. This indicates that a regulatory mechanism is operational at the level of PYK, but the advantage for bloodstream form glycolysis of being controlled at this level by Fru(2,6)P_2 is at present not clear. It is probably only aimed at maintaining a maximal flux through the pathway at all times.

The short-stumpy bloodstream form

These life-cycle stages have, in addition to a glycolytic pathway, an active mitochondrial pyruvate dehydrogenase complex and thus are able to convert pyruvate to acetate. Moreover, owing to the presence of the citric-acid-cycle enzyme 2-oxoglutarate dehydrogenase, these stages are also able to oxidise 2-oxoglutarate to succinate [36]. In pleomorphic *T. rhodesiense*, comprising a high percentage of stumpy forms, the major end-products of glucose metabolism were pyruvate (75%) with glycerol (5%), acetate (9%), succinate (1%) and CO_2 (3%).

The insect stage

Procyclic culture forms, which are supposed to be similar, if not identical, to the invertebrate stages that are found in the insect midgut, have a poor ability to convert glucose to pyruvate. This is mainly due to the low amounts of several of the glycolytic enzymes, particularly hexokinase and PYK [37, 38]. They are capable of metabolising glucose, fructose, mannose and glycerol, but not other mono- or disaccharides [39]. While the major end-product of glucose metabolism is CO_2, significant amounts of acetate, succinate and alanine are also produced. The ratio of the latter three end-products varies with both the rate and the conditions of growth [40], and with the strain under study. Early reports deal with parasites that have been cultured in biphasic medium, but with the advent of better culture procedures this kind of medium has been replaced by an all-liquid culture medium such as the SDM-79 [41]. Proline and the citric-acid cycle intermediates 2-oxoglutarate and succinate also support respiration. Respiration with proline as substrate is up to two times higher than with the other substrates, and proline apparently is the preferred substrate. In the chemostat, in the presence of both glucose and proline and at slow rates of growth, procyclics have a clear preference for glucose as the carbon and energy source. At high growth rates, or when glucose becomes a limiting factor, the cells switch to the consumption of proline as the sole energy and carbon source [40]. Under these conditions the main end-product is carbon dioxide, but also equimolar amounts of succinate and acetate are produced, together with some alanine. While under these conditions glucose is not consumed, at least two of the glycolytic enzymes (i.e. glyceraldehyde-3-phosphate dehydrogenase and enolase) increase their activity proportionally with proline consumption. This suggests that under these conditions cells may have actively engaged in gluconeogenesis. Under anaerobic conditions procyclic stages of *T. rhodesiense* are capable of utilizing glucose and glycerol, provided CO_2 is present [39]. Most of the carbon is now recovered as succinate and a smaller amount as acetate. Anaerobic data for *T. brucei* procyclic stages are not available, but should not be too different from those obtained for *T. rhodesiense*.

PYK (see above), although present in the insect stages of *T. brucei*, has dropped in activity by 20 fold relative to the bloodstream form. Moreover, no $Fru(2,6)P_2$ could be found in the insect stage when incubated in the presence of glucose [32], whereas the $Fru(2,6)P_2$-synthesizing and degrading enzymes PFK2 and FBPase2 were both present in amounts similar to those found in the vertebrate form [33]. Why $Fru(2,6)P_2$ could not be detected in these forms remains to be elucidated. In the absence of $Fru(2,6)P_2$, the relatively low activity of PYK is now mainly regulated by the cytosolic phosphate potential ([ATP]/[ADP][Pi]), the availability of the citric-acid cycle intermediates oxaloacetate and acetylCoA and the cytosolic concentration of PEP. For the latter, the $S_{0.5}$ in the absence of $Fru(2,6)P_2$ is 10 times higher than in its presence [32, 34].

The glycosomal phosphoenolpyruvate carboxykinase (PEPCK) and malate dehydrogenase (MDH), which are virtually absent from bloodstream forms, but fully expressed in the procyclics, are believed to play an important role in the fixation of carbon dioxide [1, 42]. Due to the very low activity of PYK in this life-cycle stage (see above) the PEP formed in the cytosol cannot be converted to pyruvate and is forced back into the glycosome, where it serves as a substrate for the very active PEPCK (Fig. 2). In the case of another trypanosomatid, *T. cruzi*, it has been shown that the regulation of PEPCK is such that it favours carboxylation of PEP, rather than decarboxylation of oxaloacetate [43]. It is interesting to note that pyruvate carboxylase could not be detected in the insect-stage of *T. brucei* [42] in agreement with the important role of PEPCK for the formation of oxaloacetate and the low PYK activity. The oxaloacetate produced in the glycosome is reduced to malate by MDH, which is then excreted into the cytosol. There, it may be converted to pyruvate by a cytosolic malic

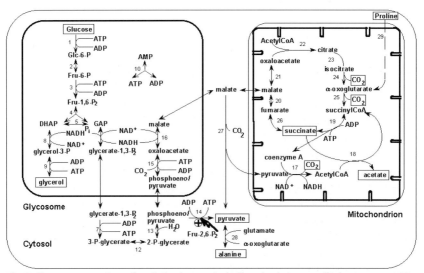

Fig. 2. Compartmentation of carbohydrate metabolism in the procyclic insect stage. For the enzymes 1 - 14: see the legend to Fig. 1. Other enzymes: 15, phosphoenolpyruvate carboxykinase; 16, malate dehydrogenase; 17, pyruvate dehydrogenase; 18, acetate: succinate CoA transferase; 19, succinylCoA synthase or succinate thiokinase; 20, fumarate hydratase; 21, mitochondrial malate dehydrogenase; 22, citrate synthase; 23, aconitase; 24, isocitrate dehydrogenase (NADP); 25, α-oxoglutarate dehydrogenase; 26, succinate dehydrogenase; 27, malic enzyme; 28, alanine aminotransferase; 29, proline oxidase

enzyme [42], or enter the mitochondrion and the citric-acid cycle directly to fulfil an anaplerotic function.

Mechanism of succinate production

Procyclic trypanosomes oxidise not only carbohydrates but also fatty acids and amino acids, such as proline, to carbon dioxide and water indicating the presence of a functional mitochondrial citric-acid cycle. Most of the enzymes of the cycle - citrate synthase [44], aconitase [45], isocitrate dehydrogenase [44, 46], succinate thiokinase [47], succinate dehydrogenase (SDH) [44, 48], fumarase [44], malate dehydrogenase ([42]; Uttaro and Opperdoes, unpublished) - have all been detected in procyclic forms of *T. brucei*. Despite the functionality of this cycle, the insect stages secrete into the extracellular medium, in addition to carbon dioxide, large amounts of succinate and acetate, as well as some alanine (see above).

Apparently, the capacity of the cycle is limited so that it is not able to cope with all the carbon units coming from glycolysis and from fatty acid and amino-acid oxidation. The formation of succinate has been interpreted by some Authors as an indication for the presence of an active fumarate reductase (FR) and indeed the enzyme has been described for the procyclic trypomastigote [49-51]. However, other

Authors have described the presence of a succinate dehydrogenase [44, 48]. It is unlikely that the cycle runs in both directions simultaneously since in anaerobic bacteria and in helminths and Crustaceae, where (part of) the cycle runs in the reductive direction, different enzymes are required. In this respect it is interesting to note that the ATP-dependent succinate thiokinase, or succinyl CoA synthase, that is thought to function in the direction of succinyl CoA breakdown, and not the GTP-dependent succinate thiokinase thought to catalyse the reverse reaction, drastically increases with transformation from bloodstream form to insect stage [47]. Moreover, in aerobic organisms SDH transfers its reducing equivalents via ubiquinone to the mitochondrial respiratory chain and ubiquinone Q9 is present in the trypanosomatids. FR uses either menaquinone (in prokaryotes), or rhodoquinone (in eukaryotes), rather than ubiquinone [52]. Neither menaquinone, nor rhodoquinone have been detected in trypanosomatids (Tielens, personal communication). So most likely, the different authors that have measured the activity of FR or SDH have measured one and the same enzyme - SDH - but assayed in two different directions.

Mechanism of acetate production

It has long remained an unsolved question by what mechanism the procyclic stages of *T. brucei*, as well as the other insect stages of trypanosomatids, produce their acetate. Since procyclics have a functional pyruvate dehydrogenase complex (Jackman and Opperdoes, unpublished) and citric-acid cycle in the mitochondrion, they must be capable of converting pyruvate to acetylCoA. Due to the limited capacity of the cycle not all acetylCoA can be oxidised and a considerable amount is excreted in the form of acetate. Recently it has been shown that both in *Leishmania* promastigotes and in *T. brucei* procyclics, acetate is produced from acetyl-CoA with the synthesis of one molecule of ATP by a cycle that comprises a novel mitochondrial acetate: succinate CoA-transferase [53] and the ATP-dependent citric-acid-cycle enzyme succinyl-CoA synthase [47]. The acetate: succinate CoA-transferase of *L. mexicana mexicana* promastigotes was shown to have a high affinity for acetyl-CoA (Km 0.1 mM), while it has a low affinity for succinate (Km 3.8 mM). This suggests that acetate production only occurs when mitochondrial succinate concentrations are high and thus the intra-mitochondrial succinate concentration would be a key-factor in switching mitochondrial metabolism from citric-acid cycle activity to the production of acetate. Interestingly, a similar pathway for the production of acetate has since long been known to be associated with hydrogenosomes, the hydrogen-producing organelles of the facultative anaerobe *Trichomonas* sp. [54]. Now that it has become likely that mitochondria and hydrogenosomes are the descendants of the same α-proteobacterial endosymbiont [55], the identification of the acetate: succinate CoA transferase/succinyl-CoA synthase cycle in the mitochondria of trypanosomatids is the first metabolic pathway common to both types of organelle.

Respiratory chain

In the insect stage the reducing equivalents that are generated by the citric-acid cycle and the pyruvate dehydrogenase complex are oxidised by the respiratory chain. There has been some doubt as to the presence of a functional complex I (NADH dehydrogenase) which was borne out by the lack of inhibition of respiration by rotenone, a potent inhibitor of NADH dehydrogenase activity. However, it has now been demonstrated that NADH dehydrogenase can be inhibited by rotenone although at much higher concentration [56]. The ubiquinone (UQ9) that is reduced by the complexes I (NADH dehydrogenase) and II (SDH) and by the FAD-linked glycerol-3-phosphate dehydrogenase, is mainly oxidised by a cytochrome-containing respiratory chain which consists of complexes similar to the complexes III and IV of the classical mammalian respiratory chain. The cytochromes a, b and c have all been detected. As a consequence, respiration in these cells is sensitive to cyanide, but due to residual alternative oxidase activity (see above) a considerable amount of cyanide-insensitive respiration may remain present and this can be inhibited by SHAM. The function for this residual alternative oxidase is not clear. It may serve for the oxidation of excess of reducing equivalents which otherwise would lead to the production of reactive oxygen species and so help in protecting the cell against oxidative stress, as has recently been shown for the plant alternative oxidase [57].

The pentose-phosphate pathway

Glucose is also metabolized by a second pathway, different from glycolysis, the pentose-phosphate pathway (PPP), also known as the hexose-monophosphate shunt (Fig. 3). While the role of the EMP has been extensively investigated in *T. brucei*, there have been only a limited number of studies on the PPP and its contribution to carbohydrate metabolism. Although there have been doubts as to the functioning of this pathway, it is now clear that it plays a crucial role in both the metabolism of *T. brucei* bloodstream forms and its protection against oxidant stress.

An important role of the pentose phosphate pathway is to maintain a pool of cellular NADPH, which serves as a hydrogen donor in reductive biosynthesis, and in the defence against oxidative stress. The PPP itself serves to convert glucose 6-phosphate (G6P) to ribose 5-phosphate (R5P), which is used in nucleotide biosynthesis. The oxidative branch

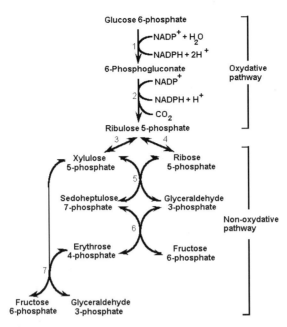

Fig. 3. The pentose-phosphate pathway. Enzymes: 1, glucose-6-phosphate dehydrogenase; 2, 6-phosphogluconate dehydrogenase; 3, ribulose 5-phosphate 3-epimerase; 4, ribulose 5-phosphate isomerase; 5, transketolase; 6, transaldolase; 7, transketolase

of the pathway converts G6P to ribulose 5-phosphate (Ru5P), a process which leads to the production of two moles of NADPH per mole of G6P consumed. The non-oxidative branch generates from various pentose-phosphates the substrates for the enzymes transketolase and transaldolase which transfer, respectively, two- and three-carbon units, between a variety of phosphorylated carbohydrates. End products of the non-oxidative branch are the glycolytic intermediates fructose 6-phosphate (F6P) and glyceraldehyde 3-phosphate (GA3P). The latter branch may also function in the other direction, where GA3P and F6P are used to generate other phosphorylated sugars.

All the enzymes of the classical PPP have been detected in the procyclic form [58]. However, in the bloodstream form two enzymes of the non-oxidative pathway, ribulose-5-phosphate epimerase and transketolase, were not detected, while transaldolase was present. The classical enzymes appear to be predominantly cytosolic, but there is some evidence to suggest that part of the G6PDH and 6PGDH may be associated with glycosomes (Heise and Opperdoes, unpublished). Rat-liver peroxisomes contain separate isoenzymes for G6PDH and 6PGDH, which are involved in the formation of NADPH required for peroxisomal alkoxyphospholipid biosynthesis. As this biosynthetic pathway is also present in glycosomes, this may be the explanation why small amounts of the oxidative branch dehydrogenases are found in these organelles.

The individual steps of carbohydrate metabolism

Glucose transport

Trypanosoma brucei utilizes glucose, fructose and mannose as substrates for the glycolytic pathway. Early experiments by Southworth and Read [59] have shown that there is a glucose transporter present in the plasma membrane of bloodstream forms that is capable of transporting all three substrates. More detailed studies have been carried out on the transporters of *T. equiperdum*, *T. rhodesiense* and *T. brucei* [60-64]. In the bloodstream form of *T. brucei* glucose crosses the plasma membrane by a mechanism of facilitated diffusion. This has been confirmed by purification and reconstitution of the transporter into liposomes [63]. The transporter has a relatively low affinity for glucose (1-2 mM), and its activity is such that at external glucose concentrations below 5 mM the rate of the glycolytic flux does not exceed the rate at which glucose enters the cell [64]. Above 5 mM one or more later steps in the glycolytic pathway, become rate limiting and glucose accumulates within the cell. This observation nicely illustrates how under conditions of a glucose homeostasis in the blood (5 mM) the number of transporter molecules in the plasma membrane of the parasite is adapted to the overall activity of the glycolytic pathway [64].

In the procyclic insect form, the glucose transporter has an affinity for glucose that is 10 - 20 fold higher than that of the transporter found in the bloodstream form. This probably reflects the low availability of glucose in the insect midgut. At low glucose concentrations Parsons and Nielsen [65] found that transport was sensitive to FCCP and KCN, suggesting possible proton-dependence and active transport, while at high substrate concentrations transport of glucose was via facilitated diffusion [66, 67]. A recent study has shown that the inhibition of transport by FCCP and KCN is dependent on the glucose concentration, suggesting that in the insect stage transport may be dependent on a proton gradient at low substrate concentrations, but not at high concentrations [68].

The glucose transporter of *T. brucei* has been cloned and sequenced [60]. This gene called THT1 makes part of a multi-gene family consisting of two isoforms THT1 and THT2 (for Trypanosome Hexose Transporter 1 and 2) encoding proteins which are 80% identical. The expression of these genes is life-cycle-stage-dependent. THT1 is expressed in the bloodstream form, while THT2 is active in the procyclics. When both genes were expressed in heterologous systems, only facilitated diffusion and not active transport was measured.

Analysis of the substrate specificity of the bloodstream transporter has revealed that the hydroxy groups at C2 and C6 of glucose are less important for the recognition of the substrate. There are some interesting differences between the *T. brucei* transporter and the human erythrocyte transporter GLUT1. The parasite transporter is 1000 times less sensitive to cytochalasin B and, unlike the mammalian transporter, it recognizes fructose with relatively high affinity (2.6 mM) [63, 69]. It does not recognize analogues of glucose altered at the C-4 position.

Pyruvate transport

Efflux of the end-product of glycolysis, pyruvate, is also mediated by a facilitated diffusion carrier [70-72]. The transporter is saturable (Km pyruvate = 2 mM) and exhibits the phenomenon of transacceleration, typical of facilitated diffusion carriers. This carrier differs from monocarboxylate transporters for pyruvate and lactate of mammals, on the basis of its sensitivity to inhibitors. It is inhibited by monocarboxylic acids, but not by lactate. Moreover, inhibition of the transporter by a specific inhibitor such as the α-cyanocinnamate derivative UK5099 led to a dramatic accumulation of pyruvate inside the cell and an acidification of the cytosol, suggesting that pyruvate is co-transported with H+ [72]. Under these conditions, respiration and glycolysis stop and the cells lyse.

Hexokinase

The first enzyme of the glycolytic pathway catalyses the transfer of the γ-phosphoryl group of ATP to a hexose molecule. The trypanosome enzyme is exclusively associated with the glycosome and therefore the phosphorylation of glucose must take place either inside the glycosome or during the transport of the hexose into the organelle. It is not known whether hexokinase itself is involved in this transport or there is a separate hexose transporter present in the glycosomal membrane. So far, no such transporter has been identified, although it was proposed by Opperdoes and Borst [6]. Hexokinase is active on glucose, mannose and fructose [29, 73], with which it has similar affinities and on which it acts with similar rates. However, mannose 6-phosphate, that is produced inside the glycosome by the action of hexokinase, is only metabolised very slowly, probably because of the low activity of mannose-6-phosphate isomerase in the glycosome (Misset and Opperdoes, unpublished), and thus accumulates inside the organelle leading to an inhibition of glycolysis [73]. Hexokinase is not inhibited by either glucose 6-phosphate or glucose 1,6-bisphosphate ([29]; Opperdoes and Van Roy, unpublished), as are the hexokinases from most other sources, but it is selectively inhibited by a number of glucose analogues ([74]; Willson and Opperdoes, unpublished). The *T. brucei* enzyme, contrary to most other hexokinases, is not specific for ATP; it also utilizes ITP, and to a lesser extent GTP, UTP and CTP [75]. Such specificity is probably not required for an enzyme that functions inside an organelle where ATP is supposed to be the major nucleotide. The enzyme has a molecular mass of 53 000 and an isoelectric point of 10. On a gel-filtration column it elutes as a hexameric protein [7, 75]. It represents 0.25% of the total cellular protein, and 6% of the glycosomal protein, and it was calculated that some 180 molecules of hexokinase are present within one organelle [7]. The gene encoding the enzyme has been cloned (Irrthum et al, unpublished). It codes for a polypeptide of 474 amino acids long, has a N-terminally located sequence that is reminiscent of a peroxisomal import signal, also called PTS-2, and has only 36% positional identity with the human enzyme that has a molecular mass of 100 000. In view of the numerous differences with the mammalian enzyme, hexokinase proves to be an excellent target for drug design.

Phosphoglucose isomerase

This enzyme that catalyses the conversion of glucose 6-phosphate into fructose 6-phosphate is mainly associated with the glycosome of the bloodstream form. It has been purified and represents 0.15% of the total protein and 4.1% of the glycosomal protein [7, 76]. The kinetics of the *T. brucei* enzyme does not significantly differ from its yeast or rabbit-muscle homologues. Affinity (fructose 6-phosphate, 0.12 mM) and inhibition constants (6-phosphogluconate, 0.14 mM) are similar. The *T. brucei* enzyme is inhibited by suramin and agaricic acid [77-79], while the mammalian counterpart is not [78]. There is a single gene that encodes a protein of 606 amino acids with a measured pI of 7.4 and a predicted molecular mass of 67280, which is 5 kDa larger than the mammalian isomerase. The protein has 54-58% positional identity with the yeast and the mammalian counterparts. The enzyme has a 38-49 amino acid long N-terminal extension, the function of which remains to be elucidated. At its C-terminus there is a tripeptide that has been shown to be capable of transporting polypeptides to peroxisomes and which is called a PTS1-type peroxisomal import signal.

In the procyclic form a small portion of the total amount of enzyme is cytosolic. Cytosolic PGI is one of the enzymes linking the EMP and PPP and is needed to reconvert F6P to G6P for repeated rounds of the PPP to generate NADPH for biosynthetic purposes and for defence against oxidative stress.

Phosphofructokinase

The *T. brucei* PFK is exclusively located in glycosomes. It represents 0.4% of the total protein and 11% of the glycosomal protein. It is a homotetrameric protein with a relative mass of 196 000 - 220 000 [7, 30] and a pI of 8.8. The enzyme is specific for ATP [29, 30]. It is an allosteric protein which is activated by its substrate Fru6P and by AMP and ADP, but not by fructose 1,6 bisphosphate. Moreover, the trypanosome PFK is not regulated by Fru(2,6)P$_2$ either, as is the case with most eukaryotic PFKs. PFK is also not regulated by any of the other compounds known to be regulators of PFK. Although strictly dependent on ATP as substrate, its subunit molecular mass of 51 000 is unique amongst the PFKs, which normally fall into two classes, the bacterial type PFKs with a subunit mass of 35 000 and the eukaryotic PFK with a mass of 75-90 kDa. This enigma has recently been solved by the cloning and sequencing of the corresponding gene. The predicted amino acid sequence of the *T. brucei* PFK is more related to the PPi-dependent PFKs from a number of eubacteria and protists, such as *Giardia*, *Naegleria* and *Entamoeba*, rather than to the ATP-dependent PFKs found in eukaryotes [80].

Aldolase

This enzyme belongs to the class I aldolases [75] that catalyse the reversible cleavage of the fructose 1,6-bisphosphate via the formation of a Schiff-base into the trioses DHAP and GA3P without the need of a divalent metal ion. In the bloodstream form it is almost completely found in the glycosome, where it represents 34% of all the protein, or 1700 molecules per organelle, and 1.2% of the total trypanosome protein. The trypanosome enzyme is a homotetrameric protein with a molecular mass of 157 000 and a pI of 9. Two almost identical tandemly repeated genes encode a polypeptide of 371 amino acids with a calculated subunit mass of 40 940 [81, 82]. The enzyme has an extension at its N-terminus of about 10 amino acids that codes for a type 2 peroxisomal targeting signal. The enzyme is between 45 and 47% identical to other eukaryotic aldolases and displays the typical properties of other class I aldolases [75, 83]. *T. brucei* aldolase has been studied in detail [83] and it has been overexpressed to high levels [84]. A low degree of phosphorylation of the enzyme at the level of a serine residue has been reported, but the function of this phosphorylation remains to be determined [85].

Triose-phosphate isomerase (TIM)

This enzyme catalyses the reversible conversion of the triosephosphates DHAP and GA3P. In the bloodstream form the enzyme is mainly found in the glycosome with 155 subunits per organelle, where it represents 1% of the protein, or less than 0.04% of the total trypanosome protein [7]. TIM is a dimer of 52 000 Mr of two identical subunits of 250 amino acids having about 50% positional identity with the TIMs from other organisms [86]. Only the dimer is active since each subunit contributes to the formation of the enzyme's active site. Its measured pI is 9.8, indicating that the enzyme carries many positively charged amino acids on its surface. The kinetic properties of the enzyme have been studied but these are not too dissimilar from other TIMs that have been described [87]. The trypanosome enzyme is inhibited by sulphate, phosphate, arsenate, and by 2-phosphoglycolate. The trypanosome TIM was the first parasite enzyme to be crystallised and its three dimensional structure was subsequently solved [88]. The TIM dimer of the trypanosome is less stable than its homologues from other organisms and easily dissociates and loses its activity upon dilution [89]. Kuntz et al. [90] have tried to exploit this reduced affinity of the subunits for each other by synthesizing cyclic peptide inhibitors mimicking part of a subunit interface loop that either would interfere specifically with the formation of the active dimer, or would promote its dissociation. Although this specific approach has failed because the enzyme precipitated in the presence of such peptides [91], Gao et al. [92], using the trypanocidal drug suramin as a tool, have shown that this is indeed a valid approach. Although suramin inhibits the active enzyme to the same extent as its rabbit-muscle homologue [87], it binds tightly to the monomers of *T. brucei* TIM and so interferes with the formation of an active dimer. Suramin that cannot be used for the treatment of Chagas' disease or leishmaniasis, binds less tightly to the subunits of the homologous enzymes from *T. cruzi* and *Leishmania* and to human TIM, and thus hardly interferes with the formation of these active dimers. Fairlamb and Bowman [93] found that trypanosomes that were exposed to suramin for prolonged times exhibited a significant reduction of their glycolytic rate. It may be that this specific effect of suramin, resulting on one hand from the high positive charge of the trypanosome TIM and on the other hand from the fact that the enzyme has to operate at relatively high protein concentrations, is the explanation for the mode of action of this important trypanocide.

Glyceraldehyde-3-phosphate dehydrogenase (GAPDH)

This enzyme catalyses the reversible oxidation and phosphorylation of glyceraldehyde-phosphate to 1,3-bisphosphoglycerate. All Trypanosomatidae studied, including *T. brucei*, have two separate isoenzymes, one in the cytosol and one in the glycosome. The glycosomal enzyme represents 0.5% of all the cellular protein in the bloodstream form and 14% of the glycosomal protein. The enzyme is homotetrameric with a molecular mass of 139 000 and a pI of 9.3 [7]. Two tandemly linked identical genes encode a polypeptide of 358 amino acids with a calculated mass of 38.9 kDa [94] and 52-57% identity with other GAPDH proteins. The polypeptide has several specific insertions that are conserved in all glycosomal GAPDHs and a C-terminal extension carrying a PTS1 import signal [95]. The enzyme is inhibited by the epoxide-containing GAPDH inhibitor pentalenolactone and by suramin, gossypol and agaricic acid [96]. Although its substrate-binding site is well conserved, it has been possible to synthesise epoxide or alpha-enone containing analogues that selectively and irreversibly inhibit not only the enzyme but also the growth of the trypanosome in an *in vitro* culture system [79]. The binding site for the NAD cofactor has some interesting differences that lead to a reduced affinity for this cofactor. The enzyme has been crystallised and its three-dimensional structure solved [97]. By the method of modelling and rational drug design a number of highly selective adenosine analogues have been synthesized that tightly fit the NAD binding pocket of the trypanosome enzyme, but do not bind to the corresponding pocket of the mammalian homologue [98, 99]. Some of these compounds have a high selectivity for the glycosomal enzyme and inhibit its activity at sub-micromolar concentrations (Gelb et al, unpublished).

The cytosolic isoenzyme differs in subunit mass (36 kDa) and pI (6.2) from the glycosomal enzyme [38] and lacks the typical insertions and extension of the glycosomal enzyme. Comparison of the sequences of the two enzymes has revealed that they are only distantly related [100] and that one of them, most likely the cytosolic enzyme which is absent from bodonine and the euglenozoan organisms, must have entered the ancestral trypanosomatid by an event of horizontal gene transfer [101]. The function of this cytosolic isoenzyme is not clear. It could be involved, together with the cytosolic PGK, in the equilibration of the cytosolic redox potential with the phosphate potential, or be involved in the conversion of GA3P produced in the pentose-phosphate pathway to glycerate 3-phosphate.

Glycerol-3-phosphate dehydrogenase (G3PDH)

This NAD-dependent enzyme catalyses the reversible oxidation of DHAP into glycerol-3-phosphate. In *T. brucei* the enzyme (0.25% of the total protein), which has a pI = 10, is exclusively present in glycosomes where it represents 7% of the protein [7]. The enzyme is inhibited by agaricic acid [77] and by the trypanocidal drugs suramin and cymelarsan (Denis and Baltz, personal communication). There are two tandemly linked genes that code for a polypeptide of 353 amino acids with at its C-terminus a PTS1 targeting signal [102, 103]. Interestingly, a comparison of the *T. brucei* sequence with that of other GPDH sequences revealed that the identity of the trypanosome enzyme with that of eukaryotes ranged from 23 - 25%, while the identity with a number of prokaryotic GPDs ranged from 32 - 36%. Moreover, the trypanosome GPDH had typical prokaryotic signatures as well [102]. Therefore, it was suggested that the glycosomal GPDH has entered the trypanosome by an event of horizontal gene transfer.

Glycerol kinase

Glycerol kinase (GK) has been purified and its kinetic properties have been examined [104]. It is a monomeric protein with molecular mass of approximately 53,000. It has a broad pH, optimum between pH 7 and 9.5, and catalyses a random bi bi reaction, with some cooperativity in substrate binding at high pH. Glycerol-3-phosphate up to 10 mM displays virtually no product inhibition of the forward reaction, but ADP is a weak inhibitor. The reverse reaction, essential for anaerobic glycolysis, proceeds at an extremely low rate and only in the presence of an efficient ATP trap. This is in agreement with the idea that the enzyme reaction is reversed by mass action only.

Phosphoglycerate kinase

This enzyme is involved in the formation of ATP from 1,3-bisphosphoglycerate and ADP by substrate-linked phosphorylation. Three isoenzymes have been detected in *T. brucei*: a glycosomal one of 47 kDa, which is directly involved in glycolysis; a second glycosomal isoenzyme of 56 kDa whose function is unknown, and a cytosolic isoenzyme of 45 kDa [7, 105, 106]. The cytosolic isoenzyme is absent from the bloodstream form, while the 47 kDa glycosomal isoenzyme is abundant and represents 4.4% of the glycosomal protein [7]. The latter is not expressed in the procyclic insect stage, but the cytosolic enzyme is. The 56 kDa isoenzyme is always expressed, but present in barely detectable quantities [107, 108]. PGK is monomeric and the glycosomal enzyme has a pI of 9.3, while the cytosolic one has a pI of 6.3. The three PGK genes are tandemly linked and exhibit a high degree of sequence identity. The glycosomal polypeptide of 47 kDa has a 20 amino acid long C-terminal extension which is responsible for the targeting of the protein towards the glycosome, while the 56 kDa glycosomal isoenzyme has an internal targeting signal [109]. The glycosomal and cytosolic enzymes have an almost absolute requirement for ATP, contrary to the enzymes from yeast and rabbit muscle, which accept GTP and ITP as well. The two major trypanosome isoenzymes are inhibited by suramin, but the glycosomal enzyme, with its higher positive charge is much more sensitive to this inhibitor [106]. The glycosomal isoenzyme has been crystallised after removal, by genetic engineering, of the C-terminal extension and its structure solved [110].

Phosphoglycerate mutase and enolase

Little is known about these two enzymes from *T. brucei*. They are both in the cytosol [6, 111], but for the mutase some activity has been observed in the glycosomal fraction as well (Opperdoes, unpublished). The enzymes have not been purified and their genes have not yet been cloned.

Two classes of phosphoglycerate mutase have been described, one that uses the cofactor 2,3-bisphosphoglycerate and the other that does not. It is not known to which class the *T. brucei* enzyme belongs.

Pyruvate kinase (PYK)

This enzyme catalyses the synthesis of ATP and pyruvate from PEP and ADP. The enzyme is cytosolic, it is a homotetramer of subunits of a mass of 54 378 and the overall identity of the trypanosome PYK with other PYKs ranges from 41-51% [112]. The allosteric enzyme –appears to be a crucial regulator of the glycolytic flux and its most striking regulator is Fru(2,6)P_2 [32-35]. The enzymes from *T. brucei* and *Leishmania mexicana* have been cloned and sequenced and overexpressed in *E. coli* [112, 113]. In addition to Fru(2,6)P_2 the enzyme is activated by Fru(1,6)P_2, glucose(1,6)P_2 and ribulose bisphosphate [34] and its activity is modulated by adenine nucleotides and inorganic phosphate, while it is insensitive to regulation by most of the metabolites that modulate PYK activity from other sources. A model of the structure of *T. brucei* PYK, based on the atomic coordinates of the cat-muscle enzyme has revealed a potential binding site for Fru(2,6)P_2. Many residues in this site are conserved within the trypanosomatids, but not in the other PYKs, and thus this site would form an interesting target for drug design.

6-phosphogluconate dehydrogenase (6PGDH)

6PGDH has been suggested as a potential target for chemotherapy [114]. In other eukaryotic organisms, deletion of the gene encoding 6PGDH is a lethal event [115, 116]. Since this is not the case for the first enzyme of the pathway (G6PDH), this indicates that the PPP itself is dispensable, but that accumulation of 6-phosphogluconate is toxic to cells. This is corroborated by the observation that the lethal effect of 6PGDH deficiency can be relieved by loss of G6PDH in *Drosophila melanogaster* [117]. Since 6-phosphogluconate (6PG) is a potent inhibitor of phosphoglucose isomerase (PGI), including the *T. brucei* enzyme [78], inhibition of 6PGDH is thought to lead to accumulation of 6PG and hence the inhibition of both pathways of glucose metabolism [116].

The gene that codes the *T. brucei* 6PGDH has been cloned and the enzyme purified and crystallized [114, 118, 119]. It is remarkably different from other 6PGDHs in that it shares less than 40% amino acid sequence identity with all other available bacterial and eukaryotic 6PGDH sequences, while these others share at least 50% sequence identity (Opperdoes, unpublished). This makes it more likely to find exploitable differences between the *T. brucei* and the host 6PGDH. The trypanocidal drug suramin inhibits 6PGDH. Also the trivalent aromatic arsenoxides inactivate the enzyme with extreme potency, although it has been ruled out that this enzyme represents a primary target for trypanocidal arsenicals [120].

Phosphoenolpyruvate carboxykinase (PEPCK)

Phosphoenolpyruvate carboxykinase isolated from glycosomes from procyclics [121] is specific for ATP rather than GTP, as are the mammalian homologues. The enzyme is a homodimer with a subunit molecular mass of 59 kDa. Studies on the corresponding gene have shown that the protein has 525 amino acids and ends in a tripeptide (SRL) which is a potential targeting signal for import into the glycosome [122]. Like other glycosomal proteins, the enzyme has a high isoelectric point (pH 8.9). It is strictly dependent on adenosine nucleotides for activity, as well as on the presence of Mn^{2+}. Quinolinic acid, a structural analogue of oxaloacetate, completely inhibits the decarboxylation reaction at a 1 mM concentration. Kinetic data suggest that under physiological conditions PEPCK in *T. brucei* is bidirectional and that its activity is regulated primarily by mass action.

Malate dehydrogenase and malic enzyme

In *T. brucei* there are three different isoenzymes of MDH. In the bloodstream form 90% of all MDH is cytosolic. The remaining 10% is present in the mitochondrion. In procyclics there is a mitochondrial MDH (10%), a cytosolic MDH (5%), identical to the major cytosolic MDH of the bloodform and a glycosomal MDH (85%) (Uttaro and Opperdoes, unpublished). In both life-cycle stages the cytosolic and mitochondrial isoenzymes, in conjunction with aspartate aminotransferase, must have a function in the transport of reducing equivalents from cytosol to mitochondrion and vice versa. In the procyclic stage the cytosolic pyruvate kinase is almost completely inactive because of the absence of its allosteric regulator fructose 2,6-bisphosphate. So all the PEP generated in the glycolytic pathway is forced back into the glycosome where PEPCK and MDH function in concert in the reoxidation of glycosomal NADH and the formation of one mole of glycosomal ATP per mole of NAD regenerated. Part of the glycosomal malate so produced is converted in the cytosol to pyruvate by malic enzyme which is mainly cytosolic [42].

In an isoenzyme study of 29 stocks of *T. evansi* from Egypt, Sudan and Indonesia a perfect correlation was found between the presence of the type VII banding pattern of malic enzyme with a high resistance to suramin [123]. However, the function of malic enzyme in this bloodstream trypanosome so closely related to *T. brucei* is not known.

Control of glycolysis and drug development

Recent studies have addressed the question which steps in the EMP control the glycolytic flux in the bloodstream form of *T. brucei*. Such information is relevant for the development of new anti-trypanosomal drugs because the selectivity of a drug may be enhanced by choosing an enzyme which has a high control in the parasite, but a low control in the host. Since for *T. brucei* sufficient kinetic data about the individual enzymes of the glycolytic pathway are now available, the answer to this question can be calculated. Interestingly, and contrary to what had been claimed before, control is not limited to the glucose transporter alone [61-64], but is spread over the transporter and several of the glycolytic enzymes [9]. Even conditions can be found where the transporter does not, or only partially, control the glycolytic flux. Control depends considerably on the external glucose concentration and on the individual enzyme activities. Under the conditions studied control shifts from the transport of glucose, on one hand, to aldolase, GAPDH, PGK and G3PDH on the other hand, with the increase in the glucose concentration. Under physiological conditions most of the control seems associated with the transporter. The other kinases, HK, PFK and PYK, which are thought to control glycolysis in other organisms, have no significant control under any of the investigated conditions [9].

Most studies aiming at the design of new anti-glycolytic trypanocides have focused on the individual enzymes without taking into consideration whether the inhibition of such an enzyme would lead to the inhibition of the glycolytic flux. Until now efforts have focused on the enzymes aldolase [124, 125], GAPDH [79, 97-99, 124, 126], and PGK [110]. Although the studies by Bakker et al. [9] have shown that the individual enzymes of glycolysis exert less control than the transporter step, this does not imply that all efforts should now be shifted to the synthesis of inhibitors of glucose transport. In the case of the enzymes GAPDH and PGK it could be calculated that a difference in effectiveness for inhibition of the glycolytic flux could be overcome by the design of inhibitors that had a two-fold higher affinity for their respective enzymes.

Conclusion

The complete dependence of the bloodstream form trypanosome on glucose as a source of carbon and energy has made glycolysis an interesting target for drug intervention. No other pathway in the trypanosome has been studied in such detail as the glycolytic pathway. Its peculiar characteristics such as its compartmentation inside glycosomes and the invariably high positive charge of the enzymes in it [127], the fact that pyruvate rather than lactate is excreted as the end-product of the pathway and the reoxidation of glycosomal NADH via the trypanosome alternative oxidase, all have attracted considerable attention. Most of its enzymes have been overexpressed and studied in great detail and their properties have been compared with their mammalian homologues, in order to identify differences that could be exploited for the design of new and better drugs. One of the drawbacks of the glycolytic pathway as target for drug design is the fact that it is probably one of the oldest pathways in Nature. It is present in almost every cell and in general its enzymes are very well conserved. Despite the great evolutionary distance between the trypanosome and its mammalian host, most of the glycolytic enzymes still share some 50% positional identity at the amino acid level, and thus share the same three-dimensional structure. Moreover, active-site residues of the glycolytic enzymes are even better conserved and this does certainly not facilitate the use of substrate analogues in the development of specific inhibitors. Moreover, most substrate analogues would act as competitive inhibitors, which would require them to be of very high affinity in order to exert any significant effect on the functioning of the overall pathway. Therefore, the development of irreversible inhibitors has received a lot of attention [79] and also much attention has been given to differences found on the surface of these enzymes, such as their high overall positively charge [127, 128], modified interface loops and unique amino acid insertions. The obvious differences with the host, such as the trypanosome GAPDH binding pocket for NAD [98], the peculiar nature of trypanosome PFK [80], the regulation of PYK by $Fru(2,6)P_2$, the pyruvate transporter [72] and the trypanosome alternative oxidase, all constitute the more attractive targets for the development of a new and specific chemotherapy. On the other hand, the recent mathematical modelling of the pathway [9, 129] has identified the steps that exert the highest control in the pathway, such as the glucose transporter, aldolase, GAPDH and G3PDH and PGK. These steps, at least from theoretical considerations, would best translate an inhibitory effect on an individual enzyme into an inhibition of the overall flux of the pathway. An entirely different type of approach, not discussed in this review, would be the interference with the compartmentation of glycolysis inside the glycosome itself. Any inhibition of, or interference with, receptors for peroxisome targeting signals would directly affect the importation of several glycolytic enzymes at the same time and eventually slow down or even interrupt the glycolytic flux.

Many different approaches have been taken so far. It can only be hoped that one of them, either alone or in combination, will lead to the development of a more effective and safer treatment of human African trypanosomiasis.

Acknowledgements

I thank my colleague Dr. Paul Michels for the critical reading of the manuscript. The writing of this paper was supported by a grant from the Interuniversity Poles of Attraction Programme - Belgian State, Prime Minister's Office - Federal Office for Scientific, Technical and Cultural Affairs.

References

1. Opperdoes FR (1987) Compartmentation of carbohydrate metabolism in trypanosomes. Annu Rev Microbiol 41:127-151
2. Opperdoes FR (1988) Glycosomes may provide clues to the import of peroxisomal proteins. TIBS 13:255-260
3. Opperdoes FR, Michels PAM (1993) The glycosomes of the Kinetoplastida. Biochimie 75:231-234
4. Hannaert V, Michels PAM (1994) Structure, function and biogenesis of glycosomes in Kinetoplastida. J Bioenerg Biomembr 26:205-212
5. Michels PAM, Hannaert V (1994) The evolution of Kinetoplastid glycosomes. J Bioenerg Biomembr 26:213-219
6. Opperdoes FR, Borst P (1977) Localization of nine glycolytic enzymes in a microbody-like organelle in *Trypanosoma brucei*: the glycosome. FEBS Lett 80:360-364
7. Misset O, Bos OJM, Opperdoes FR (1986) Glycolytic enzymes of *Trypanosoma brucei*. Simultaneous purification, intraglycosomal concentrations and physical properties. Eur J Biochem 157:441-453
8. Michels PAM (1988) Compartmentation of glycolysis in trypanosomes: a potential target for new trypanocidal drugs. Biology of the Cell 64:157-164
9. Bakker BM, Michels PAM, Opperdoes FR, Westerhoff H (1998) What controls glycolysis in bloodstream form *Trypanosoma brucei*? J Biol Chem (submitted)
10. Siedow JN, Umbach AL (1995) Plant mitochondrial electron transfer and molecular biology. Plant Cell 7:821-831
11. Lambowitz AM, Sabourin JR, Bertrand H, Nickels R, McIntosh L (1989) Immunological identification of the alternative oxidase of *Neurospora crassa* mitochondria. Mol Cell Biol 9:1362-1364
12. Day DA, Whelan J, Millar AH, Siedow JN, Wiskich JT (1995) Regulation of the alternative oxidase in plants and fungi. Aust J Plant Physiol 22:497-509
13. Sanchez-Moreno M, Lasztity D, Coppens I, Opperdoes FR (1992) Characterization of carbohydrate metabolism and demonstration of glycosomes in a *Phytomonas* sp. isolated from *Euphorbia characias*. Mol Biochem Parasitol 54:185-200
14. Van Hellemond JJ, Simons B, Milenaar FF, Tielens AG (1998) A gene encoding the plant-like alternative oxidase is present in *Phytomonas* but absent in *Leishmania* spp. J Eukaryot Microbiol 45:426-430
15. Hammond DJ, Bowman IBR (1980a) Studies on glycerol kinase and its role in ATP synthesis in *Trypanosoma brucei*. Mol Biochem Parasitol 2:77-91
16. Hammond DJ, Bowman IBR (1980b) *Trypanosoma brucei*: the effect of glycerol on the anaerobic metabolism of glucose. Mol Biochem Parasitol 2:63-75
17. Fairlamb AH, Opperdoes FR (1986) Carbohydrate metabolism in African trypanosomes, with special reference to the glycosome. In: Morgan MJ (ed) Carbohydrate metabolism of cultured cells. Plenum Press Corporation, pp 183-224
18. Opperdoes FR, Borst P, Fonck K (1976) The potential use of inhibitors of glycerol-3-phosphate oxidase for chemotherapy of African trypanosomiasis. FEBS Letters 62:169-172
19. Eisenthal R, Panes A (1985) The aerobic/anaerobic transition of glucose metabolism in *Trypanosoma brucei*. FEBS Lett 181:23-27
20. Fairlamb AH, Opperdoes FR, Borst P (1977) New approach to screening drugs for activity against African trypanosomes. Nature 265:270-271
21. Clarkson AB Jr, Brohn FH (1976) Trypanosomiasis: an approach to chemotherapy by the inhibition of carbohydrate catabolism. Science 194:204-206
22. Van der Meer C, Versluijs-Broers JA (1979) *Trypanosoma brucei*: trypanocidal effect of salicylhydroxamic acid plus glycerol in infected rats. Exp Parasitol 48:126-134
23. Van der Meer C, Versluijs-Broers JA (1986) *Trypanosoma brucei* and *T. vivax*: salicylhydroxamic acid and glycerol treatment of acute and chronically infected rats. Exp Parasitol 62:98-113

24. Grady RW, Bienen EJ, Clarkson AB Jr (1986a) p-Alkyloxybenzhydroxamic acids, effective inhibitors of the trypanosome glycerol-3-phosphate oxidase. Mol Biochem Parasitol 19:231-240
25. Grady RW, Bienen EJ, Clarkson AB Jr (1986b) Esters of 3,4-dihydroxybenzoic acid, highly effective inhibitors of the sn-glycerol-3-phosphate oxidase of *Trypanosoma brucei brucei*. Mol Biochem Parasitol 21:55-63
26. Grady RW, Bienen EJ, Dieck HA, Saric M, Clarkson AB Jr (1993) N-n-alkyl-3,4-dihydroxybenzamides as inhibitors of the trypanosome alternative oxidase: activity *in vitro* and *in vivo*. Antimicrob Agents Chemother 37:1082-1085
27. Minagawa N, Yabu Y, Kita K, Nagai K, Ohta N, Meguro K, Sakajo S, Yoshimoto A (1997) An antibiotic, ascofuranone, specifically inhibits respiration and *in vitro* growth of long slender bloodstream forms of *Trypanosoma brucei brucei*. Parasitology 84:271-280
28. Chaudhuri M, Hill GC (1996) Cloning, sequencing, and functional activity of the *Trypanosoma brucei brucei* alternative oxidase. Mol Biochem Parasitol 83:125-129
29. Nwagwu M, Opperdoes FR (1982) Regulation of glycolysis in *Trypanosoma brucei*: hexokinase and phosphofructokinase activity. Acta Trop 39:61-72
30. Cronin CN, Tipton KF (1985) Purification and regulatory properties of phosphofructokinase from *Trypanosoma (Trypanozoon) brucei brucei*. Biochem J 227:113-124
31. Cronin CN, Tipton KF (1987) Kinetic studies on the reaction catalysed by phosphofructokinase from *Trypanosoma brucei*. Biochem J 245:13-18
32. Van Schaftingen E, Opperdoes FR, Hers HG (1985) Stimulation of *Trypanosoma brucei* pyruvate kinase by fructose 2,6-bisphosphate. Eur J Biochem 153:403-406
33. Van Schaftingen E, Opperdoes FR, Hers HG (1987) Effects of various metabolic conditions and of the trivalent arsenical melarsen oxide on the intracellular levels of fructose 2,6-bisphosphate and of glycolytic intermediates in *Trypanosoma brucei*. Eur J Biochem 166:653-661
34. Callens M, Kuntz DA, Opperdoes FR (1991a) Characterization of pyruvate kinase of *Trypanosoma brucei* and its role in the regulation of carbohydrate metabolism. Mol Biochem Parasitol 47:19-30
35. Callens M, Opperdoes FR (1992) Some kinetic properties of pyruvate kinase from *Trypanosoma brucei*. Mol Biochem Parasitol 50:235-244
36. Bowman IBR, Flynn IW (1976) Oxidative metabolism of trypanosomes. In: Lumsden WHR, Evans DA (eds) Biology of the Kinetoplastida. Vol. I, Academic Press, New York, pp 435-476
37. Hart DT, Misset O, Edwards SW, Opperdoes FR (1984) A comparison of the glycosomes (microbodies) isolated from *Trypanosoma brucei* bloodstream form and cultured procyclic trypomastigotes. Mol Biochem Parasitol 12:25-35
38. Misset O, Van Beeumen J, Lambeir A-M, van der Meer R, Opperdoes, FR (1987) Glyceraldehyde-phosphate dehydrogenase from *Trypanosoma brucei*: comparison of the glycosomal and cytosolic isoenzymes. Eur J Biochem 162:501-507
39. Ryley JF (1962) Studies on the metabolism of the protozoa. 9. Comparative studies on the metabolism of the bloodstream and culture forms of *Trypanosoma rhodesiense*. Biochem J 85:211-223
40. Ter Kuile BH (1997) Adaptation of metabolic enzyme activities of *Trypanosoma brucei* promastigotes to growth rate and carbon regimen. J Bacteriol 179: 4699-4705
41. Brun R, Schönenberger M (1979) Cultivation and in vitro cloning of procyclic culture forms of *Trypanosoma brucei* in a semi-defined medium. Acta Trop 36: 289-292
42. Opperdoes FR, Cottem D (1982) Involvement of the glycosome of *Trypanosoma brucei* in carbon dioxide fixation. FEBS Lett 143:60-64
43. Cymeryng C, Cazzulo JJ, Cannata JJ (1995) Phosphoenolpyruvate carboxykinase from *Trypanosoma cruzi*. Purification and physicochemical and kinetic properties. Mol Biochem Parasitol 73:91-101
44. Durieux PO, Schutz P, Brun R, Kohler P (1991) Alterations in Krebs cycle enzyme activities and carbohydrate catabolism in two strains of *Trypanosoma brucei* during *in vitro* differentiation of their bloodstream to procyclic stages. Mol Biochem Parasitol 45:19-27
45. Overath P, Czichos J, Haas C (1986) The effect of citrate/cis-aconitate on oxidative metabolism during transformation of *Trypanosoma brucei*. Eur J Biochem 160:175-182
46. Opperdoes FR, Borst P, Bakker S, Leene W (1977) Localization of glycerol-3-phosphate oxidase in the mitochondrion and NAD^+-linked glycerol-3-phosphate dehydrogenase in the microbodies of the bloodstream form of *Trypanosoma brucei*. Eur J Biochem 76:29-39
47. Jenkins TM, Eisenthal R, Weitzman PD (1988) Two distinct succinate thiokinases in both bloodstream and procyclic forms of *Trypanosoma brucei*. Biochem Biophys Res Commun 151:257-261
48. Opperdoes FR, Markos A, Steiger RF (1981) Localization of malate dehydrogenase, adenylate kinase and glycolytic enzymes in glycosomes and the threonine pathway in the mitochondrion of cultured procyclic trypomastigotes of *Trypanosoma brucei*. Mol Biochem Parasitol 4:291-309
49. Mracek J, Snyder SJ, Chavez UB, Turrens JF (1991) A soluble fumarate reductase in *Trypanosoma brucei* procyclic trypomastigotes. J Protozool 38:554-558
50. Turrens JF (1989) The role of succinate in the respiratory chain of *Trypanosoma brucei* procyclic trypomastigotes. Biochem J 259:363-368

51. Turrens JF, Watts BP Jr, Zhong L, Docampo R (1996) Inhibition of *Trypanosoma cruzi* and *T. brucei* NADH fumarate reductase by benznidazole and anthelmintic imidazole derivatives. Mol Biochem Parasitol 82:125-129
52. Van Hellemond JJ, Klockiewicz M, Gaasenbeek, CPH, Roos MH, Tielens AGM (1995) Rhodoquinone and complex II of the electron transport chain in anaerobically functioning eukaryotes. J Biol Chem 270:31065-31070
53. Van Hellemond JJ, Opperdoes FR, Tielens AGM (1998) Trypanosomatidae produce acetate via a mitochondrial acetate: succinate CoA transferase. Proc Natl Acad Sci USA 95:3036-3041
54. Lindmark DG (1976) Acetate production by *Tritrichomonas foetus*. In: Van den Bossche H, (ed) Biochemistry of parasites and host-parasite relationships. Elsevier/North Holland Biomedical Press, Amsterdam, pp 15-21
55. Müller M (1997) Evolutionary origins of trichomonad hydrogenosomes. Parasitol Today 13:166-167
56. Beattie DS, Howton MM (1996) The presence of rotenone-sensitive NADH dehydrogenase in the long slender bloodstream and the procyclic forms of *Trypanosoma brucei* brucei. Eur J Biochem 241:888-894
57. Popov VN, Simonian RA, Skulachev VP, Starkov AA (1997) Inhibition of the alternative oxidase stimulates H_2O_2 production in plant mitochondria. FEBS Lett 415:87-90
58. Cronin CN, Nolan DP, Voorheis HP (1989) The enzymes of the classical pentose phosphate pathway display differential activities in procyclic and bloodstream forms of *Trypanosoma brucei*. FEBS Lett 244:26-30
59. Southworth GC, Read CP (1970) Specificity of sugar transport in *Trypanosoma gambiense*. J Protozool 17:396-399
60. Bringaud F, Baltz T (1992) A potential hexose transporter gene expressed predominantly in the bloodstream form of *Trypanosoma brucei*. Mol Biochem Parasitol 52:111-121
61. Eisenthal R, Game S, Holman GD (1989) Specificity and kinetics of hexose transport in *Trypanosoma brucei*. Biochim Biophys Acta 985:81-89
62. Gruenberg J, Sharma PR, Deshusses J (1978) D-glucose transport in *Trypanosoma brucei*. D-glucose transport is the rate-limiting step of its metabolism. Eur J Biochem 89:461-469
63. Seyfang A, Duszenko M (1991) Specificity for glucose transport in *Trypanosoma brucei*. Effective inhibition by phloretin and cytochalasin B. Eur J Biochem 202:191-196
64. Ter Kuile BH, Opperdoes FR (1991) Glucose uptake by *Trypanosoma brucei*. J Biol Chem 266:857-862
65. Parsons M, Nielsen B (1990) Active transport of 2-deoxy-D-glucose in *Trypanosoma brucei* procyclic forms. Mol Biochem Parasitol 42:197-203
66. Ter Kuile BH, Opperdoes FR (1992) Mutual adjustment of glucose uptake and metabolism in *Trypanosoma brucei* grown in a chemostat. J Bacteriol 174:1273-1279
67. Wille U, Seyfang A, Duszenko M (1996) Glucose uptake occurs by facilitated diffusion in procyclic forms of *Trypanosoma brucei*. Eur J Biochem 236:228-233
68. Barrett MP, Tetaud E, Seyfang A (1995) Functional expression and characterization of the *Trypanosoma brucei* procyclic glucose transporter, THT2. Biochem J 312:687-691
69. Kasahara M, Hinkle PC (1977) Reconstitution and purification of the D-glucose transporter from human erythrocytes. J Biol Chem 252:7384-7390
70. Barnard JP, Reynafarje B, Pedersen PL (1993) Glucose catabolism in African trypanosomes. Evidence that the terminal step is catalyzed by a pyruvate transporter capable of facilitating uptake of toxic analogs. J Biol Chem 268:3654-3661
71. Wiemer EAC, Ter Kuile B, Michels PAM, Opperdoes FR (1992) Pyruvate transport across the membrane of the bloodstream form of *Trypanosoma brucei* is mediated by a facilitated diffusion carrier. Biochem Biophys Res Comm 184:1028-1034
72. Wiemer EAC, Michels PAM, Opperdoes FR (1995b) Inhibition of pyruvate transport across the plasma membrane of the bloodstream form of *Trypanosoma brucei* and its metabolic implications. Biochem J 312:479-484
73. Hara T, Kanbara H, Nakao M, Fukuma T (1997) Rapid uptake and phosphorylation of D-mannose, and limited D-mannose 6-phosphate isomerization in the glycolytic pathway of bloodstream forms of *Trypanosoma brucei gambiense*. Kurume Med J 44:105-113
74. Trinquier M, Périé J, Callens M, Opperdoes F, Willson M (1995) Specific inhibitors for the glycolytic enzymes of *Trypanosoma brucei*. Bioorg Med Chem 3:1423-1427
75. Misset O, Opperdoes FR (1984) Simultaneous purification of hexokinase, class-I fructose-bisphosphate aldolase, triosephosphate isomerase and phosphoglycerate kinase from *Trypanosoma brucei*. Eur J Biochem 144:475-483
76. Hannon RH, Parr CW (1978) The phosphoglucose isomerases of the bloodstream forms of *Trypanosoma brucei* and *Trypanosoma vivax*. Comp Biochem Physiol B60:177-181
77. Kuntz D, Kooystra U, Opperdoes FR (1988) Agaricic acid inhibits the glycolytic enzymes of *Trypanosoma brucei*. Arch Intern Phys Biochim 96:B99
78. Marchand M, Kooystra U, Wierenga RK, Lambeir A-M, Van Beeumen J, Opperdoes FR, Michels PAM (1989) Glucosephosphate isomerase from *Trypanosoma brucei*. Cloning and characterization of the gene and analysis of the enzyme. Eur J Biochem 184:455-464

79. Willson M, Lauth N, Périé J, Callens M, Opperdoes FR (1994) Inhibition of glyceraldehyde-3-phosphate dehydrogenase by phosphorylated epoxides and alpha-enones. Biochemistry 33:214-220
80. Michels PAM, Chevalier N, Opperdoes FR, Rider MH, Rigden DJ (1997) The glycosomal ATP-dependent phosphofructokinase of *Trypanosoma brucei* must have evolved from an ancestral pyrophosphate-dependent enzyme. Eur J Biochem 250:698-704
81. Clayton CE (1985) Structure and regulated expression of genes encoding fructose biphosphate aldolase in *Trypanosoma brucei*. EMBO J 4:2997-3003
82. Marchand M, Poliszczak A, Gibson WC, Wierenga RK, Opperdoes FR, Michels PAM (1988) Characterization of the genes for fructose-bisphosphate aldolase in *Trypanosoma brucei*. Mol Biochem Parasitol 29:65-76
83. Callens M, Kuntz DA, Opperdoes FR (1991b) Kinetic properties of fructose bisphosphate aldolase from *Trypanosoma brucei* compared to aldolase from rabbit muscle and *Staphylococcus aureus*. Mol Biochem Parasitol 47:1-9
84. Chevalier N, Callens M, Michels PAM (1995) High-level expression of *Trypanosoma brucei* fructose-1,6-bisphosphate aldolase in *Escherichia coli* and purification of the enzyme. Protein Expression and Purification 6:39-44
85. Clayton CE, Fox JA (1989) Phosphorylation of fructose bisphosphate aldolase in *Trypanosoma brucei*. Mol Biochem Parasitol 33:73-79
86. Swinkels BW, Gibson, WC, Osinga KA, Kramer R, Veeneman, GH, van Boom JH, Borst P (1986) Characterization of the gene for the microbody (glycosomal) triosephosphate isomerase of *Trypanosoma brucei*. EMBO J 5:1291-1298
87. Lambeir AM, Opperdoes FR, Wierenga RK (1987) Kinetic properties of triose-phosphate isomerase from *Trypanosoma brucei brucei*. Eur J Biochem 168:69-74
88. Wierenga RK, Noble MEM, Vriend G, Nauche S, Hol WGJ (1991) Refined 1.83 Å structure of trypanosomal triosephosphate isomerase crystallized in the presence of 2.4 M-ammonium sulphate. J Mol Biol 220:995-1015
89. Borchert TV, Pratt K, Zeelen JP, Callens M, Noble MEM, Opperdoes FR, Michels PAM, Wierenga RK (1993) Overexpression of trypanosomal triosephosphate isomerase in *Escherichia coli* and characterisation of a dimer-interface mutant. Eur J Biochem 211:703-710
90. Kuntz DA, Osowski R, Schudok M, Wierenga RK, Müller K, Kessler H, Opperdoes FR (1992) Inhibition of triosephosphate isomerase from *Trypanosoma brucei* with cyclic hexapeptides. Eur J Biochem 207:441-447
91. Callens M, Van Roy J, Zeelen JP, Borchert TV, Nalis D, Wierenga RK, Opperdoes FR (1993) Selective interaction of glycosomal enzymes from *Trypanosoma brucei* with hydrophobic cyclic hexapeptides. Biochem Biophys Res Commun 195:667-672
92. Gao XG, Garza-Ramos G, Saavedra-Lira E, Cabrera N, Tuena de Gomez-Puyou M, Perez-Montfort R, Gomez-Puyou A (1998) Reactivation of triosephosphate isomerase from three trypanosomatids and human: effect of suramin. Biochem J 332:91-96
93. Fairlamb AH, Bowman IB (1980) Uptake of the trypanocidal drug suramin by bloodstream forms of *Trypanosoma brucei* and its effect on respiration and growth rate in vivo. Mol Biochem Parasitol 1:315-333
94. Michels PAM, Poliszczack A, Osinga KA, Misset O, Van Beeumen J, Wierenga RK, Borst P, Opperdoes FR (1986) Two tandemly linked identical genes code for the glycosomal glyceraldehyde-phosphate dehydrogenase in *Trypanosoma brucei*. EMBO J 5:1049-1056
95. Hannaert V, Opperdoes FR, Michels PAM (1998) Comparison and evolutionary analysis of the glycosomal glyceraldehyde-3-phosphate dehydrogenase from different Kinetoplastida. J Mol Evol 47:728-738
96. Lambeir A-M, Loiseau A, Kuntz DA, Vellieux FMD, Michels PAM, Opperdoes FR (1991) The cytosolic and glycosomal glyceraldehyde 3-phosphate dehydrogenase isoenzymes from bloodstream form *Trypanosoma brucei*: Kinetic properties and comparison with homologous enzymes. Eur J Biochem 198:429-435
97. Vellieux FMD, Hajdu J, Verlinde CLMJ, Groendijk H, Read RJ, Greenhough TJ, Campbell JW, Kalk KH, Littlechild JA, Watson HC, Hol WGJ (1993) Structure of glycosomal glyceraldehyde-3-phosphate dehydrogenase from *Trypanosoma brucei* determined from Laue data. Proc Natl Acad Sci USA 90:2355-2359
98. Verlinde CLMJ, Callens M, Van Calenbergh S, Van Aerschot A, Herdewijn P, Hannaert V, Michels PAM, Opperdoes FR, Hol WGJ (1994) Selective inhibition of trypanosomal glyceraldehyde-3-phosphate dehydrogenase by protein structure based design: toward new drugs for the treatment of sleeping sickness. J Med Chem 37:3605-3613
99. Van Calenbergh S, Verlinde CL, Soenens J, De Bruyn A, Callens M, Blaton NM, Peeters OM, Rozenski J, Hol WGJ, Herdewijn P (1995) Synthesis and structure activity relationships of analogs of 2'-deoxy-2'-(3-methylbenzamido)adenosine, a selective inhibitor of trypanosomal glycosomal glyceraldehyde-3-phosphate dehydrogenase. J Med Chem 38:3838-3849
100. Michels PAM, Marchand M, Kohl L, Allert S, Wierenga RK, Opperdoes FR (1991) The cytosolic and glycosomal isoenzymes of glyceraldehyde-3-phosphate dehydrogenase in *Trypanosoma brucei* have a distant evolutionary relationship. Eur J Biochem 198:421-428
101. Wiemer EAC, Hannaert V, Van den IJssel PRLA, Van Roy J, Opperdoes FR, Michels PAM (1995a) Molecular analysis of glyceraldehyde-3-phosphate dehydrogenase in *Trypanoplasma*

borelli. An evolutionary scenario of subcellular compartmentation in Kinetoplastida. J Mol Evol 40: 443-454
102. Kohl L, Drmota T, Do Thi C, Callens M, Van Beeumen J, Opperdoes FR, Michels PAM (1996) Glycerol-3-phosphate dehydrogenase of *Trypanosoma brucei* and *Leishmania mexicana*. Cloning and characterization of the genes and overexpression of the trypanosomal protein in *Escherichia coli*. Mol Biochem Parasitol 76:159-173
103. Stebeck CE, Frevert U, Mommsen TP, Vassella E, Roditi I, Pearson TW (1996) Molecular characterization of glycosomal $NAD^{(+)}$-dependent glycerol-3-phosphate dehydrogenase from *Trypanosoma brucei rhodesiense*. Mol Biochem Parasitol 76: 145-158
104. Krakow JL, Wang CC (1990) Purification and characterization of glycerol kinase from *Trypanosoma brucei*. Mol Biochem Parasitol 43:17-25
105. Osinga KA, Swinkels BW, Gibson WC, Borst P, Veeneman GH, Van Boom JH, Michels PAM, Opperdoes FR (1985) Topogenesis of microbody enzymes: a sequence comparison of the genes for the glycosomal (microbody) and cytosolic phosphoglycerate kinases of *Trypanosoma brucei*. EMBO J 4:3811-3817
106. Misset O, Opperdoes FR (1987) The phosphoglycerate kinases from *Trypanosoma brucei*: a comparison of the glycosomal and the cytosolic isoenzyme and their sensitivity towards Suramin. Eur J Biochem 162:493-500
107. Swinkels BW, Loiseau A, Opperdoes FR, Veenhuis M, Borst P (1992) A phosphoglycerate kinase related gene, conserved between *Trypanosoma brucei* and *Crithidia fasciculata*, encodes a glycosomal (microbody) protein. Mol Biochem Parasitol 50:69-78
108. Alexander K, Parsons M (1993) Characterization of a divergent glycosomal microbody phosphoglycerate kinase from *Trypanosoma brucei*. Mol Biochem Parasitol 60:265-272
109. Peterson GC, Sommer JM, Klosterman S, Wang CC, Parsons M (1997) *Trypanosoma brucei*: identification of an internal region of phosphoglycerate kinase required for targeting to glycosomal microbodies. Exp Parasitol 85: 16-23
110. Bernstein BE, Michels PAM, Hol WGJ (1997) Synergistic effects of substrate induced conformational changes in phosphoglycerate kinase activation. Nature 385:275-278
111. Visser N, Opperdoes FR (1980) Glycolysis in *Trypanosoma brucei*. Eur J Biochem 103:623-632
112. Allert S, Ernest I, Poliszczak A, Opperdoes FR, Michels PAM (1991) Molecular cloning and analysis of two tandemly-linked genes for pyruvate kinase of *Trypanosoma brucei*. Eur J Biochem 200:19-27
113. Ernest I, Callens M, Opperdoes FR, Michels PAM (1994) Pyruvate kinase of *Leishmania mexicana mexicana*. Cloning and analysis of the gene, overexpression in *Escherichia coli* and characterization of the enzyme. Mol Biochem Parasitol 64:43-54
114. Barrett MP, Le Page RWF (1993) A 6-phosphogluconate dehydrogenase gene from *Trypanosoma brucei*. Mol Biochem Parasitol 57: 89-100
115. Lobo Z, Maitra PK (1982) Pentose phosphate pathway mutants of yeast. Mol Gen Genet 185:367-368
116. Rosemeyer MA (1987) The biochemistry of glucose-6-phosphate dehydrogenase, 6-phosphogluconate dehydrogenase and glutathione reductase. Cell Biochem Funct 5: 79-95
117. Gvodzdev VA (1976) Role of the pentose phosphate pathway in metabolism of *Drosophila melanogaster* elucidated by mutations affecting glucose-6-phosphate and 6-phosphogluconate dehydrogenases. FEBS Lett 64:85-88
118. Barrett MP, Phillips C, Adams MJ, Le Page RW (1994) Overexpression in *Escherichia coli* and purification of the 6-phosphogluconate dehydrogenase of *Trypanosoma brucei*. Protein Expression and Purification 5:44-49
119. Phillips C, Barrett MP, Gover S, Le Page RW, Adams MJ (1993) Preliminary crystallographic study of 6-phosphogluconate dehydrogenase from *Trypanosoma brucei*. J Mol Biol 233:317-321
120. Hanau S, Rippa M, Bertelli M, Dallochio, F, Barrett, MP (1996) 6-Phosphogluconate dehydrogenase from *Trypanosoma brucei*: kinetic analysis and inhibition by trypanocidal drugs. Eur J Biochem 240:592-599
121. Hunt M, Kohler P (1995) Purification and characterization of phospho enol pyruvate carboxykinase from *Trypanosoma brucei*. Biochim Biophys Acta 1249:15-22
122. Sommer JM, Nguyen TT, Wang CC (1994) Phosphoenolpyruvate carboxykinase of *Trypanosoma brucei* is targeted to the glycosomes by a C-terminal sequence. FEBS Lett 350: 125-129
123. Boid R, Jones TW, Payne RC (1989) Malic enzyme type VII isoenzyme as an indicator of suramin resistance in *Trypanosoma*. Exp Parasitol 69:317-323
124. Périé J, Rivière-Alric I, Blonski C, Gefflaud T, Lauth de Viguerie N, Trinquier M, Willson M, Opperdoes FR, Callens M (1993) Inhibition of the glycolytic enzymes in the trypanosome: an approach in the development of new leads in the therapy of parasitic diseases. Pharmac Ther 60:347-365
125. Samson I, Rozenski J, Samyn B, Van Aerschot A, Van Beeumen J, Herdewijn P (1997) Screening a random pentapeptide library, composed of 14 D-amino acids, against the COOH-terminal sequence of fructose-1,6-bisphosphate aldolase from *Trypanosoma brucei*. J Biol Chem 272: 11378-11383
126. Kim H, Feil IK, Verlinde CLMJ, Petra PH, Hol WGJ (1995) Crystal structure of glycosomal glyceraldehyde-3-phosphate dehydrogenase from *Leishmania mexicana*: implications for

structure-based drug design and a new position for the inorganic phosphate binding site. Biochemistry 34: 14975-14987
127. Wierenga RK, Swinkels BW, Michels PAM, Osinga K, Misset O, Van Beeumen J, Gibson WC, Postma JPM, Borst P, Opperdoes FR, Hol WGJ (1987) Common elements on the surface of glycolytic enzymes from *Trypanosoma brucei* may serve as topogenic signals for import into glycosomes. EMBO J 6:215-221
128. Willson M, Callens M, Kuntz D, Périé J, Opperdoes FR (1993) Synthesis and activity of anti-hot-spot inhibitors specific for the glycolytic enzymes from *Trypanosoma brucei*. Mol Biochem Parasitol 59:201-210
129. Bakker BM, Michels PAM, Opperdoes FR, Westerhoff HV (1997) Glycolysis in bloodstream form *Trypanosoma brucei* can be understood in terms of the kinetics of the glycolytic enzymes. J Biol Chem 272:3207-3215

CHAPTER 5

Polyamine metabolism

JC Breton, B Bouteille

Introduction

The polyamines encountered in the parasites responsible for African trypanosomiasis include putrescine or 1,4-diaminopropane and its mono N-propylated form, spermidine [1, 2]. These amines not only play a role in the parasite's growth and differentiation, but also contribute to the formation of a trypanosome-specific molecule, trypanothione, which represents for the trypanosome the equivalent of glutathione in eukaryotic cells. The enzymatic systems which participate in the synthesis of these polyamines are hence preferential targets for research of inhibitors which could be put to use as antiparasitic drug treatment. Perhaps it is such a potential clinical application that has encouraged the numerous general reviews of the metabolism of these polyamines [2-9]; see also Fig. 1).

Our discussion here will overview first spermidine and then trypanothione synthetic pathways, concentrating on the steps where inhibition may be most promising.

Polyamine synthetic pathway: from ornithine and methionine to spermidine

Spermidine results from decarboxylation of ornithine (creating putrescine) followed by an aminopropylation reaction involving decarboxylated S-adenosylmethionine. We will review these different steps in succession.

Putrescine

Formation of putrescine

The enzyme ornithine decarboxylase (ODC, E.C. 4.1.1.17) or L-ornithine carboxylyase transforms L-ornithine (L-Orn) into putrescine by decarboxylation utilizing pyridoxal phosphate (PLP) as coenzyme.

$$\text{L-Orn} \xrightarrow{ODC} \text{Putrescine} + CO_2$$

As ODC is an essential enzyme for polyamine synthesis and cell growth, it has been well studied in mammals, microbes, and plants (see [10]). Its full amino acid sequence has been deciphered from cDNA nucleotide sequences matching the mRNA ODC coding regions in mice [11] and in trypanosomes [12, 13]. The enzyme is a homodimer. Each subunit is composed of 423 amino acids [14] and the active site, located at the dimer interface, is composed of the N terminal domain of one subunit and the C terminal domain of the other [15, 16]. At the N-terminal end of the subunit, lysine (Lys 69) is linked by its terminal amino group to PLP forming a Schiff base. This Lys is situated in a PFYAVKC sequence found in all eukaryotic ODCs. The substrate, ornithine, by Schiff base reaction binds to PLP in place of Lys to undergo decarboxylation. This mechanism occurs via formation of a quinoid intermediate that was recently demonstrated by Brooks and Phillips [17]. Additionally, glutamic acid (Glu 274) stabilizes the positive charge of the PLP's pyridine nitrogen, while the 5'-phosphate of PLP interacts with a glycine-rich loop (Gly 235-237) and with arginine (Arg 277) by ionic bonding [18]. At the C terminal end of the other ODC subunit, are located the amino acid residues which interact with the substrate. Studies utilizing difluoromethylornithine (DFMO), a false substrate which serves as an enzyme inhibitor, have revealed a covalent bond with Cys 360 [18, 19] which is situated in a WGPTCDGL(I)D sequence common to all eukaryotes [19]. The roles of Lys 69 and Cys 360 have been confirmed by use of mutants where these amino acids were replaced by alanine [15, 20]. Comparison of certain ODC sequences in *Trypanosoma brucei* and mice have shown a 60% homology with notable conservation of the sequence involving the dimer interface. This homology has enabled creation of crossed species heterodimers [21] in order to elucidate structure-activity correspondences without the ability to crystallize ODC for X-ray diffraction study.

In mammals, ODC is characterized by its particularly rapid turnover, the most rapid among known enzymes, with a half-life of 10 to 30 min. Such rapid disappearance is due to antizyme, a protein inhibitor of ODC which dissociates the enzyme's dimers and binds to the protein chain. The enzyme protein is then degraded at the 26 S proteasome by a process that is ATP-dependent, but not requiring ubiquitin [22]. Polyamines exert feedback-control on ODC synthesis [23] and on its degradation by induction of antizyme [14]. Truncation of the C terminal portion of mouse ODC renders it more stable, suggesting the presence of a degradation signal in this region [14]. The ODC in mammals, like most proteins with a short half-life, possess a region termed PEST, rich in proline (P), glutamic acid (E), serine (S), and threonine (T). Mammalian ODC contains 2 PEST regions with one localized in the C terminal region at a site distinct from the antizyme-binding region [24]. It appears that the C-terminal PEST region is requisite for basal degradation while the antizyme's binding region is necessary for degradation under polyamine control [24].

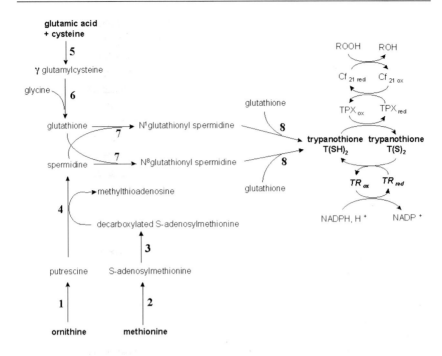

Fig. 1. Polyamine metabolism in *T. brucei* (1, ornithine decarboxylase; 2, S-adenosylmethionine synthetase; 3, S-adenosylmethionine decarboxylase; 4, spermidine synthase; 5, γ-glutamylcysteine synthetase; 6, glutathione synthetase; 7, glutathionylspermidine synthetase; 8, trypanothione synthetase; TR, trypanothione reductase; TPX, tryparedoxine or Cf16)

Comparison of the ODC amino acid composition in *T. brucei* with that of the mouse [12, 24] yields the following differences:
– 423 amino acids composing the ODC of the trypanosome, 461 for that of the mouse;
– 61.5% homology between the 376 amino acids of the enzymes' central region;
– 20 additional amino acids located at the N terminal end in the trypanosomal ODC;
– 36 fewer amino acids in the C terminal end of *T. brucei* ODC than in the mouse.

While the mouse ODC is rapidly degraded, (< 1h), that of *T. brucei* is more stable (> 6h). If one expresses the ODC gene of the parasite and that of the mouse in ovarian cells of the CHO hamster deficient in ODC, one observes that the trypanosomal ODC is stable while that of the mouse is unstable. Thus, degradation rate must be at least partially attributable to the intrinsic properties of the protein structure of the enzyme itself [25]. While the mouse ODC is capable of binding antizyme and thereby undergo inhibition, the *T. brucei* ODC doesn't bind, and thus cannot be inhibited by antizyme. Since polyamine feedback inhibition is largely via antizyme, this may explain the absence of ODC regulation by the intracellular level of polyamines observed in *T. brucei* [24]. Additionally, the truncation of the C terminal tail and the absence of PEST sequence in the C terminal region favor a longer ODC half-life in the parasite by also avoiding the mechanism for its basal degradation. Recombinant *T. brucei* ODC with a C terminal tail from mammalian ODC, if expressed in mammalian cells, confers a rapid half-life on this protein, thereby demonstrating the key role played by this C terminal region [14]. These mechanisms are reviewed by Fairlamb and Le Quesne [9].

Ornithine decarboxylase inhibitors

The first demonstration of the activity of ODC on mammalian cell growth was obtained through use of a competitive inhibitor, a methylated substrate analog, α-methylornithine [26]. As this drug product was not sufficiently active, researchers at Merell Dow synthesized the bifluorinate derivative, α-difluoromethylornithine (DFMO) [27], which was found to be effective. DFMO serves as an ODC substrate and reacts with PLP; after its decarboxylation, the imine intermediate inactivates the enzyme by forming a covalent bond with a nucleophilic residue of the enzyme [28]. This mechanism was further worked out by Poulin et al. [19] using [5-^{14}C]-DFMO; they demonstrated that the covalent bond is made with the Cys 360 in mouse ODC. DFMO also appears to have activity on the ODC of *T. brucei* [29]. *In vitro* evidence of this parasite ODC inhibition can be surmised from depletion of the intracellular pool of polyamines in *T. brucei* after DFMO introduction [30]; furthermore, this effect is reversible *in vitro* by introducing putrescine to the media [12]. This polyamine inhibition is manifested in *T. brucei* by change from the organism's long slender forms, capable of rapid movement and multiplication and having mitochondria without cristae, to short stumpy forms presenting slow movement, incapable of multiplying, and presenting mitochondria with numerous cristae [13, 30-32]. These short forms are also not capable of modifying their variable surface glycoprotein and, because of this fact, are much more susceptible to the host's immune defenses. It has been shown, *in vitro* as well as *in vivo*, a general reduction in protein synthesis and variable surface antigen as measured by diminution in incorporated myristic acid [33, 34]. Another natural consequence of inhibition of polyamine synthesis is a reduction in trypanothione levels which, as we will see later, additionally favors the trypanosome clearance by DFMO [35].

Despite all these points of impact on *T. brucei*'s metabolism, DFMO has no activity with respect to *T. rhodesiense* [36].

S-adenosylmethionine (AdoMet)

Formation of S-adenosylmethionine

This molecule has a dual role: it participates in transmethylation reactions, but also acts in the transfer of an aminopropyl group.

The sulfur atom of methionine (Met) attacks the 5' carbon of ATP, rapidly hydrolysing it into mineral phosphate and pyrophosphate, with accompanying transfer of an adenosine group:

$$\text{Met} + \text{ATP} \rightarrow \text{AdoMet} + \text{Pi} + \text{PP}$$

The enzyme which catalyzes this reaction is an adenosyl-methionine transferase (S-adenosylmethionine synthetase or AdoMet synthetase, E.C. 2.5.1.6). Active over a wide pH range (7.5 to 10.0), it has linear kinetics with respect to ATP for concentrations from 0.01 to 0.25 mM with a K_M of $53 \pm 6\,\mu M$, but for concentrations between 0.5 and 5.0 mM the K_M changes (to 1.35 ± 0.21 mM). One observes a biphasic curve according to the Lineweaver-Burk curve when the Met concentration ranges from 10 µM to 50 mM. These facts suggest the existence of two isoforms of trypanosome AdoMet synthetase [37], as have been previously described in mammalian cells [38, 39]. The molecular weight of the enzyme, determined by agarose gel filtration, is 145 kDa. AdoMet, the reaction product, is a potent inhibitor of this enzyme with a K_i of 0.24 mM, an effect 8 to 80-fold greater than that found for the mammalian, yeast or *Escherichia coli* enzyme [37].

S-adenosylmethionine synthetase inhibitors

Analogs of L-Met (D-Met, selenoMet and DL-ethionine) are competitive inhibitors with a K_i of 0.4 to 0.7 mM. *cis* AMB (L-2-amino-4-methoxy-*cis*-but-3-enoic acid), while capable of inhibiting AdoMet synthetase in mammalian tumor cells, is less effective than the Met analogs on the trypanosomal enzyme ($K_i = 1.40$ mM) and serves as additional proof of the difference between these two enzymes [37].

During DFMO treatment the pool of putrescine is depleted, and decarboxylated AdoMet (dcAdoMet), lacking the substrate for donating its aminopropyl radical, increases signficantly in concentration in mammalian [40] as well as trypanosomal cells. However, in trypanosomes, the response to DFMO is paradoxical. The high value of K_i for AdoMet synthetase in the trypanosome explains in part the increases AdoMet levels seen in the presence of DFMO [37]. It is also necessary to take into account the role played by AdoMet in the course of transmethylation reactions.

Other transmethylation enzymes and reaction products have been studied in the presence of DFMO by Yarlett and Bacchi [41] and more recently by Bacchi et al. [42]; the latter group, utilizing sinefungin (a transmethylation inhibitor) have demonstrated that sinefungin increases the half-life and decreases AdoMet turnover. This demonstrates that the trypanosome metabolizes Met along transmethylation pathways.

Decarboxylated S-adenosylmethionine (dcAdoMet)

Synthesis of dcAdoMet

S-adenosylmethionine decarboxylase (AdoMetDC, E.C.4.1.1.5.0) decarboxylates AdoMet and forms 5'-deoxyadenosyl-(5')-3-methylthiopropylamine (dcAdoMet):

$$\text{Ado} - \overset{\oplus}{\underset{CH_3}{S}} - CH_2 - CH_2 - \underset{COOH}{CH} - NH_2 \xrightarrow{AdoMetDC} \text{Ado} - \overset{\oplus}{\underset{CH_3}{S}} - CH_2 - CH_2 - CH_2 - NH_2 + CO_2$$

AdoMet → dcAdoMet

This enzyme has been described in both eukaryotes and prokaryotes [4, 43, 44]. While most amino acid decarboxylases have PLP as a cofactor, AdoMet DC has a pyruvate covalently bound to the enzyme's catalytic site and belongs to the group of "Pyruvoyl-dependent enzymes" [45]. This pyruvoyl prosthetic group is found in all species' AdoMet DC [46].

In mammals, AdoMetDC is synthesized in the form of a 38 kDa proenzyme [23]. During enzyme maturation, cleavage at a Ser residue (Ser 68 in man) liberates two chains:
– a large subunit of 32 kDa with Ser at the N terminal end which is transformed into pyruvic acid serving as cofactor, and also, as in ODC, a PEST region;
– a small subunit of 6 kDa whose function is poorly understood, but which contributes to the activity of the molecule [44].

The two subunits form an $\alpha_2\beta_2$ heterotetramer in eukaryotes and an $\alpha_4\beta_4$ octameric structure in *E. coli* [47]. The AdoMetDC amino acid composition, determined from cDNA sequences, has been compared by Stanley and Shantz [44] among man, *Saccharomyces cerevisiae*, *Solanum tuberosum*, and *E. coli* with that of *Leishmania donovani* cells and *T. brucei brucei*, whose reference sequences are published in a database (EMBL/GenBank/DDBJ). The structure of the human gene for AdoMetDC has also now been determined [48].

The enzyme extract of *T. brucei brucei* has been purified via a MGBG-Sepharose column [49] and its properties compared to those of the human enzyme. SDS-PAGE electrophoresis of the purified enzyme produces a band corresponding to 34 kDa which differs slightly with the 32 kDa of human AdoMetDC, and a 26 kDa band which reflects only protein degradation of the 34 kDa band. This trypanosomal enzyme

does not crossreact with antiserum prepared against the human enzyme. Its activity is stimulated 7-fold by putrescine and less strongly by other diamines such as 1,3-diaminopropane (2.5-fold), or 1,5-diaminopentane, or cadaverine (4.5-fold). Tekwani et al. [49] have considered the enzyme to be putrescine-dependent. Heby and Persson [23] have shown that the enzyme activity increases during cell growth because of two changes:
– increased enzyme protein synthesis via increased mRNA and increased number of ribosomes per polysome producing improved translational efficiency;
– prolonged enzyme half-life.

Putrescine increases the catalytic activity of AdoMetDC by decreasing the K_M for the substrate AdoMet, and by facilitating the conversion of the proenzyme into active enzyme [44]. Spermidine at elevated concentrations inhibits this conversion and increases the rate of the enzyme's degradation [23].

S-adenosylmethionine decarboxylase inhibitors

A large number of molecules have been synthesized with the purpose of inhibiting AdoMetDC [50, 51]. These can be divided into the following groupings.

Competitive inhibitors

Methylglyoxal bis guanylhydrazone (MGBG) is an example of a reversible enzyme inhibitor, capable of binding to the pyruvic group of AdoMetDC:

Synthesized in Germany in 1898 by Thiele and Dralle, its activity against animal tumors was recognized in 1950 [52]; its *in vitro* trypanocidal effect was subsequently appreciated [53, 54]. This inhibitor has a K_i of 32 mM for the AdoMetDC of *T. brucei brucei*, higher than the K_i of 0.3 mM determined for AdoMetDC extracted from rat prostate [55]. MGBG also interferes with the parasite's incorporation of nucleosides [54] and demonstrates cytotoxicity. However, utilized at the dose as high as 25 mg/kg for 3 days, it does not cure mice infected with *T. brucei brucei* [56] and its inability to cross the blood-brain barrier also has limited its clinical interest.

There is greater interest in MGBG derivatives with greater AdoMetDC specificity, which do not inhibit diaminooxidase and hence are less toxic to the host, such as those belonging to the amidinoarylidene- and (hetero)-arylidene-guanylhydrazones series [57]. A particular bis amidinobenzylidene guanylhydrazone (CGP 40215A) has been demonstrated to be particularly active *in vitro* [58]:

The following Table 1 indicates the minimal inhibitory concentration (MIC) and the 50% inhibitory concentration (IC$_{50}$) of CGP 40215A for different trypanosomes *in vitro*.

Table 1

	MIC (µM)	IC$_{50}$ (µM)
T. brucei rhodesiense	0.093	0.0045
T. brucei gambiense	1.31	0.33
T. brucei brucei	0.18	0.052
T. congolense	0.37	0.08

CGP 40215A has been studied *in vitro* in the mouse with success against 19 isolates of *T. brucei* and one of *T. congolense*; significantly, some of these isolates were resistant to standard trypanocidal agents. Trypanosomiasis with CNS involvement has been successfully treated with a combination of DFMO (14 days of DFMO at 2% *per os* equaling 5 g/kg/day) and CGP 40215A administered at a dose of 5, 10 or 25 mg/kg/day for 3 or 7 days by intraperitoneal injection or with a subcutaneous implanted miniosmotic pump [56].

Pentamidine, diamidino-4,4'-diphenoxypentane, also acts on the trypanosomal AdoMetDC [55] by competitive inhibition. This inhibition may be observed at low concentration (10 µM), but one must reach pentamidine concentrations of 10 mM to obtain complete enzyme inhibition. However, in trypanosome-infected rats treated with pentamidine (4 mg/kg), when the parasites were isolated and their polyamine level was determined by HPLC [59], no modification was found in their synthesis of putrescine, spermidine, glutathionylspermidine and trypanothione. Its effect on AdoMetDC is hence not the primary target of pentamidine's action.

Irreversible inhibitors

One of the common characteristics of these numerous inhibitors is a general similarity with the structure of AdoMet.

Attracting particular attention is 5'-{[(Z)-4-amino-2-butenyl] methylamino}-5'-deoxyadenosine or (Z)AbeAdo or MDL 73811. A selective and potent inhibitor of rat liver AdoMetDC, it has been shown to inhibit the *T. brucei brucei* enzyme [60] by its action on the active site as a competing substrate, since high concentrations of AdoMet protect the enzyme. In the rat infected with *T. brucei brucei*, there is rapid inhibition of the trypanosomal enzyme with a resulting reduction in spermidine and an increase in putrescine. Cure is achieved with intraperitoneal injection twice daily for 4 days using a dose of 20 mg/kg. The mouse infected by a resistant strain of *T. brucei rhodesiense* can be successfully treated by a combination of MDL 73811 at a dose of 50 mg/kg per day for 5 days and DFMO, or by MDL 73811 used alone when injected by microosmotic pump at the same dose for 7 days [60]. This has been confirmed by others authors [61], using the mouse and *T. brucei rhodesiense*; they have also found that the effective dose is between 50 and 100 mg/kg per day for 7 days using an osmotic pump and that the dose can be reduced to 25 – 50 mg/kg depending on the isolates used when associated with DFMO (0.25 to 1% concentration in drinking water for 7 days). Using this molecule, one observes rapid parasite clearance from the blood circulation. Inhibition of AdoMetDC activity is complete within 10 min of MDL 73811 injection and is maintained for at least 4 hours [62]. Rapidly, one finds an elevated concentration (1.9 ± 0.1 mM) of (Z)AbeAdo in the trypanosome, suggesting its rapid uptake by the parasite's purine nucleoside transporters [62]; this also can explain the greater trypanosome sensitivity to this molecule *in vivo* as the host's cells do not have this transporter [63]. Perhaps the marked increase in AdoMet that occurs with AdoMetDC inhibition correlates with degree of antitrypanosomal activity, suggesting that AdoMet itself has trypanocidal properties [62] which may be linked to its enhancing transmethylation of proteins and lipids [64]. Moreover, inhibition of ODC further reduces AdoMetDC activity, yielding higher specific elevations of AdoMet; this finding has been put to use in therapeutic combinations of MDL 73811 and DFMO.

Other AdoMet analogs have been synthesized [65] and tested on *E. coli* and *T. brucei brucei*.

Among these, AdoMac or S-(5'-deoxy-5'-adenosyl)-1-amino-4-methylthio-2-cyclopentene has the greatest effectiveness with a IC_{50} of 5.2 µM on *T. brucei brucei* in culture. This compound is equally active against three drug-resistant *T. brucei rhodesiense* isolates (IC_{50} of 1, 16, and 25 µM). The specific molecular configuration necessary to achieve AdoMetDC inhibition has been worked out by Wu and Woster [66].

AdoMao or S-(5'-deoxy-5'-adenosyl)-1-aminoxy-4-(methylsulfonio)-2-cyclopentene also has potent trypanocidal activity *in vitro* [67].

5'-deoxy-5'-[(2-hydrazinoethyl)methylamino]adenosine or MHZEA has been found, among the other 14 molecules tested, to be even more inhibitory than MDL 73811 with a K_i of 0.4 µM [68]. Furthermore, it displays much greater effect on the parasite's enzyme than on human cells enzyme.

Spermidine

Formation of spermidine

In a reaction catalyzed by spermidine synthase (E.C. 2.5.1.16), dcAdoMet donates its aminopropyl group to putrescine, leading to the formation of spermidine and methylthioadenosine (MTA):

$$\text{dcAdoMet} + \text{Putrescine} \xrightarrow{\text{Spermidine synthase}} \text{Spermidine} + \text{MTA}$$

Spermidine synthase has been isolated and characterized in animal tissues [69, 70] and humans [71]. The structure and method of cloning human spermidine synthase have been described by Eloranta et al. [72]. It is composed of two identical subunits. The amino acid sequence is very different from that found in the bacterial enzyme. Utilizing cDNA, human spermidine synthase has been found to have a molecular weight of 33 827 and is composed of 302 amino acids. Shiramata et al. [73] have proposed a model of the enzyme's active site comprised of a large hydrophobic cavity adjacent to a negatively charged site where the protonated amino group of putrescine binds while its unbound amino group settles into another hydrophobic cavity of the enzyme protein to undergo aminopropylation by dcAdoMet.

This enzyme has been described by Bitonti et al. [74] in *T. brucei brucei* noting a molecular weight around 74 kDa, a K_M of 0.2 mM for putrescine and of 0.1 µM for dcAdoMet.

Spermidine synthase inhibitors

Spermidine synthase activity in inhibited *in vitro* by dicyclohexylamine (IC_{50} = 3 µM) and cyclohexylamine, two molecules who serve as competitive inhibitors of the putrescine molecule. However, in the trypanosome-infected mouse, administration of dicyclohexylamine does not prolong survival.

Having a nucleoside structure that is an analog of two substrates, S-adenosyl-1,8-diamino-3-thiooctane is a very potent spermidine synthase inhibitor (IC_{50} = 2.5 µM) showing stereospecificity (the R diastereoisomer has greater activity than the S form) [75].

Many other molecules have also shown promising inhibitory activity on the enzyme isolated from various animal tissues, notably:
– trans-4-methylcyclohexylamine with an IC_{50} of 1.7 µM against pig spermidine synthase [76];
– and, among nucleoside derivatives:
· S-adenosyl-1,12-diamino-3-thio-9-azadodecane [77];
· adenosine-5'-thioethers derivatives and their methylsulfonium salts have been shown to have activity against the enzyme isolated from rat prostate [78];
· adenosylspermidine which is the most potent of the group [79].

Another method of inhibition can be via analogs of spermidine's reaction product. The 5'-deoxy-5'-methylthioadenosine (MTA) also produced in the reaction catalyzed by spermidine synthesis is rapidly cleaved by MTA phosphorylase into adenine (reutilized by the trypanosome) and methylthioribosylphosphate (that ultimately liberates a Met which also may be reutilized by the parasite).

An MTA analog, 5'-deoxy-5'-hydroxyethylthioadenosine (HETA) is a preferential substrate for MTA phosphorylase. Administration of HETA prevents the action of the phosphorylase on MTA, its natural substrate, and thereby deprives the parasite of adenine which it is incapable of synthesizing for itself. HETA is also rapidly taken up by the parasite because of its purine nucleoside transporter [80]. *In vivo*, HETA and 3 O-acetylated derivatives have been tested in mice infected with *T. brucei brucei* or *T. brucei rhodesiense*. Greater than 60% cure was obtained with HETA at a dose of 200 mg/kg using continuous infusion by osmotic micropump for 7 days. The di- and triacetylated derivatives are also effective and have greater activity than the tri O-propionyled forms [81]; these derivatives may yield new therapeutic possibilities.

Spermidine/spermine-N^1-acetyltransferase (SSAT), an enzyme which helps modulate intracellular polyamine levels in mammals, can be targeted by the bis-alkylpolyamines; these agents have anti-tumor effect by inducing SSAT synthesis and downregulating ODC and AdoMetDC synthesis [82].

During these studies, some authors have investigated the antiparasitic activity of these agents, in addition to the known anti-tumor effect. Certain molecules have indeed shown activity against the trypanosome, notably the bis-benzoyl derivative MDL 27695 (I), active on isolates of *T. brucei brucei* and *T. brucei rhodesiense* (having IC_{50} ranging from 12.3 to 15.1 µM), or the bis methylenecycloheptane derivative (II) whose high potency exceeds that of the preceding derivative (IC_{50} = 1.125 to 0.98 µM), and is of the same order as melarsen oxide. This toxicity may be attributable to the trypanosome's tumor-like rapidity of metabolism [83].

$R_1\underset{H_2}{\overset{\oplus}{N}}$~~~$\underset{H_2}{\overset{\oplus}{N}}$~~()5~$\underset{H_2}{\overset{\oplus}{N}}$~~~$\underset{H_2}{\overset{\oplus}{N}}$~$R_2$

(I) $R_1 = R_2 =$ benzyl (II) $R_1 = R_2 =$ cycloheptylmethyl

Trypanothione synthetic pathway

Trypanothione (TS$_2$), a name given in 1985 by Fairlamb et al. [84] to N^1,N^8-bis(glutathionyl)spermidine, is associated with the enzyme trypanothione reductase (TR) and plays an essential role in maintenance of the intracellular redox system linked to thiols, defense against oxidative stress due to peroxides and free radicals, and detoxication of foreign molecules. The protozoan trypanothione/trypanothione reductase system of the genera *Trypanosoma* and *Leishmania* replaces the glutathione/glutathione reductase system of their mammalian hosts. This biochemical difference between parasite and host represents an attractive target for the development of new drug molecules. Trypanothione genesis requires glutathione synthesis by enzymes common with those of mammalian cells, and then conjugation of glutathione and spermidine, a step specific to the trypanosome.

Trypanothione synthesis

First step: glutathione synthesis

$$\text{L-Glu} + \text{L-Cys} + \text{ATP} \xrightarrow{\gamma\text{-GCS}} \text{L-}\gamma\text{-glutamyl-L-Cys} + \text{ADP} + \text{P}_i$$

The enzyme, γ-glutamylcysteine synthetase (γ-GCS), can be specifically inhibited by buthionine sulfoximine (BSO). Trials using BSO against *T. brucei* indicate that the trypanosome is more sensitive to glutathione depletion than mammalian cells are [85]. In the trypanosome-infected mouse, treatment with BSO reduces the parasitemia 50% within 5 hours. Parasites which are isolated at this point are extremely fragile and, by 16 to 18 hours after treatment, have disappeared from blood samples. However, these aparasitemic mice undergo relapse within several days, perhaps because of CNS sequestration of some parasites which can then reinfect the bloodstream [85]. Indeed, it has been shown that the BSO concentration in the CNS is lower than that found in other tissues [86]. γ-GCS is similarly inhibited by cystamine (87% inhibition at 0.5 mM). The enzyme has been cloned and corresponds to a 679 amino acid protein with a molecular weight of 77.4 kDa [87].

Glutathione synthetase then forms the tripeptide in the presence of ATP:

$$\text{L-}\gamma\text{-glutamyl-L-Cys + Gly + ATP} \xrightarrow{\textit{Glutathione synthetase}} \text{glutathion + ADP + P}_i$$

Second step: condensation with spermidine

This reaction creates an amide bond between the Gly of glutathione and a primary amino group of spermidine to form, in the presence of ATP, a N^1 or N^8 glutathionylspermidine which reacts with a new molecule of glutathione to form trypanothione. Among the two possible glutathionylspermidines, it is the N^8 form which is preferentially transformed into trypanothione; thus, the N^1 form may accumulate in the milieu [88].

The enzyme catalyzing the latter reaction, trypanothione synthetase, has been isolated and purified from *Crithida fasciculata* cultures [89]; it is different from glutathionylspermidine synthetase which catalyzes the first part of the reaction [90]. Its activity on various spermidine analogs has been measured: putrescine, 1,3-diaminopropane, and 1,8-diaminooctane are poor substrates, while 1-propyl-1,3-diaminopropane and spermine compete effectively with spermidine [89]. One should note that aminopropyl-cadaverine can replace spermidine to form a trypanothione analog: N^1,N^9-bis(glutathionyl)aminopropylcadaverine or homotrypanothione in *T. cruzi* whose epimastigote form can then synthesize the two derivatives, trypanothione and homotrypanothione [91].

Glutathionylspermidine synthetase, like trypanothione synthetase, is an ATP-dependent carbon-nitrogen ligase. ATP forms an acylphosphate at the site of the Gly carboxyl; this activation induces nucleophilic attack by one of spermidine's amines with formation of an amide bond.

Substitution of the substrate's carboxyl group with a phosphonic or phosphinic acid group inhibits the activity of enzymes of this class. Trials using these derivatives have also been conducted on trypanosomal enzymes. Verbruggen et al. [92] have demonstrated that there is weak inhibitory effect for those glutathione derivatives having a phosphonic or phosphinic pseudopeptide or those obtained by substituting valine (Val) or leucine (Leu) in place of L-Cys. The most active structure found was L-γ-Glu-L-Leu-Gly^P, having a K_i of 60 ± 9 μM.

More recently, Chen et al. [93] have synthesized phosphonic analogs of glutathionylspermidine [94]. The derivative, corresponding to the following structure,

$$\gamma\text{-Glu-Ala}-N(H)-CH_2-P(=O)(O^-)-O-(CH_2)_3-N(H)-(CH_2)_4-NH_2$$

is a potent inhibitor of glutathionylspermidine synthetase (K_i = 6 μM). The same derivative in the phosphinic series is also a strong inhibitor (K_i = 3.2 μM) [93].

Study of glutathionylspermidine's specificity with respect to the peptide portion of the substrate is possible by synthesis of glutathione analogs. It appears that recognition of the γ-glutamyl radical is essential as the least modification produces a loss of activity [95]. Alkylation of the thiol by a methyl, ethyl, or propyl radical produces a substrate equivalent or even superior to glutathione; conversely, S-butylation yields a very poor substrate.

Gly appears important for activity as its replacement by Ala renders the new peptide an inhibitor: γ-L-Glu-L-Val-L-Ala acts as a competitive inhibitor (K_i = 1.35 ± 0.05 mM), and γ-L-Glu-L-Leu-L-Ala functions as a non competitive inhibitor (K_i = 0.50 ± 0.01 mM).

Trypanothione reductase (TR)

Enzyme characteristics

The intracellular thiol-disulfur equilibrium is assured in the majority of aerobic organisms by glutathione and glutathione reductase (GR) [96]. In Trypanosomatidae, as seen above, glutathione is a key intermediary of trypanothione, but GR is absent, replaced by the specific trypanothione reductase (TR, E.C. 1.6.4.8). Thus, while host and parasite have very similar enzyme systems, there is a difference at the molecular level which could serve as a potential therapeutic target. TR and GR, both members of the family of disulfur oxidoreductases (as dihydrolipoamide dehydrogenase), reduce the disulfur substrate into dithiol homodimers using FAD as cofactor and NADPH as an electron donor [97-99]:

$$TS_2 + NADPH + H^+ \xrightleftharpoons{TR} T(SH)_2 + NADP^+$$

This same reaction also occurs with monoglutathionylspermidine (G$_{SP}$):

$$(G_{SP})_{2OX} + NADPH + H^+ \xrightleftharpoons{TR} 2\, G_{SP} + NADP^+$$

TR has been purified from *C. fasciculata* [100] and subsequently from *T. cruzi* [101]. TR is very similar to GR with close molecular weights (50 kDa for the TR monomer, 55 kDa for GR). These enzymes have a disulfur active site containing the cofactor FAD and both use NADPH as reducing agent [102]. Shames et al. [103] have cloned and sequenced the TR gene and Sullivan et al. [104] have expressed the TR gene of *T. congolense* in *E. coli*. The TR enzyme of *T.cruzi* has also been cloned and sequenced [105]. The three-dimensional structure of the enzymes' active sites have also been compared [106, 107] and the role of certain TR and GR amino acids have been determined. Directed mutagenesis has enabled the role of certain sites to be elucidated [108] and has even yielded a mutated TR with GR-type activity [109]. Despite the over 50% homology between the GR structures of *T. congolense* and those of human red blood cells, the two enzymes have a mutually exclusive specificity for their substrate [103]. The substrates of TR are trypanothione and monoglutathionylspermidine. Marsh and Bradley [110] have shown that the enzyme has a slight preference for spermidine over other symmetrical polyamines of the same length and for N-acetyl-cysteinylglycyl-N-glutathionylspermidines.

Trypanothione reductase inhibitors

Given the TR enzyme's specificity to the parasite, it represents a natural target molecule for inhibitor development [107].

Covalent inhibitors
Arsenicals

The trivalent arsenicals, such as melarsen oxide, react *in vitro* and *in vivo* with trypanothione to form a stable compound of the dithioarsane type (Mel T) [111]. This compound, Mel T, is a very effective competitive inhibitor of *T. brucei* TR. A disulfur bridge at the active site, after reduction by NADPH, necessarily participates in the reaction mechanism (involving C_{52}, C_{53} in *T. cruzi*) [112]. This reduction is necessary for the inhibition of TR. Modified melarsen oxide where the arsenic (As) is combined with two molecules of cysteinamine (Mel Cy or Cymelarsan®) is as effective as the unmodified version of melarsen oxide, while melarsoprol (Mel B or Arsobal®), where melarsen oxide is combined with dimercaptopropanol, is a much less potent inhibitor [113] as shown in the following Table 2.

Table 2

	Relative inhibition of TR (%)
NADPH + melarsen oxide	97.4
NADPH + Mel Cy	96.9
NADPH + melarsoprol	10.1

Cunningham et al. [113] describe a two-step mechanism of inhibition:
– first, after reducing the disulfur bridge existing between the Cys of the active site with NADPH, the trivalent arsenical compound reacts preferentially with the distant sulfhydryl of FAD to form a monothioarsane, an enzyme-inhibiting covalent complex. At this point, the reaction is reversible either by dilution or by sequestering the melarsen oxide by dimercatopropanol in the form of melarsoprol;
– second, the compound undergoes an internal rearrangement to form a more stable dithioarsane. In the presence of a NADPH excess, this enzyme-inhibitor complex accepts two electrons on FAD.

The covalent inhibition of TR by melarsen oxide requires that all the dihydrotrypanothion T(SH)2 be blocked in the form of Mel T (see Fig. 2). However, Fairlamb et al. [111] have shown that at the point of parasite death and lysis, only 10% of the T(SH)2 is found in the form of Mel T, which suggests this melarsen oxide must inhibit other trypanothione-dependent enzymes. It has been demonstrated elsewhere that melarsoprol inhibits the parasite's adenosine transporters and contributes to a reduction of FAD and NADPH synthesis; it is in this manner that trypanothione's metabolism is likely impaired [99].

Nitrosoureas

These derivatives, clinically used as cancer chemotherapy, react with the thiols of certain enzymes. 1,3-bis(2-chloroethyl)-1-nitrosourea (BCNU) is an inhibitor of GR, lipoamide dehydrogenase and of TR [114, 115]. When it is presented to the TR of *T. cruzi*, the enzyme's activity is only 5% of its normal level after a 3 hour incubation. However, the toxicity and general mutagenic potential of the nitrosoureas prevent their utilization as antiparasitic therapy.

Competitive inhibitors

TR's specific recognition of its substrate relies upon a hydrophobic binding region (Leu 17, Try 21, Tyr 110, Met 113 and Phe 114) [109]. Dynamic molecular studies have suggested that to function, inhibitors need to bind to this zone. Correct binding structures may include drugs of the phenothiazine and tricyclic antidepressant classes.

Among 30 tested structures of such a type [116], clomipramine (K_i = 7 µM), amitriptyline and trifluoperazine (K_i = 22 µM) are effective competitive inhibitors of TR without adversely affecting GR, but their CNS effects complicate their therapeutic use.

The 2-aminodiphenylsulfurs, similar to the phenothiazines but without significant neuroleptic activity, are synthesized in mono or dimeric form, with lateral amino chains for binding with the carboxylic terminals of the glutamic residues close to the hydrophobic pocket of TR. The best inhibitor is a bis aminodiphenylsulfur with an IC_{50} of 0.55 µM [117, 118].

The quinacrine mepacrine is also a TR inhibitor with a K_i of 15 µM [107].

"Subversive" substrates or turncoat inhibition

These are compounds inducing electron transfer reactions with a NADPH-dependent enzyme which normally catalyzes the reduction of a disulfide substrate. Their name comes from the fact that they tranform a reductase, responsible for maintainance of the physiological state of cellular thiols, into an oxidase producing the superoxide oxygen $O_2^{\bullet-}$ and other oxygenated reactive species. These inhibitors increase cellular stress oxidation because of several synergistic effects: reducing agents of the NADPH type are lost, disulfide reduction is inhibited and reactive forms of oxygen are created.

Having postulated that TR has the capacity to reduce functional groups other than disulfides, Henderson et al. [119] have synthesized naphtoquinones having either two lateral carboxylic chains, or having one or two chains terminated by a guanidine residue. These naphtoquinones oxidize several molar equivalents of NADPH, which suggests the existence of a redox cycle with reoxidation by the oxygen undergoing concomitant superoxide $O_2^{\bullet-}$ formation as follows:

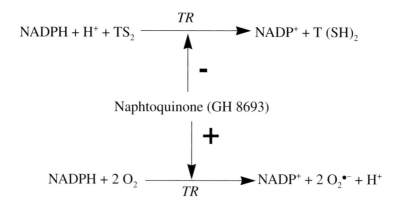

2,3-bis[3-(2-amidinohydrazono)-butyl]-1,4-naphtoquinone or GH 8693 is both a TR inhibitor ($IC_{50} = 1$ μM) and a substrate for NADPH ($K_M = 14$ μM) [114]. The formation of $T(SH)_2$ is inhibited and superoxide ion $O_2^{\bullet-}$, very toxic to the parasite, is liberated. The parasite's antioxidant enzyme is thus made pro-oxidant [112]. This effect is not observed with GR and this induction of oxidative stress could become a promising strategy in the battle against the trypanosome.

In the nitrofuran series, chinifur or 2-5'-{nitro(furo-2'-yl)-ethene-1-yl}-4(N,N-diethylamino)-1methyl-but-1-yl-aminocarbonyl-4-quinoline is one of the most selective TR inhibitors ($K_i = 4.5$ μM), with slight effect on GR ($K_i = 100$ μM) [120].

Tryparedoxine

These reduction pathways involve NADPH and trypanothione but also bring into play two parasite proteins [121, 122]. One, a 16 kDa protein, has WCPPC peptide motif which resembles the WCG(A)PC sequence of thioredoxine: this molecule has been named tryparedoxine (TPX or Cf_{16}). The other, at 21 kDa (Cf_{21}), belongs to the peroxyredoxine family, and has highly specific peroxidative action. The metabolic cascade is as follows:

$$\text{NADPH, H}^+ \searrow \nearrow TR_{OX} \searrow \nearrow T(SH)_2 \searrow \nearrow TPX_{OX} \searrow \nearrow Cf_{21\,red} \searrow \nearrow ROOH$$
$$NADP^+ \nearrow \searrow TR_{red} \quad TS_2 \nearrow \searrow TPX_{red} \quad Cf_{21\,ox} \nearrow \searrow ROH$$

Thus, TPX and protein Cf_{21} represent additional new target sites for trypanocidal molecules.

Perspectives

As we have just seen, the search for molecules active against the trypanosome has been, most often, the result of *in vitro* study of molecules' action in an enzyme system isolated from the parasite; in this manner study of structure-activity relationships can determine which chemical products demonstrate the greatest effectiveness and specificity. However, for this inhibitory action to be observed *in vivo*, it is necessary for the exogenous molecule to be able to cross the trypanosome membrane. This key requirement has also attracted the interest of researchers.

Study of the mechanism of resistance to arsenicals in certain isolates of *T. b. rhodesiense* has highlighted the role of transporters for purine bases, bases that parasite cannot synthesize and hence must extract from its host. These transporters are also utilized by molecules with *in vivo* therapeutic activity [123]. Carter and Fairlamb [124] have identified two transporters, P1 and P2, capable of transporting adenine/adenosine structures. The uptake of 1 µM of adenosine by P1 can be inhibited in a dose-dependent and saturable manner by inosine. The same is true for the P2 transporter, except that is adenine the one to inhibit its uptake of adenosine. Adenine and inosine, utilized in saturating concentrations, completely inhibit the transport of adenosine [123]. The P1 transporter has activity for adenosine and inosine and the P2 transporter for adenine and adenosine, but also for dipyridamole and melarsen oxide derivatives (e.g. melarsoprol). The absence or conformational modification of P2 renders the trypanosomal isolate resistant to arsenical drugs. The melarseno-resistant clones of *T. brucei brucei,* hence, probably have reduced intracellular concentration of the drug because of transporter defect. These clones also demonstrate *in vivo*, to variable degrees, pentamidine resistance which also seems linked to an alteration of P2 [123]. After extensive screening, the necessary structures for passage across the P2 receptor have been determined [125] and comprise type 1 substances for adenine derivatives, type 2 substances for benzamidine derivatives, and type 3 substances for melamine derivatives. The derivatives of type 2 and 3 are not substrates for mammalian adenosine transporters, which confers good selectivity on products of this type (e.g. pentamidine and melarsoprol). Of those polyamines substituted by benzylamine having trypanocidal activity, if benzylamine is itself replaced by melamine or if this radical attaches to a median polyamine nitrogen, it is possible to enable these molecules to cross the trypanosome membrane using the P2 transporter [125, 126].

Nonetheless, there is not a tight correlation between the effectiveness of the binding with the transporter and the antiparasitic activity, the latter also requiring an effective interaction with the enzymatic targets of the parasite.

References

1. Bacchi CJ, Vergara C, Garofalo J, Lipschik GY, Hutner SH (1979) Synthesis and content of polyamines in bloodstream *Trypanosoma brucei*. J Protozool 26:484-488
2. Bacchi CJ (1981) Content, synthesis, and function of polyamines in Trypanosomatids: relationship to chemotherapy. J Protozool 28:20-27
3. Bacchi CJ, McCann PP (1987) Parasitic protozoa and polyamines. In: McCann PP, Pegg AE, Sjoerdsma A (eds) Inhibition of polyamine metabolism. Biological significance and basis for new therapies. Academic Press Inc, Orlando, pp 317-344
4. Pegg AE (1987) The use of inhibitors to study the biochemistry and molecular biology of polyamine biosynthesis and uptake. In: Mc Cann PP, Pegg AE, Sjoerdsma A (eds) Inhibition of polyamine metabolism. Biological significance and basis for new therapies. Academic Press Inc, Orlando, pp 107-119
5. Giffin BF (1988) The role of polyamines in the growth and transformation of the African trypanosome. In: Zappia V, Pegg AE (eds). Progress in polyamine research: novel biochemical, pharmacological and clinical aspects. Adv Exp Med Biol 250:651-665
6. Balaña Fouce R, Alunda JM (1990) Antiprotozoal therapy with polyamine synthesis inhibitors. Rev Ibér Parasitol 50:161-177
7. Shukla OP (1990) Polyamine metabolism as a target for chemotherapy of parasite infection. J Sci Ind Res 49:263-282
8. Bacchi CJ, Yarlett N (1993) Effects of antagonists of polyamine metabolism in African trypanosomes. Acta Trop 54:225-236
9. Fairlamb AH, Le Quesne SA (1997) Polyamine metabolism in trypanosomes. In: Hide G, Mottram JC, Coombs GH, Holmes PH (eds). Trypanosomiasis and Leishmaniasis. CAB International, Wallingford, UK, pp 149-161
10. Tabor CW, Tabor H (1984a) Polyamines. Annu Rev Biochem 53:749-790
11. Gupta M, Coffino P (1985) Mouse ornithine decarboxylase complete amino acid sequence deduced from cDNA. J Biol Chem 260:2941-2944
12. Phillips MA, Coffino P, Wang CC (1987) Cloning and sequencing of the ornithine decarboxylase gene from *Trypanosoma brucei*. Implications for enzyme turnover inhibition and selective difluoromethylornithine. J Biol Chem 262:8721-8727
13. Wang CC (1991) A novel suicide inhibitor strategy for antiparasitic drug development. J Cell Biochem 45:49-53
14. Svensson F, Ceriani C, Lövkvist Wallström E, Kockum I, Algranati ID, Heby O, Persson L (1997) Cloning of a trypanosomatid gene coding for an ornithine decarboxylase that is metabolically unstable even though it lacks the C-terminal degradation domain. Proc Natl Acad Sci USA 94:397-402
15. Tobias KE, Kahana C (1993) Intersubunit location of the active site of mammalian ornithine decarboxylase as determined by hybridization of site-directed mutants. Biochemistry 32:5842-5847
16. Coleman CS, Stanley BA, Viswanath R, Pegg AE (1994) Rapid exchange of subunits of mammalian ornithine decarboxylase. J Biol Chem 269:3155-3158
17. Brooks HB, Phillips MA (1997) Characterization of the reaction mechanism for *Trypanosoma brucei* ornithine decarboxylase by multiwavelength stopped-flow spectroscopy. Biochemistry 36:15147-15155
18. Osterman AL, Brooks HB, Rizo J, Phillips M.A (1997) Role of Arg-277 in the binding of pyridoxal-5'-phosphate to *Trypanosoma brucei* ornithine decarboxylase. Biochemistry 36:4558-4567
19. Poulin R, Lu L, Ackermann B, Bey P, Pegg AE (1992) Mechanism of the irreversible inactivation of mouse ornithine decarboxylase by α-difluoromethylornithine. Characterization of sequences at the inhibitor and coenzyme binding site. J Biol Chem 267:150-158
20. Osterman AL, Kinch LN, Grishin NV, Phillips MA (1995) Acidic residues important for substrate binding and cofactor reactivity in eukariotic ornithine decarboxylase identified by alanine scanning mutagenesis. J Biol Chem 270:11797-11802
21. Osterman AL, Grishin NV, Kinch LN, Phillips MA (1994) Formation of functional cross-species heterodimers of ornithine decarboxylase. Biochemistry 33:13662-13667
22. Hayashi S (1995) Antizyme-dependent degradation of ornithine decarboxylase. Essays Biochem 30:37-47
23. Heby O, Persson L (1990) Molecular genetics of polyamine synthesis in eukariotic cells. Trends Biochem Sci 15:153-158
24. Li X, Coffino P (1992) Regulated degradation of ornithine decarboxylase requires interaction with polyamine-inducible protein antizyme. Mol Cell Biol 12:3556-3562
25. Ghoda L, Phillips MA, Bass KE, Wang CC, Coffino P (1990) Trypanosome ornithine decarboxylase is stable because it lacks sequences found in the carboxyl terminus of the mouse enzyme which target the latter for intracellular degradation. J Biol Chem 265:11823-11826

26. Mamont PS, Böhlen P, McCann PP, Bey P, Schuber F, Tardif C (1976) α-methylornithine, a potent competitive inhibitor of ornithine decarboxylase, blocks proliferation of rat hepatoma cells in culture. Proc Natl Acad Sci USA 73:1626-1630
27. Bey P, Danzin C, Jung M (1987) Inhibition of basic amino acid decarboxylases involved in polyamine biosynthesis. In: McCann PP, Pegg AE, Sjoerdsma A (eds) Inhibition of polyamine metabolism. Biological significance and basis for new therapies. Academic Press Inc, Orlando, pp 1-32
28. Metcalf BW, Bey P, Danzin C, Jung MJ, Casara P, Vevert JP (1978) Catalytic irreversible inhibition of mammalian ornithine decarboxylase (E.C. 4.1.1.17) by substrate and product analogues. J Am Chem Soc 100:2551-2553
29. Bitonti AJ, Bacchi CJ, McCann PP, Sjoerdsma A (1985) Catalytic irreversible inhibition of *Trypanosoma brucei brucei* ornithine decarboxylase by substrate and product analogs and their effects on murine trypanosomiasis. Biochem Pharmacol 34:1773-1777
30. Giffin BF, McCann PP, Bitonti AJ, Bacchi CJ (1986) Polyamine depletion following exposure to DL-α-difluoromethylornithine both *in vivo* and *in vitro* initiates morphological alterations and mitochondrial activation in a monomorphic strain of *Trypanosoma brucei brucei*. J Protozool 33:238-243
31. De Gee ALW, Carstens PHB, McCann PP, Mansfield JM (1984) Morphological changes in *Trypanosoma brucei rhodesiense* following inhibition of polyamine biosynthesis *in vivo*. Tissue Cell 16:731-738
32. Giffin B, McCann PP (1989) Physiological activation of the mitochondrion and the transformation capacity of DFMO induced intermediate and short stumpy bloodstream form trypanosomes. Am J Trop Med Hyg 40:487-493
33. Bitonti AJ, Cross-Doersen DE, McCann PP (1988) Effects of α-difluoromethylornithine on protein synthesis and synthesis of the variant-specific glycoprotein (VSG) in *Trypanosoma brucei brucei*. Biochem J 250:295-298
34. McCann PP, Bacchi CJ, Bitonti AJ (1988) Inhibition of ornithine or arginine decarboxylase as in experimental approach to African or American trypanosomiasis. In: Zappia V, Pegg AE (eds) Progress in polyamine research: novel biochemical, pharmacological and clinical aspects. Adv Exp Med Biol 250:727-735
35. Fairlamb AH, Henderson GB, Bacchi CJ, Cerami A (1987) *In vivo* effects of difluoromethylornithine on trypanothione and polyamine levels in bloodstream forms of *Trypanosoma brucei*. Molec Biochem Parasitol 24:185-191
36. Bacchi CJ, Garofalo J, Ciminelli M, Rattendi D, Goldberg B, McCann PP, Yarlett N (1993) Resistance to DL-α-difluoromethylornithine by clinical isolates of *Trypanosoma brucei rhodesiense*. Role of S-adenosylmethionine. Biochem Pharmacol 46:471-481
37. Yarlett N, Garofalo J, Goldberg B, Ciminelli MA, Ruggerio V, Sufrin JR, Bacchi CJ (1993) S-adenosylmethionine synthetase in bloodstream *Trypanosoma brucei*. Biochem Biophys Acta 1181:68-76
38. Liau MC, Lin GW, Hurlbert RB (1977) Partial purification and characterization of tumor and liver S-adenosylmethionine synthetases. Cancer Res 37:427-435
39. Cabrero C, Puerta J, Alemany S (1987) Purification and comparison of two forms of S-adenosyl-L-methionine synthetase from rat liver. Eur J Biochem 170:299-304
40. Mamont PS, Danzin C, Wagner J, Siat M, Joder-Ohlenbusch AM, Claverie N (1982) Accumulation of decarboxylated-S-adenosyl-L-methionine in mammalian cells as a consequence of the inhibition of putrescine biosynthesis. Eur J Biochem 123:499-504
41. Yarlett N, Bacchi CJ (1988) Effect of DL-α-difluoro-methylornitine on methionine cycle intermediates in *Trypanosoma brucei brucei*. Molec Biochem Parasitol 27:1-10
42. Bacchi CJ, Goldberg B, Garofalo-Hannan J, Rattendi D, Lyte P, Yarlett N (1995) Fate of soluble methionine in African trypanosomes: effect of metabolic inhibitors. Biochem J 309:737-743
43. Tabor CW, Tabor H (1984b) Methionine adenosyltransferase (5-adenosylmethionine synthetase) and S-adenosylmethionine decarboxylase. Adv Enzymol 56:251-282
44. Stanley B, Shantz LM (1994) S-adenosylmethionine decarboxylase structure-function relationships. Biochem Soc Trans 22:863-869
45. Van Poelje PD EE, Snell (1990) Pyruvoyl-dependent enzymes. Annu Rev Biochem 59:29-59
46. Wickner RB, Tabor CW, Tabor H (1970) Purification of adenosylmethionine decarboxylase from *Escherichia coli*: evidence for covalently bound pyruvate. J Biol Chem 245:2132-2139
47. Anton DL, Kutney R (1987) *Escherichia coli* S-adenosyl methionine decarboxylase. Subunit structure, reductive amination, and NH2-terminal sequences. Biol Chem 262:2817-2822
48. Maric SC, Crozat A, Jänne OA (1992) Structure and organization of the human S-adenosylmethionine decarboxylase gene. J Biol Chem 267:18915-18923
49. Tekwani BL, Bacchi CJ, Pegg AE (1992a) Putrescine activated S-adenosylmethionine decarboxylase from *Trypanosoma brucei brucei*. Mol Cell Biochem 117:53-61
50. Secrist III JA (1987) New substrate analogues as inhibitors of S-adenosylmethionine decarboxylase. Nucleosides & nucleotides 6:73-83
51. Marton LJ, Pegg AE (1995) Polyamines as targets for therapeutic intervention. Annu Rev Pharmacol Toxicol 35:55-91
52. Warrell RP Jr, Burchenal JM (1983) Methylglyoxal bis(guanylhydrazone) (Methyl-GAG): current status and future prospect. J Clin Oncol 1:52-63

53. Chang KP, Steiger RF (1976) Anti-trypanosome activity of methylglyoxal bis(guanylhydrazone). J Parasitol 62 (suppl):32
54. Chang KP, Steiger RF, Dave C, Cheng YC (1978) Effects of methylglyoxal bis(guanylhydrazone) on Trypanosomatid flagellates: inhibition of growth and nucleoside incorporation in *Trypanosoma brucei*. J Protozool 25:145-149
55. Bitonti AJ, Dumont JA, McCann PP (1986) Characterization of *Trypanosoma brucei brucei* S-adenosyl-L-methionine decarboxylase and its inhibition by Berenyl, pentamidine and methylglyoxal bis(guanylhydrazone). Biochem J 237:685-689
56. Bacchi CJ, Brun R, Croft SL, Alicea K, Bülher Y (1996) *In vivo* trypanocidal activities of new S-adenosylmethionine decarboxylase inhibitors. Antimicrob Agents Chemother 40:1448-1453
57. Stanek J, Caravatti G, Capraro HG, Furet Mett H, Schneider P, Regenass U (1993) S-adenosylmethionine decarboxylase inhibitors: new aryl and heteroaryl analogues of methylglyoxal bis(guanylhydrazone). J Med Chem 36:46-54
58. Brun R, Bühler Y, Mett H, Stanek J, Kaminsky N (1994) S-adenosylmethionine decarboxylase inhibitors as agents against African trypanosomes. In: 3rd Acrival meeting on chemotherapy of Trypanosomatidae, June 27-28, Bâle, Switzerland
59. Berger BJ, Carter NS, Fairlamb AH (1993) Polyamine and pentamidine metabolism in African trypanosomes. Acta Trop 54:215-224
60. Bitonti AJ, Byers TL, Bush TL, Casara PJ, Bacchi CJ, Clarkson AB Jr, McCann PP, Sjoerdsma A (1990) Cure of *Trypanosoma brucei brucei* and *Trypanosoma brucei rhodesiense* infections of mice with an irreversible inhibitor of S-adenosylmethionine decarboxylase. Antimicrob Agents Chemother 34:1485-1490
61. Bacchi CJ, Nathan HC, Yarlett N, Goldberg B, McCann PP, Bitonti AJ, Sjoerdsma A (1992) Cure of murine *Trypanosoma brucei rhodesiense* infections with an S-adenosylmethionine decarboxylase inhibitor. Antimicrob Agents Chemother 36:2736-2740
62. Byers TL, Bush TL, McCann PP, Bitonti AJ (1991) Antitrypanosomal effects on polyamine biosynthesis inhibitors correlate with increases in *Trypanosoma brucei brucei* S-adenosylmethionine. Biochem J 274:527-533
63. Byers TL, Casara P, Bitonti AJ (1992) Uptake of the antitrypanosomal drug 5'-{[(Z)-4amino-2butenyl]methylamino}-5'-deoxyadenosine (MDL 73811) by the purine transport system of *Trypanosoma brucei brucei*. Biochem J 283:755-758
64. Goldberg B, Rattendi D, Yarlett N, Lloyd D, Bacchi C (1997a) Effects of carboxymethylation and polyamine synthesis inhibitors on methylation of *Trypanosoma brucei* cellular proteins and lipids. J Euk Microbiol 44:352-358
65. Guo JQ, Wu YQ, Farmer WL, Douglas KA, Woster PM, Garofalo J, Bacchi CJ (1993) Restricted rotation analogs of S-adenosylmethionine: synthesis, evaluation as inhibitors of S-adenosylmethionine decarboxylase, and potential use as selective antitrypanosomal agent. Bioorg Med Chem Letters 3:147-152
66. Wu Y, Woster PM (1992) S-(5'-deoxy-5'-adenosyl)-1-ammonio-4-(methylsulfonio)-2-cyclopentene: a potent enzyme-activated irreversible inhibitor of S-adenosylmethionine decarboxylase. J Med Chem 35:3196-3201
67. Guo J, Wu YQ, Rattendi D, Bacchi CJ, Woster PM (1995) S-(5'-deoxy-5'-adenosyl)-1-aminoxy-4-(methylsulfonio)-2-cyclopentene (Ado Mao): an irreversible inhibitor of S-adenosylmethionine decarboxylase with potent *in vitro* antitrypanosomal activity. J Med Chem 38:1770-1777
68. Tekwani BL, Bacchi CJ, Secrist III JA, Pegg AE (1992b) Irreversible inhibition of S-adenosylmethionine decarboxylase of *Trypanosoma brucei brucei* by S-adenosylmethionine analogues. Biochem Pharmacol 44:905-911
69. Samejima K, Yamanoha B (1982) Purification of spermidine synthase from rat ventral prostate by affinity chromatography on immobilized S-adenosyl-(5')-3-thiopropylamine. Arch Biochem Biophys 216:213-222
70. Yamanoha B, Samejima K, Nakajima T, Yasuhara T (1984) Differences between homogeneous spermidine synthase isolated from rat and pig liver. J Biochem 96:1273-1281
71. Eloranta T, Kajander O, Kauppinen L, Hyvönen T, Linnala-Kankkunen A, Kalkkinen N, Kulomaa M, Almonen L, Jänne J (1988) Approaching the structure of mammalian propylamine transferases and their genes. In: Progress in polyamine research. Adv Exp Med Biol 250:117-126
72. Eloranta TO, Wahlfors J, Almonen L, Hyvönen T, Jänne J (1990) Cloning and primary structure of human spermidine synthase. In: Goldenberg SH, Algranati ID (eds) The biology and chemistry of polyamines. ICSU Symposium Series, vol 12, IRL Press, pp 91-98
73. Shiramata A, Morohoshi T, Fukai M, Akatsu S, Samejima K (1991) Putrescine or spermidine binding site of aminopropyltransferase and competitive inhibitors. Biochem Pharmacol 41:205-212
74. Bitonti A, Kelly SE, McCann PP (1984) Characterization of spermidine synthase from *Trypanosoma brucei brucei*. Molec Biochem Parasitol 13:21-28
75. Liu C, Coward JK (1991) Stereospecific synthesis of (R)-and (S)-S-adenosyl-1,8-diamino-3-thiooctane, a potent inhibitor of polyamine biosynthesis. Comparison of asymmetric induction vs enantiomeris synthesis. J Med Chem 34:2094-2101

76. Shiramata A, Morohoshi T, Samejima K (1988) Trans-4-methylcyclohexylamine, a potent new inhibitor of spermine synthase. Chem Pharm Bull 36:3220-3222
77. Woster PM, Black AH, Duff KJ, Coward JK, Pegg AE (1989) Synthesis and biological evaluation of S-adenosyl-1,12-diamino-3-thio-9-azadodecane, a multisubstrate adduct inhibitor of spermidine synthase. J Med Chem 32:1300-1307
78. Tang KC, Pegg AE, Coward JK (1980) Specific and potent inhibition of spermidine synthase by the transition-state analog, S-adenosyl-3-thio-1,8-diaminooctane. Biochem Biophys Res Commun 96:1371-1377
79. Lakanen JR, Pegg AE, Coward JK (1995) Synthesis and biochemical evaluation of adenosylspermidine, a nucleoside-polyamine adduct inhibitor of spermidine synthase. J Med Chem 38:2714-2727
80. Goldberg B, Yarlett N, Sufrin J, Lloyd D, Bacchi CJ (1997b) A unique transporter of S-adenosylmethionine in African trypanosomes. FASEB J 11:256-260
81. Bacchi CJ, Sanabria K, Spiess AJ, Vargas M, Marasco CJ Jr, Jimenez LM, Goldberg B, Sufrin JR (1997) In vivo efficacies of 5'-methylthioadenosine analogs as trypanocides. Antimicrob Agents Chemother 41:2108-2112
82. Casero RA Jr, Pegg AE (1993) Spermidine/spermine N1-acetyltransferase - the turning point in polyamine metabolism. FASEB J 7:653-661
83. Bellevue III FH, Boahbedason M, Wu R, Woster PM (1996) Structural comparison of alkylpolyamine analogues with potent in vitro antitumor or antiparasitic activity. Bioorg Med Chem Letters 6:2765-2770
84. Fairlamb AH, Blackburn P, Ulrich P, Chait BT, Cerami A (1985) Trypanothione: novel bis(glutathionyl)spermidine cofactor for glutathione reductase in Trypanosomatids. Science 227:1485-1487
85. Arrick BA, Griffith OW, Cerami AS (1981) Inhibition of glutathione synthesis as a chemotherapeutic strategy for trypanosomiasis. J Exp Med 153:720-725
86. Griffith O, Meister A (1979) Glutathione: interorgan translocation, turnover, and metabolism. Proc Natl Acad Sci USA 76:5606-5610
87. Lueder DV, Phillips MA (1996) Characterization of *Trypanosoma brucei* γ-glutamylcysteine synthetase, an essential enzyme in the biosynthesis of trypanonothione (diglutathionylspermidine). J Biol Chem 271:17485-17490
88. Fairlamb AH, Henderson GB, Cerami A (1986) The biosynthesis of trypanothione and N^1-glutathionylspermidine in *Crithidia fasciculata*. Molec Biochem Parasitol 21:247-257
89. Henderson GB, Yamaguchi M, Novoa L, Fairlamb AH, Cerami A (1990) Biosynthesis of the trypanosomatid metabolite trypanothione: purification and characterization of trypanothione synthetase from *Crithidia fasciculata*. Biochemistry 29:3924-3929
90. Smith K, Nadeau K, Bradley M, Walsh C, Fairlamb AH (1992) Purification of glutathionylspermidine and trypanothione synthetases from *Crithidia fasciculata*. Protein Sci 1:874-883
91. Hunter KJ, Le Quesne SA, Fairlamb AH (1994) Identification and biosynthesis of N1, N9-bis(glutathionyl)aminopropylcadaverine (homotrypanothione) in *Trypanosoma cruzi*. Eur J Biochem 226:1019-1027
92. Verbruggen C, De Craecker S, Rajan P, Jiao XY, Borloo M, Smith K, Fairlamb AH, Haemers A (1996) Phosphonic acid and phosphinic acid tripeptides as inhibitors of glutathionylspermidine synthetase. Bioorg Med Chem Letters 6:253-258
93. Chen S, Lin CH, Walsh C, Coward JK (1997) Novel inhibitors of trypanothione biosynthesis: synthesis and evaluation of phosphinate analog of glutathionylspermidine (GSP), a potent, slow-binding inhibitor of GSP synthetase. Bioorg Med Chem Letters 7:505-510
94. Kwon DS, Lin CH, Chen S, Coward JK, Walsh CT, Bollinger JM Jr (1997) Dissection of glutathionylspermidine synthetase/amidase from *Escherichia coli* into autonomously folding and functional synthetase and amidase domains. J Biol Chem 272:2429-2436
95. De Craecker S, Verbruggen C, Rajan PK, Smith K, Haemers A, Fairlamb AH (1997) Characterization of the peptide substrate specificity of glutathionylspermidine synthetase from *Crithidia fasciculata*. Molec Biochem Parasitol 84:25-32
96. Meister A, Anderson ME (1983) Glutathione. Annu Rev Biochem 52:711-760
97. Murgolo NJ, Cerami A, Henderson GB (1989) Trypanothione reductase. Ann New York Acad Sci 569:193-200
98. Walsh C, Bradley M, Nadeau K (1991) Molecular studies on trypanothione reductase, a target for antiparasitic drugs. Trends Biol Sci 16:305-309
99. Fairlamb AH, Cerami A (1992) Metabolism and functions of trypanothione in the kinetoplastida. Annu Rev Microbiol 46:695-729
100. Shames SL, Fairlamb AH, Cerami A, Walsh CT (1986) Purification and characterization of trypanothione reductase from *Crithidia fasciculata*, a newly discovered member of the family of disulfide containing flavoprotein reductases. Biochemistry 25:3519-3526
101. Krauth-Siegel RL, Enders B, Henderson GB, Fairlamb AH, Schirmer RH (1987) Trypanothione reductase from *T. cruzi*, purification and characterization of the cristalline-enzyme. Eur J Biochem 164:123-128
102. Sullivan FX, Krauth-Siegel RL, Pai EF, Walsh CT (1990) Molecular approaches in analysis of the substrate specificity of trypanothione reductase, a flavoprotein from trypanosomatid parasites. UCLA Symp Mol Cell Biol New Ser 110 (Protein Pharm Eng):119-134

103. Shames SL, Kimmel BE, Peoples OP, Agabian N, Walsh CT (1988) Trypanothione reductase of *T. congolense*: gene isolation, primary sequence determination, and comparison to glutathion reductase. Biochemistry 27:5014-5019
104. Sullivan FX, Shames SL, Walsh CT (1989) Expression of *Trypanosoma congolense* trypanothione reductase in *Escherichia coli*: overproduction/purification and characterization. Biochemistry 25:3519-3526
105. Sullivan FX, Walsh C (1991) Cloning, sequencing overproduction and purification of trypanothione reductase from *Trypanosoma cruzi*. Molec Biochem Parasitol 44:145-148
106. Aboagye-Kwarteng J, Smith K, Fairlamb AH (1992) Molecular characterization of the trypanothione reductase gene from *Crithidia fasciculata* and *Trypanosoma brucei*: comparison with other flavoprotein disulfide oxidoreductase with respect to substrate specificity and catalytic domain. Mol Microbiol 6:3089-3099
107. Krauth-Siegel RL, Schöneck R (1995) Trypanothione reductase and lipoamide dehydrogenase as targets for a structure-based drug design. FASEB J 9:1138-1146
108. Borges A, Cunningham ML, Tovar J, Fairlamb AH (1995) Site-directed mutagenesis of the redox-active cysteines of *Trypanosoma cruzi* trypanothione reductase. Eur J Biochem 228:745-752
109. Sullivan FX, Sobolov SB, Bradley M, Walsh CT (1991) Mutational analysis of parasite trypanothione reductase activity in a triple mutant. Biochemistry 30: 2761-2767
110. Marsh IR, Bradley M (1997) Substrate specificity of trypanothione reductase. Eur J Biochem 243:690-694
111. Fairlamb AH, Henderson GB, Cerami A (1989) Trypanothione is the primary target for arsenical drugs against African trypanosomes. Proc Natl Acad Sci USA 86:2607-2611
112. Shirmer RH, Müller JG, Krauth-Siegel RL (1995) Disulfide-reductase inhibitors as chemotherapeutic agents: the design of drugs for trypanosomiasis and malaria. Angew Chem Int Ed Engl 34:141-154
113. Cunningham ML, Zvelebil MJJ, Fairlamb AH (1994) Mechanism of inhibition of trypanothione reductase and glutathione reductase by trivalent organic arsenicals. Eur J Biochem 221:285-295
114. Jockers-Scherübl MC, Schirmer RH, Krauth-Siegel RL (1989) Trypanothione reductase from *Trypanosoma cruzi*. Catalytic properties of the enzyme and inhibition studies with trypanocidal compounds. Eur J Biochem 180:267-272
115. Becker K, Schirmer RH (1995) 1,3-bis(2-chloroethyl)-1-nitrosourea as thiol-carbamoylating agent in biological systems. Methods Enzymol 251:173-188
116. Benson TJ, McKie JH, Garforth J, Borges A, Fairlamb AH, Douglas KT (1992) Rationally designed selective inhibitors of trypanothione reductase. Biochem J 286:9-11
117. Fernandez-Gomez R, Moutiez M, Aumercier M, Bethegnies G, Luyckx M, Ouaissi A, Tartar A, Sergheraert C (1995) 2-Aminodiphenylsulfides as new inhibitors of trypanothione reductase. Int J Antimicrob Agents 6:111-118
118. Girault S, Baillet S, Horvath D, Lucas V, Davioud-Charvet E, Tartar A, Sergheraert C (1997) New potent inhibitors of trypanothione reductase from *Trypanosoma cruzi* in the 2-aminodiphenylsulfide series. Eur J Med Chem 32:39-52
119. Henderson GB, Ulrich P, Fairlamb AH, Rosenberg I, Pereira M, Sela M, Cerami A (1988) "Subversive" substrates for the enzyme trypanothione disulfide reductase: alternative approach to chemotherapy of Chagas disease. Proc Natl Acad Sci USA 85:5374-5378
120. Cenas N, Bironaite D, Dickancaite E, Anusevicius Z, Sarlanskas J, Blanchard JS (1994) Chinifur, a selective inhibitor and "subversive substrate" for *Trypanosoma congolense* trypanothione reductase. Biochem Biophys Res Commun 204:224-229
121. Gommel DU, Nogoceke E, Morr M, Kiess M, Kalisz HM, Flomé L (1997) Catalytic characteristics of tryparedoxin. Eur J Biochem 248:913-918
122. Nogoceke E, Gommel DU, Kiess M, Kalisz HM, Flomé L (1997) A unique cascade of oxidoreductases catalyses trypanothione-mediated peroxide metabolism in *Crithidia fasciculata*. Biol Chem Hoppe Seylers 378:827-836
123. Carter NS, Berger BJ, Fairlamb AH (1995) Uptake of diamidine drugs by the P2 nucleoside transporter in melarsen-sensitive and -resistant *Trypanosoma brucei brucei*. J Biol Chem 270:28153-28157
124. Carter NS, Fairlamb AH (1993) Arsenical-resistant trypanosomes lack an unusual adenosine transporter. Nature 361:173-176
125. Tye CK, Kasinathan, Barrett MP, Brun R, Doyle VE, Fairlamb AH, Weaver R, Gilbert IH (1998) An approcah to use an unusual adenosine transporter to selectively deliver polyamine analogues to trypanosomes. Bioorg Med Chem Letters 8:811-816
126. Barrett MP, Fairlamb AH (1999) The biochemical basis of arsenical-diamidine crossresistance in African trypanosomes. Parasitol Today 15:136-140

CHAPTER 6

Cytokines and the blood-brain barrier in human and experimental African trypanosomiasis

VW Pentreath

Introduction

Amongst the extensive immune and pathologic symptoms occurring in African sleeping sickness, damage to capillary and other tissue compartment barrier systems are widespread. At a general level this is manifest by petechial haemorrhages and blood vessel fragility which occurs in many tissues throughout the body. However, in the context of the nervous system, especially the central nervous system (CNS), these changes take on extra significance. This is because a range of substances, including cytokines and mediator substances may then exchange between blood, cerebrospinal fluid (CSF), other tissue compartments and the nervous tissue, thus perturbing vital functions normally protected by the blood-brain barrier. Lymphocyte infiltration and other neuroimmune changes may be initiated, exacerbated or otherwise modified to further aggravate the pathological changes. In this chapter the nature and consequences of the damage to the barrier systems in human and experimental African trypanosomiasis with special respect to the nervous system are summarised.

Altered capillary permeability and the involvement of cytokines and mediator substances

Over thirty years ago Goodwin and his co-workers studied the vascular lesions in rabbits infected with *Trypanosoma brucei brucei* [1]. These were analysed following intravenous injection of particulate material, using a valuable preparation consisting of rabbit ear-chambers [2]. The studies were extended by Edeghere [3]. A major conclusion was that the increased permeability was caused by substances released in the tissues, such as histamine, 5-hydroxytryptamine (5-HT) or bradykinin, all of which were known to cause weakness in the junctions between endothelial cells. The substances were collectively called autocoids, a term then applied to a range of amines, lipids and peptides which were found in tissues and known to have widespread pharmacological activities (see Boreham [4]). It was hypothesised that these substances were responsible for the permeability changes. Several causes for the autocoid release were suggested. These include substances released from trypanosomes such as proteases (reviewed by Tizard et al. [5]), immune complexes interacting with the cells or a consequence of complement activation [4].

Tissue barrier damage is also evident in the kidney glomeruli, causing proteinuria, negative water balance and serious imbalances in serum calcium, albumin, urea and creatinine [4, 6].

Since these studies the potential for perturbation of tissue barrier systems by endogenous substances has been vastly increased by the understanding of cytokines, prostaglandins and other immune mediators. Many of these are known to potently modify tissue permeability by direct and indirect actions, including the bidirectional movement of immune cells from blood to tissues and vice versa. Only very limited studies have yet been made on the likely alterations of these substances in African trypanosomiasis. Tumor necrosis factor (TNF)-α, for example, is markedly elevated in the serum of patients infected with *T. b. gambiense* [7]. The level of TNF-α correlates with the disease severity. TNF-α is involved in multiple immunological and physiological functions, including the responses to a range of bacterial, viral and parasitic infections and systemic inflammatory reactions (e.g. Cerami and Beutler [8]). There is now significant information concerning this and several other cytokines, including interferon (IFN)-γ [9, 10] in the immune responses to trypanosomes, which are discussed in other chapters of this volume. Some of this information is relevant to barrier systems within nervous tissue, and this is described in the following sections.

Sites and causes of barrier damage and movement of parasites from blood to brain

Trypanosomes cause breakdown of several barrier systems in the brains of infected animals and from the limited studies available, also in man. The changes are brought about by the different strains of *T. brucei*, with *T. b. brucei* in the experimental animals and *T. b. gambiense* in the chronic human form best documented. The parasites occupy the meninges fairly soon after infection, and this produces a meningitis which is of variable severity. Parasites then damage the choroid plexus and the blood-brain barrier. It is not understood whether this damage is a direct physical effect due to parasite numbers and movement, or whether substances released locally in response to the parasites or by the parasites themselves induce detachment of cells indirectly.

The question of trypanosome presence in the parenchyma is also not always clear. Parasite influx does not however appear to a widespread or wholesale phenomenon, comparable to that which occurs in lymphoid or connective tissue, except in some terminal disease states. In the rat model *T. b. brucei* selectively penetrates some areas where the blood-brain barrier is reduced (e.g. pineal, area postrema and circumventricular organs) quite early following infection, and it has been argued that this may underlie the neuro-endocrine dysfunctions and disturbed endogenous rhythms [11, 12]. The pituitary is also penetrated by the parasite. For example in both rodents and sheep the microvasculature is badly damaged and parasites occupy the extracellular spaces [11, 13]. In contrast *T. congolense* causes degenerative changes in the pituitary of cattle, but does not occupy the nervous tissue [14]. Scattered trypanosomes have been observed in the brains of rats and mice in several other studies [15-17]. However, other studies suggest that the brain parenchyma may in fact be a relatively inhospitable environment for the parasite. Direct inoculation of trypanosomes into the brain of mice, or providing them with free access to the parenchyma by mechanical damage to the blood-brain barrier leads to an initial dissemination, but this is soon followed by reduction and ultimately clearance from the nervous tissue [18, 19]. Also CSF does not provide a good medium for parasite growth ([20]; see Fig. 1). It is also not clear why those areas where the blood-brain barrier is reduced may encourage local parasite penetration. It seems very improbable that this is due to the physical properties of the barrier, because the term reduced or leaky refers to the access of substances with low molecular weights, whereas trypanosomes (diameter approximately 2 μm) could only have easier access if the capillaries were seriously damaged. On the other hand the relationship between parasite spread and reduced blood-brain barrier may be a consequence of diffusion of signalling substances which promote parasite entry and growth. A likely substance is IFN-γ which has been shown to enhance growth of *T. b. brucei* in lymph tissue and possible also in dorsal root ganglia (which also are penetrated by the parasite; see Bakhiet [9]). This chemotactic cytokine may be released from $CD8^+$ T-cells which infiltrate the same brain areas [21].

The choroid and CSF

Injury to the choroid plexus appears to be a key stage in the neuropathogenesis, with numerous parasites filling the stromal spaces and the epithelium obviously damaged. This has been described in several studies with *T. b. gambiense*, *T. b. rhodesiense* and *T. b. brucei* in rodents and primates [15, 22-25, 51, 58, 59]. The parasites then have ready access to the CSF and subsequently, via the subarachnoidal spaces to the extensions of the perivascular spaces (Virchow-Robin spaces) which extend into the brain. The breaching of the choroid appears to be a key determinant in the disease progression from meningitis to encephalitis, iniating the perivascular infiltrations (cuffing) in the brain [29, 57, 59]. Indeed the presence of parasites in the CSF is the accepted cardinal diagnostic criterion for the meningoencephalitis (late-stage) phase of the disease. Some interesting properties of the CSF regarding parasite survival have been found [20]. Different types of artificial human CSF containing all the major components (including amino acids, electrolytes and carbohydrates) of healthy fluid was very similar to the CSF of patients without evidence of CNS involvement (i.e without evidence of parasite presence or choroid damage) in terms of parasite growth and survival. These fluids provided a moderately satisfactory environment for supporting short-term survival of the trypanosomes, although they were not as favourable as blood or serum. The normal CSF did not appear to contain endogenous unknown substances which possessed trypanocidal or growth promoting activities. On the order hand CSF from late-stage *T.b. gambiense* sickness patients (i.e with choroid damage and parasites present in the CSF) was much less supportive of parasite survival (Fig. 1). It was suggested that the limited survival of trypanosomes could correlate with the increased numbers of lymphocytes, antibodies and perhaps other unknown suppressive factors in the late-stage CSF [20]. These findings argue against the early suggestions of Peruzzi [26] who thought that in the late stage of the disease substances gaining access from the blood conferred enhanced survival in the CSF. In another study by Waitumbi et al. [27] on monkeys a recurring increase in the CSF/serum albumin index was demonstrated to be caused by *T. b. rhodesiense* infections. This may also be of great potential importance because it indicates that the blood-CSF and/or blood-brain barrier impairment is not a permanent feature. A recurring pattern of barrier damage and repair, perhaps associated with waves of parasitaemia and cytokine production is thus possible. In the human patient these issues require much further research, including regular measurements on the same infected individual, which unfortunately requires very difficult procedures.

The blood-brain barrier

Disruption of the blood-brain barrier is a common feature of many forms of neurological damage and brain inflammation. It has a complex aetiology but hypoxic and radical mediated damage or toxic substances (chemicals or substances resulting from parasite viral, or bacterial infections, for example endotoxins) cause cytokine/mediator production which in turn alter the endothelial integrity and its control by astrocytes. It also has potentially serious consequences, for example by allowing access of substances from

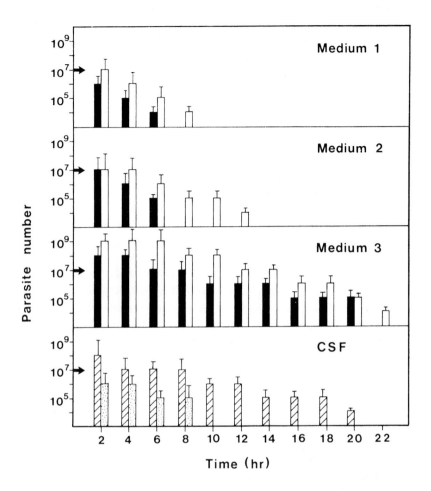

Fig. 1. Properties of artificial and human CSF for the survival of *T. b. brucei*. Flasks of CSF were inoculated with 10^7 parasites at the start of the experiment (*arrows*) and the parasite numbers per ml of CSF measured at 2 hr intervals. Each histogram is the mean ± SD [N=20 for the artificial media (media 1-3), N=12 for the abnormal CSF and N=14 for the normal CSF (*bottom*, CSF)]. The artificial CSF were based on the detailed analyses described in Fishman. Medium 1 contained essential electrolytes and glucose, medium 2 contained in addition all the major amino acids and medium 3 contained in addition other substances present in trace quantities in healthy CSF (*solid bars* in 1-3). To test for the effects of other possible non-identified factors present in trace quantities, and for comparison with survival in serum, fetal calf serum (FCS) was added to another set of samples (*open bars* in 1-3). The CSF was collected from patients at the Projet de Recherche Clinique sur la Trypanosomiase, Daloa, Côte d'Ivoire. The cross-hatched bars are for "normal" CSF (i.e. for patients with no evidence of CNS involvement), the stipled bars are for the CSF samples from patients diagnosed as having the late-stage disease. Note the very limited survival time of the parasites in the latter. From [20], reproduced with permission from the Liverpool School of Tropical Medicine

the blood to the parenchyma. Although it seems a likely site of damage in the chronic human disease state because of the oedema and some other pathological features, including the serum CSF ratios of albumin [28, 29], clear information is only available for the experimental rat model. Infected animals develop a progressive barrier damage in chronic *T. b. brucei* infections as demonstrated by the movement of the fluorescent fluid-phase marker sulphorhodamine B (molecular weight = 574) into the brain tissue [17]. Bilateral opening of the barrier was consistently observed in the late stages of the disease model (days 35 and 50 post infection).The thalamus and hypothalamus were first affected, but dye efflux continued to extend to most brain regions with dye penetration throughout both white and grey matter. At the same time occasional parasites were observed in the parenchyma, but their numbers were very few and showed no obvious correlation with the distribution of dye. There was also considerable variation between different experimental animals in the extent of barrier damage in the different brain regions. As discussed above it appears most unlikely that parasite penetration is a physical cause of the barrier damage.

Breakdown of the blood-brain barrier could be of great significance in relation to drug treatment in the late disease stages. It is possible it may facilitate drug access and could therefore be advantageous in drug treatment, but it may also be a potential cause of aggravated pathology. Difluoromethylornithine (DFMO) has potent anti-inflammatory effects, including down-regulation of astrocyte reactivity and protection against blood-brain barrier breakdown and vasogenic oedema resulting from experimental ischaemia [31, 32], as well as trypanocidal activity. Part of its chemotherapeutic effectiveness may therefore be due to alterations in cytokines associated with the anti-inflammatory actions. On the other hand another frequently used drug, melarsoprol, can produce extremely severe side effects and cause reactive arsenical encephalopathy, which is fatal in an estimated 10% of patients. It has been shown in the rodent model that the fatal reactions may correlate with subcurative therapy, which leaves parasites remaining in the brain [33]. It is possible that the arsenicals, which are themselves neurotoxic, may exacerbate damage to the barrier leading to an aggravated inflammatory response in a similar way to the mortalities associated with some drug therapies for some forms of bacterial meningitis (see Quagliarello and Scheld, [34]).

Cerebral oedema

Cerebral oedema is an increase in brain volume (brain swelling) due to an increase in water content (see Katzman and Pappius, [35]). A common form is vasogenic oedema which is generally caused by damage to vascular elements within the CNS [36]. It is characterised by breakdown of the blood-brain barrier and build-up of extravasated fluid similar to plasma which accumulates in the extracellular spaces in the parenchyma. It is prevalent in the white matter because it is thought that the barriers between the parallel-running myelinated fibres are less resistant to the accumulation of extracellular fluid than the tightly interwoven process of the grey matter. Brain oedema during experimental trypanosomiasis has been studied in the rat model by measuring brain weight, density and electrolyte concentrations [17]. Increases in the wet brain weights and to a lesser extent the dry brain weights were found in chronic infections. Brain sodium and potassium contents were also significantly elevated. The pathological changes correlated with the breakdown of the blood-brain barrier as demonstrated by fluorescent dye efflux. Taken together the data were interpreted as providing strong evidence that the late stages of experimental trypanosomiasis were accompanied by a mild-to-moderate vasogenic oedema [17]. The increases in the dry brain weights of the infected animals were thought to be due to the extravasated substance accumulating in the parenchyma, similar to that demonstrated for other pathological states (see, for example, Kuroiwa et al. [37]). Relatively little is known about these possible changes in patients with CNS involvement, although brain swelling was a common feature amongst fatal cases who died as a result of acute arsenical encephalopathy [29].

Substances initiating brain pathology

Important questions concern the nature of the biologically active products which are responsible for triggering the damage to the barrier systems in the CNS. The physical movement of trypanosomes into the parenchyma would not appear to be of primary importance for the reasons discussed above. There are several other likely causes. African trypanosomes and their dead remnants contain a range of potentially pathogenic substances including the variable antigenic coat, phospholiphase A, proteases (especially cysteine), saturated fatty acids and aromatic amino acid derivatives (e.g. tryptophol) which have been reviewed by Tizard et al. [5]. Other causative substances may have non-parasite origins. One class of these for which there is growing evidence are endotoxins from Gram-negative bacteria. In relation to this it has frequently been noted that many of the clinical and laboratory criteria are similar in trypanosomiasis and endotoxaemia. Amongst these are the polyclonal enhancement and subsequent immunosuppression, the complement and kinin activation, hypergammaglobulinaemia, and also the fever, headache and some features of the altered endogenous rhythms. Small doses of purified endotoxin given before trypanosome infection increases resistance of mice to *T. b. rhodesiense*, *T. congolense* and *T. duttoni* [38], and of rats to *T. lewisi* [39]. However, when the endotoxin injections are given simultaneously with the trypanosomes, the course of the infection is not altered [38], or may be aggravated [39]. It was hypothesized by Greenwood [40] that many aspects of the pathology in sleeping sickness may be due to non-specific endotoxin-like substances.

Studies with the mouse model and *T. b. brucei* infections have demonstrated that serum endotoxin levels become significantly elevated soon after infection and that the levels are maintained high throughout the disease course. Some evidence was also obtained for the presence of endotoxin activity in the parasite [41, 42]. In patients with African sleeping sickness serum levels of endotoxin are also markedly elevated, and of great potential importance was the finding that the levels in the CSF of late-stage patients with clear-cut neurological involvement (i.e. trypanosomes, high protein levels and large numbers of lymphocytes in the CSF, together with other neuropsychiatric symptoms) were also high. There was a high degree of correlation of endotoxin levels in the serum and CSF [43]. Many of the increased values could be clinically interpreted as representing moderate to severe endotoxaemia/bacteraemia, according to the accepted practice. The levels of endotoxin were reduced in both fluid compartments following chemotherapy. Subsequent analysis has demonstrated that the levels of antibodies against several Gram-negative bacteria are also markedly elevated in the serum and CSF of the late-stage patients, again with a high degree of correlation in the two compartments [44]. These findings confirm the nature of the bacterial origin of the endotoxins, and also that the elevated levels are a chronic feature of the disease. The sources of the bacterial endotoxins are not yet clear, although several sources appear likely. In the murine model systemic bacteraemia was present in more than 50% of the experimental chronic infections with *T. b. brucei* [41]. Damage to the hepatic portal system, with an inability of the Kupffer cells to cope with the materials released from parasites as well as endogenous endotoxin from the gut is another obvious possibility. Damage to the barrier systems in the gut due to ischaemia, physical damage by parasite and mediator substances released in the epithelium by the parasite in combination with the immunosuppression are other likely causes. However, regardless of the sources of the endotoxin is clear that they are potentially a major contributing factor to many aspects of the disease pathology. Endotoxins are not able to cross an intact blood-brain barrier, but are responsible for inducing its breakdown (e.g. Tunkel et al. [45]; Burroughs et al. [46]). They can also evoke cytokine/mediator production (e.g. prostaglandins) which can enter the parenchyma from the circulation or which may themselves damage the barrier [47]. The breakdown of the barrier in the late disease stages, with high levels of endotoxin and endotoxin antibodies in the CSF and with direct access of endotoxin to the neurons and glial cells will very likely exacerbate the disease pathology. In relation to this it has been demonstrated *in vivo* that endotoxins synergise with trypanosomes to cause enhanced reactive changes and prostaglandin production by astrocytes [48, 49]. Once again, further work is required to unravel this important component of African sleeping sickness.

Cytokines/mediator substances and the CNS in trypanosomiasis

The levels of cytokines and mediator substances are seriously perturbed in trypanosomiasis. The involvement of the alterations in the immune responses and the immunopathology of the nervous system are discussed in other chapters in this volume. However, in the context of the damage to the barrier system in the brain, these substances have extra significance because of the roles they may play in perturbing the barriers, and also because they may flux into the extracellular fluid pools and CSF of the brain to modify other physiological events.

Investigations of the possibilities have been made by analysis of the CSF of patients with *T. b. gambiense* from the Côte d'Ivoire [50]. In the late-stage infections there were no significant increases in the levels of interleukin (IL)-1, but there were some very large increases in prostaglandin (PG) D_2. High levels of PGE_2 and PGF_2 have also been observed [51]. The levels of prostaglandins were reduced following chemotherapy. *In vitro* studies with murine astrocytes indicate that these cells are a major source of the prostaglandins [48]. Upregulation of mRNA for several cytokines (including IL-1 and TNF-α) and accompanying astrocyte activation has been strongly implicated as an important event in the CNS involvement in the experimental murine model [52, 53]. Much further work is required to clarify the roles of these substances in the CNS pathology, but the studies have clearly established that major changes in their levels occur in sleeping sickness.

The implications for human African trypanosomiasis

The pathology of the CNS damage in sleeping sickness is exceedingly complex and has multiple components. However, there are several emerging features from the foregoing account which can be constructively discussed in relation to information established from other CNS infections and neurological disorders.

The breaching of choroid is an important event because the parasites, endotoxins and cytokines will then have access to the CSF and Virchow-Robin spaces. The blood-brain barrier is susceptible to damage from either the blood capillaries (caused by for example radical - and/or cytokine-mediated ischaemia) or from the parenchyma (e.g. by altered astrocyte secretions). In bacterial meningitis (which is a frequent complication of sleeping sickness) fever and inflammation break out when the concentration of bacterial products reach certain critical levels, causing production of cytokines such as IL-1 (see Tuomanen, [54]). The cytokines are produced on the parenchymal side of the blood-brain barrier and cause the influx of leucocytes from the circulation. Cytokines such as IL-1 and TNF-α can disrupt the integrity of the barrier via activation of cyclooxygenase and the production of eicosanoids by the endothelial cells (e.g. de Vries

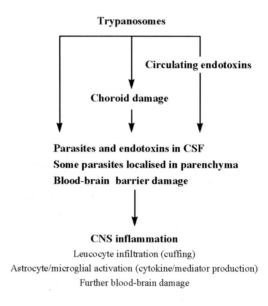

Fig. 2. The likely cascade of events in the CNS initiated by trypanosomes leading to barrier damage and other pathological changes in African sleeping sickness

et al. [55]). Several cell types including microglia and astrocytes may be responsible for the cytokine production. However, astrocytes in particular are known to play a primary role in the control of the blood-brain barrier and be capable of producing a range of inflammatory cytokines and mediators when they are activated. Their activation in sleeping sickness to express a number of inflammatory cytokines has been clearly demonstrated [52, 53]. Other aspects of their activation by trypanosomes and the synergistic effects of endotoxins have been shown *in vitro* [49]. It seems likely that these cells play key roles in mediating the inflammatory changes in the CNS of late-stage sleeping sickness [56]. A likely sequence of events and consequences of the damage to the CNS barrier systems in African sleeping sickness are summarised in Fig. 2. Only by continuing research will the many facets of the neuropathology be understood.

References

1. Goodwin LG, Hook SVM (1968) Vascular lesions in rabbits infected with *Trypanosoma (Trypanozoon) brucei*. Br J Pharmacol Chemother 32:305-311
2. Goodwin LG (1971) Pathological effects of *Trypanosoma brucei* on small blood vessels in rabbit ear-chambers. Trans R Soc Trop Med Hyg 65:82-89
3. Edeghere HUFI (1979) Vascular changes in experimental African trypanosomiasis. Parasitology 79:IX
4. Boreham PFL (1985) Autocoids: their release and possible role in the pathogenesis of African trypanosomiasis. In: Tizard I (ed) Immunology and Pathogenesis of Trypanosomiasis. CRC Press Inc, Boca Raton, Florida, pp 45-66
5. Tizard I, Nielsen KH, Seed JR, Hall JE (1978) Biologically active products from African trypanosomes. Microbiol Rev 42:661-681
6. Goodwin LG (1970) The pathology of African trypanosomiasis. Trans R Soc Trop Med Hyg 64:797-804
7. Okomo-Assoumou MC, Daulouède S, Lemesre J-L, N'zila-Mouanda A, Vincendeau P (1995) Correlation of high serum levels of tumour necrosis factor-α with disease severity in human African trypanosomiasis. Am J Trop Med Hyg 53:539-543
8. Cerami A, Beutler B (1988) The role of cachectin/TNF in endotoxic shock and cachexia. Immunol Today 9:28-31
9. Bakhiet AMO (1993) Immunopathogenesis of experimental African trypanosomiasis, interactions between *Trypanosoma brucei brucei*, $CD8^+$ T cells and interferon gamma. Karolinksa Institute, Stockholm
10. Lucas R, Magez S, Songa B, Darji I, Hamers R, De Baetselier P (1993) A role for TNF during African trypanosomiasis - involvement in parasite control, immunosuppression and pathology. Res Immunol 144:370-375
11. Schultzberg M, Ambatsis M, Samuelsson E-B, Kristensson K, Van Meirvenne N (1988) Spread of *Trypanosoma brucei* to the nervous system: early attack on the circumventricular organs and sensory ganglia. J Neurosci Res 21:51-61
12. Bentivoglio M, Grassi-Zucconi G, Olsson T, Kristensson K (1994) *Trypanosoma brucei* and the nervous system. Trends Neurosci 17:325-329
13. Ikede BO, Losos GJ (1975) Pathogenesis of *Trypanosoma brucei* infection in sleep. II. Hypophyseal and other endocrine lesions. J Comp Pathol 85:37-44
14. Abebe G, Shaw MK, Eley, RM (1993) *Trypanosoma congolense* in the microvasculature of the pituitary gland of experimentally infected Boran cattle (*Bos indicus*). Vet Pathol 30:401-409
15. Van Marck EAE, Le Ray D, Beckers A, Jacob W, Wéry M, Gigase PLJ (1981) Light and electron microscope studies on extravascular *Trypanosoma brucei gambiense* in the brain of chronically infected rodents. Ann Soc Belge Méd Trop 61:57-78
16. Schmidt H, Bafort JM (1985) African trypanosomiasis: treatment-induced invasion of brain and encephalitis. Am J Trop Med Hyg 34:64-68
17. Philip KA, Dascombe MJ, Fraser PA, Pentreath VW (1994) Blood-brain barrier pathology in experimental trypanosomiasis. Ann Trop Med Parasitol 88:607-616
18. Bafort JM, Schmidt H, Molyneux DH (1987) Development of *Trypanosoma brucei* in suckling mouse brain following intracerebral injection. Trans R Soc Trop Med Hyg 81:487-490
19. Schmidt H, Bafort JM (1987) African trypanosomiasis: haematogenic brain parasitism early in experimental infection through bypassing the blood-brain barrier, with considerations of brain trypanosomiasis in man. Parasitol Res 73:15-21
20. Pentreath VW, Owolabi OA, Doua F (1992) Survival of *Trypanosoma brucei brucei* in cerebrospinal fluid. Ann Trop Med Parasitol 86:29-34
21. Schultzberg M, Olsson T, Samuelsson E-B, Maehlen J, Kristensson K (1989) Early major histocompatibility complex (MHC) class I antigen induction in hypothalamic supraoptic and paraventricular nuclei in trypanosome-infected rats. J Neuroimmunol 24:104-112
22. Stevens RR, Moulton JE (1977) Experimental meningoencephalitis in *Trypanosoma brucei* infection of deer mice (*Peromyscus maniculatus*). A light, immunofluorescent, and electron microscopic study. Acta Neuropath (Berlin) 38:173-180
23. Poltera AA, Hochmann A, Rudin W, Lambert PH (1980) *Trypanosoma brucei brucei*: a model for cerebral trypanosomiasis in mice - an immunological, histological and electronmicroscopic study. Clin Exp Immunol 40:496-507
24. Van Marck EAE, Mulumba P, Beckers A, Gigase P, Wéry M (1983) Studies of choriod plexus involvement in chronic infections of mice infected with *Trypanosoma brucei gambiense*. Contrib Microbiol Immunol 7:173-182

25. Rudin W, Poltera AA, Jenni L (1983) An EM study on cerebral trypanosomiasis in rodents and primates. Contrib Microbiol Immunol 7:165-172
26. Peruzzi M (1928) Pathological - anatomical and serological observations on trypanosomiasis. Final Report of the League of Nations, International Commission on Human Trypanosomiasis, Geneva, pp 245-328
27. Waitumbi JN, Sayer PD, Gould SS (1988) Evidence of blood cerebrospinal fluid barrier permeability impairment in *Trypanosoma rhodesiense* infection in vervet monkeys. Bull Soc Path Ex 81:502-505
28. Lambert PH, Berney M, Kazyumba G (1981) Immune complex in serum and in cerebrospinal fluid in African trypanosomiasis. Correlations with polyclonal B cell activation and with intracerebral immunoglobulin synthesis. J Clin Invest 67:77-85
29. Adams JH, Haller L, Boa FY, Doua F, Dago A, Konian K (1986) Human African trypanosomiasis (*T. b. gambiense*): a study of 16 fatal cases of sleeping sickness with some observations on acute arsenical encephalopathy. Neuropathol Appl Neurobiol 12:81-94
30. Fishman RA (1980) Cerebrospinal Fluid in Diseases of the Nervous System. WB Saunders, Philadelphia
31. Zoli M, Zini I, Grimaldi R, Biagini G, Agnati L (1993) Effects of polyamine synthesis blockade on neuronal loss and astroglial reaction after transient forebrain ischemia. Int J Develop Neurosci 11:175-187
32. Schmitz MP, Combs DJ, Dempsey RJ (1993) Difluoromethylornithine decreases postischemic brain oedema and blood-brain barrier breakdown. Neurosurgery 33:882-887
33. Jennings FW, Hunter CA, Kennedy PGE, Murray M (1993) Chemotherapy of *Trypanosoma brucei* infection of the central nervous system: the use of a rapid chemotherapeutic regimen and the development of post-treatment encephalopathies. Trans R Soc Trop Med Hyg 87:224-226
34. Quagliarello V, Scheld WM (1992) Bacterial meningitis: pathogenesis, pathophysiology and progress. New Eng J Med 327:864-872
35. Katzman R, Pappius HM (1973) Brain Electrolytes and Fluid Metabolism. Williams and Wilkins, Baltimore
36. Klatzo I (1967) Neuropathological aspects of brain edema. Neuropathol Exptl Neurol 26:1-14
37. Kuroiwa T, Cahn R, Juhler M, Goping G, Campbell G, Klatzo I (1985) Role of extracellular proteins in the dynamics of vasogenic brain oedema. Acta Neuropathol (Berlin) 66:3-11
38. Singer I, Kimble ET, Ritts RE (1964) Alterations in the host-parasite relationship by administration of endotoxin to mice with infections of trypanosomes. J Infect Dis 14:243-248
39. Styles TJ (1965) Effect of bacterial endotoxin on *Trypanosoma lewisi* infections in rats. J Parasitol 51:650-653
40. Greenwood BM (1974) Possible role of B cell mitogen in hypergammaglobulinaemia in malaria and trypanosomiasis. Lancet 1:435-436
41. Alafiatayo RA, Crawley B, Oppenheim BA, Pentreath VW (1993) Endotoxins and the pathogenesis of *Trypanosoma brucei brucei* infection in mice. Parasitology 107:49-53
42. Pentreath VW (1994) Endotoxins and their significance for murine trypanosomiasis. Parasitol Today 10:226-229
43. Pentreath VW, Alafiatayo RA, Crawley B, Doua F, Oppenheim BA (1996) Endotoxins in the blood and cerebrospinal fluid of patients with African sleeping sickness. Parasitology 112:67-73
44. Pentreath VW, Alafiatayo RA, Barclay GR, Crawley B, Doua F, Oppenheim BA (1997) Endotoxin antibodies in African sleeping sickness. Parasitology 114:361-365
45. Tunkel AR, Rosser SW, Hansen EJ, Scheld WM (1991) Blood-brain alterations in bacterial meningitis: development of an *in vitro* model and observations on the effects of lipopolysaccharide. In Vitro Cell Dev Biol 27A:113-120
46. Burroughs M, Carbellos C, Prasad S, Tuomanen E (1992) Bacterial components and the pathophysiology of injury to the blood-brain barrier: Does cell wall add to the effects of endotoxin in Gram-negative meningitis? J Infect Dis 165 (Suppl 1):582-585
47. Dascombe MJ, Milton AS (1979) Study on the possible entry of bacterial endotoxin and prostaglandin E_2 into the central nervous system from the blood. Brit J Pharmacol 66:565-572
48. Alafiatayo RA, Cookson MR, Pentreath VW (1994) Production of prostaglandins D_2 and E_2 by mouse fibroblasts and astrocytes in culture caused by *Trypanosoma brucei brucei* products and endotoxin. Parasitol Res 80:223-229
49. Pentreath VW, Cookson MR, Ingram GA, Mead C, Alafiatayo RA (1994b) *Trypanosoma brucei* products activate components of the reactive response in astrocytes *in vitro*. Bull Soc Path Ex 87:323-329
50. Pentreath VW, Rees K, Owolabi OA, Philip KA, Doua F (1990) The somnogenic T lymphocyte suppressor prostaglandin D_2 is selectively elevated in cerebrospinal fluid of advanced sleeping sickness patients. Trans R Soc Trop Med Hyg 84:795-799
51. Pentreath VW, Baugh PJ, Lavin DR (1994a) Sleeping sickness and the nervous system. Onderstep J Vet Res 61:369-377
52. Hunter CA, Gow JW, Kennedy PGE, Jennings FW, Murray M (1991) Immunopathology of experimental African sleeping sickness: detection of cytokine mRNA in the brains of *Trypanosoma brucei brucei* infected mice. Infect Immun 59:4636-4646

53. Hunter CA, Jennings FW, Kennedy PGE, Murray M (1992) Astrocyte activation correlates with cytokine production in central nervous system of *Trypanosoma brucei brucei* infected mice. Lab Invest 67:635-642
54. Tuomanen E (1993) Breaching the blood-brain barrier. Scientific American (Feb):56-60
55. de Vries HE, Blom-Roosemalen MCM, van Oosten M, de Boer AG, van Berkel TJC, Breimer DD, Kuiper J (1996) The influence of cytokines on the integrity of the blood-brain barrier *in vitro*. J Neuroimmunol 64:37-43
56. Pentreath VW (1989) Neurobiology of sleeping sickness. Parasitol Today 5:215-218
57. Greenwood BM, Whittle HC (1980) The pathogenesis of sleeping sickness. Trans R Soc Trop Med Hyg 74:716-725
58. Ndung'u JM, Ngure RM, Ngotho JM, Sayer PD, Omuse JK (1995) Total protein and white cell changes in the cerebrospinal fluid of vervet monkey infected with *Trypanosoma rhodesiense* and the post-treatment reaction. Jap J Protoz Res 73:484-497
59. Schmidt H (1983) The pathogenesis of trypanosomiasis of the CNS. Studies on parasitological and neurohistological findings in Trypanosoma rhodesiense infected vervet monkey. Virchows Arch (Pathol Anat) 399:333-343

CHAPTER 7

Cytokines in the pathogenesis of human African trypanosomiasis: antagonistic roles of TNF-α and IL-10

SG Rhind, PN Shek

Introduction

At the beginning of the twentieth century, human African trypanosomiasis (HAT) or sleeping sickness surpassed all other public health problems in sub-Saharan Africa. By the late 1950s, the incidence of HAT had been decreased in all endemic countries of West and Central Africa as a result of mass detection and treatment campaigns. Political unrest and civil war have disturbed the normal public health infrastructure, causing a resurgence of this devastating protozoal infection in regions of tropical and subtropical Africa. Recent medical surveys reveal shockingly high prevalence (> 700 cases per 1000) in some endemic areas of equatorial Africa [1].

The immune dysregulation associated with the pathogenesis of African trypanosomiasis has been the object of considerable research attention over the past 30 years [2-5]. Despite these efforts, HAT has remained elusive, due mainly to logistical limitations associated with conducting human studies in Africa and on account of the elaborate capacity of trypanosomes for antigenic variation [6] and immunosuppression [7]. The ability of trypanosomes for host immune evasion results in a chronic debilitating disease which is invariably fatal when untreated. Thus, it is critical to gain a better understanding of the underlying immunological mechanisms involved in its pathogenesis, leading ultimately to more effective therapeutic and prophylactic control modalities.

Advances in the area of cytokine biology are currently expanding at an unprecedented rate. To date, upwards of 200 cytokines have been identified (reviewed by Aggerwall and Purish, [8]). A growing body of evidence implicates many of these molecules as decisive factors in host–parasite immunobiology [40]. Several cytokines are capable of mediating both resistance and disease outcome, and it is now clear that the cytokine cascade has a central role in the pathophysiology of HAT [98, 99].

This chapter reviews recent information demonstrating reciprocal and antagonistic interactions between trypanosomes and host-derived cytokines, and their impact on disease. The intent of this contribution is not to provide a comprehensive survey of the literature, but to attempt a synthesis of the current status of the field with emphasis on new findings. Human studies are often limited and therefore, relevant animal studies will also be discussed to offer a broader examination of the condition.

The cytokine network in parasitic infections

In response to infectious or inflammatory agents, the immune and inflammatory systems (or more precisely – as they are inextricably linked – the immunoinflammatory system) are activated by a series of predominantly low molecular weight polypeptide mediators which regulate both the amplitude and duration of effector responses. These cell-derived mediators may act on the releasing cell (autocrine effect), cells in the local microenvironment (paracrine effect), and more distant systemic tissues (endocrine effect).

Confounding any discussion of the "cytokine network" is the tremendous redundancy of cellular sources and the range of biological activities exhibited by a particular cytokine. Cytokines are highly active at very low concentrations, combining with small numbers of high affinity receptors to induce changes in RNA and protein synthesis. The production and action of cytokines is normally transient and tightly regulated by positive and negative feedback loops. The extent of immunoregulatory influence exerted by this array of cytokines enables them to play the opposing roles of contributing to the pathophysiology of infection, but also to confer protection against many parasites, including African trypanosomes [9]. The final consequences of infection, whether resistance or disease, are dependent upon the balance of counter-regulatory cytokines elicited by a parasite.

It is impossible to accurately group cytokines according to unique tissue sources or biologic activities. For the purpose of this review, however, cytokines will be grouped according to their predominant source of production (i.e., mononuclear phagocyte-derived – monokines or T-lymphocyte derived – lymphokines) and activity patterns (i.e., immunoinflammatory or immunosuppressive).

Cytokine induction and polarization of T cell function

Many cytokines are required for the initiation and coordination of clonal proliferation and activation of specific cell populations. The particular cytokines that are released in response to a pathogen influence whether the resulting immune response will be mainly humoral, cell-mediated or inflammatory in nature. An optimal host response against different microbial agents requires highly specialized defence reactions.

Specific humoral and cellular immune responses are regulated to a large extent via the secretion of cytokines by $CD4^+$ T-helper (Th) cell subsets following antigen recognition. Parasitic disease models have provided important insight into the significance of the dichotomy in cytokine production by Th subsets [10]. The secretion of aberrant sets of cytokines may result in nonprotective immune responses and, consequently, allow disease to develop [11].

The combat of extracellular parasites involves production of soluble antibodies which first neutralize the invasion and opsonize parasites for phagocytosis. In contrast, the destruction of intracellular parasites requires activation of monocyte-macrophages. Thus, the immune response to parasitic protozoa is often characterized by a dominance of either cell-mediated or humoral effector mechanisms [12].

In general, infections by intracellular parasites (e.g., *Leishmania, Trypanosoma cruzi*) are associated with the induction of a strong Th1-type cytokine (IL-2, IFN-γ, TNF-α, TGF-β) response and are, therefore, particularly efficient activators of phagocyte and cytotoxic-dependent host responses [13]. At the other extreme, extracellular parasites (e.g., *Schistosoma mansoni, Plasmodium falciparum*) induce predominantly Th2-type cytokine (IL-4, IL-5, IL-6, IL-10) responses that favor B cell maturation and provide efficient help for the production of appropriate antibody isotypes [9]. The majority of peripheral blood Th cells, however, have an unrestricted cytokine secretion pattern (Th0 cells), simultaneously producing type-1 and type-2-associated cytokines [11].

The development of either pattern is governed by many factors, such as antigen density, the type of antigen-presenting cell, and local production of hormones [98]. Cytokines themselves have been shown to be potent modulators. Each T cell subset serves as its own autocrine growth factor and produces cytokines which cross-regulate the development and activity of each other [11]. The net result of autocrine-amplification and cross-regulation is that once a T-cell immune response begins to develop along one pathway, namely Th1 or Th2, it tends to become progressively polarized in that direction (see Fig. 1). Such polarization is a characteristic feature of persistent immune stimulation in chronic parasitic infections [14].

Pathogenesis of HAT

The vast majority of experimental studies have been conducted using murine [15] or bovine models [16] of trypanosomiasis using *Trypanosoma brucei brucei,* which is the ancestor of *T. b. rhodesiense* and *T. b. gambiense*. Although morphologically indistinguishable from the human pathogens, *T. b. brucei* is non-infective to humans. The restricted host range of *T. b. brucei* is due to its sensitivity to a circulating nonimmune factor which causes lysis of these organisms within a few minutes of exposure to normal human serum [17]. The trypanolytic component of human serum has been identified as a minor subfraction of high-density lipoprotein [18] called the trypanolytic factor [19]. Why this does not occur in human-infective trypanosomes is under investigation.

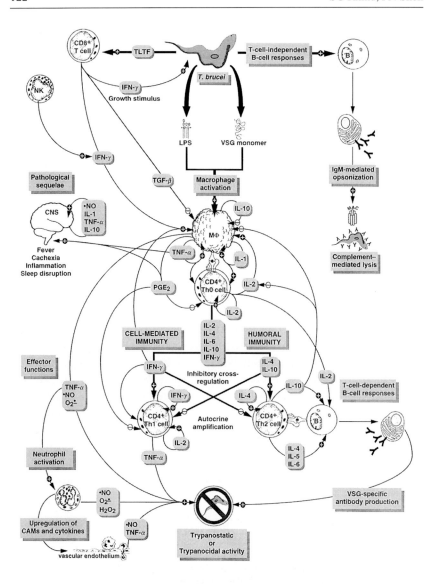

Fig. 1. Activation of the cytokine network during African trypanosomiasis infections. Trypanosomes stimulate a cascade of host-derived cytokines by INF-γ– activated macrophages (MΦ) and T helper (Th) cell subsets. Both specific and non-specific defences are activated and can lead to host protection or pathology. Antagonistic actions of TNF-α and IL-10 figure prominently in pathological sequelae (see text for details; abbreviations: ADCC = antibody-dependent cellular cytotoxicity; B = B cell; CAMs = cellular adhesion molecules; CNS = central nervous system; H_2O_2 = hydrogen peroxide; IFN = interferon; Ig = immunoglobulin; IL = interleukin; LPS = lipopolysaccharide; MAC = membrane attack complex; NO = nitric oxide; PGE_2 = prostaglandin E_2; $O_2^{\bullet-}$ = superoxide; TLTF = trypanosome-released lymphocyte triggering factor; TNF = tumor necrosis factor; VSG = variable surface glycoprotein; ⊕ = stimulation; ⊖ = inhibition

HAT, caused by the subspecies *T. b. gambiense*, is a progressive infection that leads to development of severe systemic disease, usually comprising two stages. Stage I – the hemolymphatic phase – begins when the inoculated organisms replicate extracellularly at the wound site producing a chancre which is characterized by a localized inflammatory response [20]. Subsequently, trypanosomes spread through the lymphatic vessels to the bloodstream and other body tissues. This invasion results in successive peaks of parasitemia producing fever, chronic fatigue and anemia [3]. If untreated, *T. b. gambiense* infections progress to stage II, the chronic phase, which is defined by the presence of parasites in the central nervous system (CNS) leading to a variety of CNS abnormalities [21]. There is demyelination and neuronal damage leading to meningoencephalitis, disruption of the sleep-wake cycle with daytime somnolence [22], progressive neurologic compromise and, if still untreated, coma, and death by inanition or intercurrent infection [20].

Chronic HAT infections are accompanied by numerous functional aberrations of the immune system. It is well established that trypanosomes possess unique mechanisms for evading the consequences of mammalian immune surveillance [23]. In mammals, African trypanosomes are completely covered by an outer coat of approximately 10^7 variant surface glycoproteins (VSG) [24]. Since trypanosomes live and multiply extracellularly in the blood and tissues during their infectious stage in the vertebrate host, they are easy targets for the host's humoral immune mechanisms. In a classical T-cell independent interaction, this highly immunogenic array of VSG molecules stimulates polyclonal B cell proliferation (particularly of the IgM class) [25]. Three to four days post-infection, the T cell-dependent production of high titers of VSG-specific antibodies occurs. Together, these specific and nonspecific effector mechanisms can destroy over 90% of the original infecting population via complement-mediated lysis, opsonization facilitated-phagocytosis, and antibody-dependent cell-mediated cytotoxicity (Fig. 1) [26, 27].

Totally effective humoral eradication rarely occurs, because trypanosomes evade complete destruction by antibodies by altering their surface antigens via transcriptional switching of VSG-genes [28]. The incomplete neutralization of trypanosomes perpetuates the chronic infection with only transient reduction in parasitemia. In addition, during antigenic switching, the VSG coat is shed into the blood where it is responsible for further immune dysregulation [29]. This interplay of host VSG-specific immunity and parasite antigenic variation induces the archetypical fluctuating parasitemias associated with HAT and likely contributes to the overall immunosuppression by a disturbance of cytokine networks. It appears that as the replicative cycle continues, the host immune response becomes increasingly suppressed and less sensitive [25].

Immunopathogenic and protective actions of cytokines in HAT

Immunosuppression during human and experimental African trypanosomiasis infections is a well-documented phenomenon that has hitherto lacked adequate mechanistic explanation [30-36]. In addition to their partial evasion of antibody-mediated destruction, African trypanosomes have also evolved effective subversion strategies by which they can actively disrupt normal activation and effector mechanisms, such as antigen or mitogen-stimulated T cell activation, proliferation, and cytokine production by lymphocytes and mononuclear phagocytes [6]. Although it is unclear to what extent cytokines participate directly in protective immunity, it is apparent that the elaboration of many host-derived cytokines are grossly disrupted by trypanosomal infections [98] and that some of these potent mediators contribute to the immunopathological symptoms of this disease [7, 37].

Reduced responsiveness has been attributed to parasite-induced suppressive actions of IFN-γ-activated macrophages [15], which block IL-2 production via increased prostaglandin (PG) E synthesis [32, 38] and IL-2R expression through PGE – independent mechanisms [34]. The downregulation of IL-2 production and receptor expression however, does not induce a generalized suppression of T cell activation but causes a shift in the dominant cytokine production profile [5]. The production of IFN-γ by $CD8^+$ T cells is polyclonally triggered by a trypanosome – released factor (termed trypanosome lymphocyte triggering factor, or TLTF) [37]. In contrast to many other protozoal infections including American trypanosomiasis, IFN-γ does not mediate a protective role but instead acts as a virulence factor by providing a growth promoting signal to African trypanosomes [39].

Mononuclear phagocyte-derived cytokines

Activated tissue macrophages and circulating monocytes have the capacity to secrete a large number of well-defined mediators which include cytokines, growth factors, PGs and nitric oxide (NO). These mononuclear phagocytes are responsible for the bulk of the production of proinflammatory monokines, such as TNF-α, IL-1β, and IL-6 in response to various infectious and noninfectious stimuli [40]. The processing of antigens as they are taken up by phagocytes, bound to major histocompatibility complex (MHC) II molecules and presented to Th cells provides a major pathway of monokine production. Alternatively, monocyte-macrophages may be directly triggered to produce cytokines by microbial-derived membrane components, such as endotoxin/lipopolysaccharide (LPS) through interaction with CD14 receptors [41].

TNF-α is a pleiotrophic cytokine that acts in conjunction with other monokines in the orchestration of systemic inflammatory cascades and immune stimulation [42]. Its effects were described in the last century, but the soluble factor was not isolated for some time [43]. TNF-α was originally named cachectin for its association with the negative protein balance and cachexia (wasting syndrome) that accompanies terminal malignancies and severe parasitic infections [44]. In addition to mononuclear phagocytes, TNF-α is also produced by Th1 cells, and astrocytes and microglial cells in the CNS [45]. TNF-α has a broad spectrum of immunomodulatory actions including enhanced cytotoxicity, anti-infection and growth modulation. These multiple biological activities are mediated through two distinct, high affinity receptors present on most nucleated cells [42]. A primary target of TNF-α is the vascular endothelium where it gives rise to the expression of several cellular adhesion molecules (CAMs) that serve as receptors for inflammatory cells. Exaggerated neutrophil endothelial adherence often causes endothelial injury by enhanced leukocyte-released proteases and respiratory burst activity (i.e., H_2O_2 and $O_2^{\bullet-}$ radical production) [46].

From a pathophysiologic perspective, TNF-α is one of the most intensively studied cytokines in infection, and despite its long history it is only now reaching major clinical prominence [47]. The exuberant production of this cytokine can be highly deleterious to the host and may lead to multisystem injury, systemic metabolic abnormalities, and circulatory collapse [48]. This is exemplified in septic shock [49] and cerebral malaria [50], where it is responsible for much of the pathological sequelae associated with these conditions. Plasma TNF-α levels often correlate with disease severity or mortality during experimental infections and anti-TNF-α antibodies can reduce mortality [51]. However, TNF-α has also been shown to be necessary in host defence [52].

Trypanosome-induced, host-derived TNF-α is recognized as an important pathological and protective mediator of both animal and human trypanosomiasis [53]. Secretion of TNF-α during the early stages of trypanosomal infection induces fever, elicits the hepatic acute phase response, and activates mononuclear phagocytes in an autocrine fashion [54, 55]. These non-specific immunoinflammatory responses initiated by TNF-α provide host protection, by inhibiting parasite development, limiting parasite spread, and preparing the host for a prolonged specific immune defence [56].

In the case of experimental trypanosomiasis infections, administration of soluble trypanosomal lysates induces murine TNF-α secretion, resulting in direct trypanolysis and/or inhibition of parasite growth [57, 58]. In contrast, this anti-parasitic effect is abolished by co-treatment with neutralizing anti-TNF-α antibodies [58]. While these results are indicative of a positive role for TNF-α in the control of trypanosomes, severe anemia and cachexia are major causes of death in bovine trypanosomiasis and have been linked to excessive TNF-α production [48]. Indeed, many of the clinical and pathophysiologic features of trypanosomiasis including anemia, fever, headaches and cachexia resemble those induced by *in vivo* injection of LPS or recombinant TNF-α [52, 59].

Human studies measuring TNF-α levels have demonstrated that circulating concentrations are significantly increased in *T b. gambiense*-infected patients relative to healthy African controls, and that these cytokine disruptions are linked to changes in neuroendocrine circadian rhythmicity [60, 61, 98]. Okomo-Assoumou et al. [62] reported a substantial elevation of TNF-α in the serum of *T. b. gambiense*-infected patients which were correlated with disease severity. Their work also implied that drug treatment quickly reduces elevated circulating TNF-α levels. In support of this, Rhind et al. [63] observed chronically elevated circulating levels TNF-α in patients with advanced meningoencephalitic *T. b. gambiense* infection. Furthermore, these levels were significantly reduced, but not completely abrogated, following chemotherapy with melarsoprol (see Fig. 2B).

It has been shown that increased endogenous production of NO, enzymatically derived from activated macrophages via the action of the inducible nitric oxide synthase (iNOS), has direct parasiticidal activity against numerous intra- and extracellular pathogens [64]. Specifically, NO inactivates enzymes involved in mitochondrial respiration of target cells [65]. Since TNF-α is a potent enhancer of iNOS in IFN-γ-primed monocyte-macrophages, recent interest has focused on a possible dual regulatory and effector role for NO in mediating disease protection and pathology [66, 67]. It is hypothesized that in certain conditions, such as cerebral malaria, high local concentrations of TNF-α induce excessive synthesis of NO in cerebral vascular endothelial cells, and that this NO is responsible for synaptic dysregulation and coma [68]. On the other hand, Anstey et al. [69] recently concluded that rather than being increased in cerebral malaria, NO production was inversely proportional to disease severity. Findings from animal models of African trypanosomiasis support this dualism. *In vitro* experiments using BCG-infected or IFN-γ activated murine macrophages demonstrate that NO is cytostatic or cytotoxic against *T. b. brucei* and *T. b. gambiense* [70]. Likewise, in *T. b. brucei*-infected rats Buguet et al. [71] discovered that circulating NO levels were decreased, but that NO levels in the CNS were elevated compared to uninfected controls. Results from a large group of *T. b. congolense*-infected cattle show that NO production by IFN-γ-activated peripheral blood mononuclear cells (PBMCs) is suppressed [72]. However, recent studies of *T. b. brucei*-infected mice reveal that both NO-dependent and NO-independent immuno-suppressive mechanisms are operative in the early and late stages of murine infection, respectively [100].

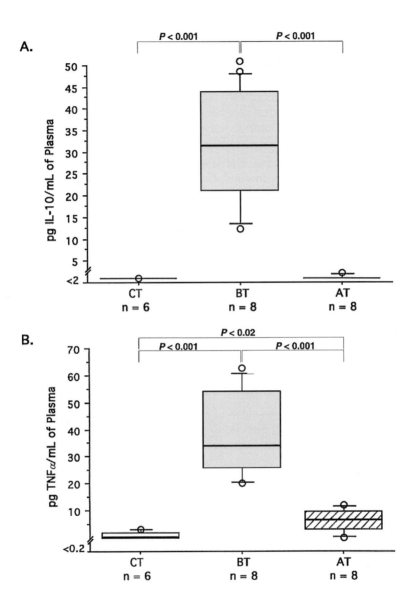

Fig. 2. Box-and-wisker plots of circulating plasma cytokine concentrations in healthy African controls (CT), unmedicated late-stage trypanosomiasis patients (BT), and patients after 6 days (AT) of combination chemotherapy. The median is indicated by a horizontal bar across each box. Boxes encompass 50% of the data and delineate the lower (25th) and upper (75th) percentiles. Wisker caps on each box represent the 10th and 90th percentile points of data. Individual data points (o) show the outliers

Microglias are emerging as key players in the immune response to CNS injury. LPS activated microglia are observed in areas of tissue damage where they produce several inflammatory mediators following clinical injury or experimentally-induced neuropathy [73]. This histopathology coupled with the ability of microglia to produce cytotoxic factors such as NO and TNF-α argues for an active role in tissue damage [74]. The brains of late-stage *T. b. brucei*-infected mice commonly exhibit upregulation of TNF-α [75]. We have found that successfully treated patients exhibit substantially reduced, although above normal, TNF-α levels one day following a three-day treatment regimen [63]. This suggests that TNF-α levels of cured patients may stabilize quickly following therapy. Conceivably then, prophylactic blockade of TNF-α activity could be used to mitigate morbidity and mortality in late-stage patients undergoing chemotherapy. On the other hand, it also implies an advantageous role for moderate TNF-α levels during the recovery from HAT. In support of this view, work with TNF-α receptor knock-out mice shows that TNF-α can be protective in the CNS, where it prolongs neuronal cell survival following excitotoxic or ischemic insults [76].

Soluble parasite-derived factors are implicated as triggers of TNF-α production during trypanosomal infection [77]. Pentreath et al. [78] confirmed that markedly elevated LPS levels were expressed in the blood and cerebrospinal fluid of late-stage trypanosomiasis patients. The source of the LPS is not clear, but could be a direct by-product of trypanolysis, a consequence of intestinal or hepatic damage, and/or secondary bacterial infection. Other microbial products, such as muramyl peptides, may also be involved in initiating cytokine production in HAT. These peptidoglycan-derived factors give rise to several cytokines including TNF-α and are well-known endogenous somnogenic agents [79]. Recently, Takahashi et al. [80] suggested that centrally produced TNF-α is a primary mediator responsible for the sleep-inducing activity of muramyl peptides both under normal and pathological conditions.

Thus, it seems that while regulated release of TNF-α may stimulate beneficial local antiparasitic host responses by enhancing cytotoxic activity, its overproduction can lead to life-threatening systemic complications. Strategies to mitigate the shock-inducing properties of TNF-α while retaining its antiparasitic effects should be adopted. For example, prior treatment with indomethacin before intravenous injection with TNF-α reduces lethality in mice by 70% [81]. Similarly, the use of bismuth subnitrate, a free radical scavenger, protected mice from TNF-α toxicity while not interfering with its antimicrobial efficacy [82].

Anti-inflammatory or immunosuppressive cytokines

One of the major findings in cytokine biology during the past decade has been the discovery of cytokines that are functional antagonists of immunostimulatory and/or proinflammatory cytokines. Two of the most important cytokines in the control of immuno-inflammatory systems are IL-10, and transforming growth factor (TGF)-β, both of which have predominantly anti-inflammatory or inhibitory effects on the synthesis of certain proinflammatory mediators.

Numerous studies have documented the inhibitory effects of IL-10 on Th1-type cytokine synthesis and monocyte-macrophage activation [83]. IL-10 is produced by murine Th2 cells, human Th0 and Th2 clones, B cells and human monocyte-macrophages [84]. The pleiotrophic actions of IL-10 on monocyte-macrophages and T cells include: inhibition of production of proinflammatory cytokines; inhibition of mitogen and antigen-driven cell proliferation; and the ability to block several accessory cell functions such as antigen presentation [83].

Specifically, IL-10 is a very potent inhibitor of IFN-γ-stimulated monocyte-macrophage effector functions, including their oxidative burst, NO synthesis and microbicidal activity [85]. This inhibition is mediated indirectly, via the action of IL-10 on cytokine production by monocyte-macrophages (i.e., IL-6, TNF-α, IL-12) and Th1 cells (i.e., IFN-γ, IL-1β). IL-10 is also a potent inhibitor of IFN-γ secretion by NK cells in response to IL-2. Also, IL-10 impairs the ability of monocytes to provide co-stimulatory signals for resting T cell proliferation, and this impairment is primarily mediated by the inhibition of IL-2 production [83].

Th2-type responses are frequently beneficial, especially for controlling intracellular pathogens. However, under certain circumstances, these host protective responses can induce immune pathology. In particular, very high levels of IL-10 have been measured in the serum of patients suffering from cerebral malaria [86]. We have recently determined that IL-10 levels are elevated in late-stage HAT patients [63]. Remarkably, following treatment with melarsoprol, these levels dropped to below the detection threshold (see Fig. 2A). Although to the best of our knowledge this is the first such report in human subjects, evidence from other protozoan illnesses [8] infer multiple possible regulatory roles for IL-10 in this disease. Taylor et al. [72], showed that IL-10 mRNA was upregulated in the lymph nodes, spleens, and PBMCs of *T. b. congolense*-infected cattle. A modulatory role for IL-10 in bovine trypanosomiasis is also supported by results showing that similar to the human, but divergent from murine systems, all subsets of bovine Th cells are able to produce and be downregulated by IL-10 [87]. In addition, recent evidence demonstrating markedly elevated levels of IL-10 protein and mRNA, in the blood and splenocyte supernatants of *T. congolense*-infected BALB/c mice, further implicates IL-10 as a central immuno-suppressive effector molecule in African trypanosomiasis [101].

The kinetics of IL-10 production differ from those of most cytokines; mRNA for IL-1, IL-2, IL-4 and IFN-γ can be identified within 15 min and peak 2–4 hours after stimulation, but IL-10 mRNA does not appear until 8 hours and is maximal only at 12–24 hours. TNF-α is a potent stimulus for IL-10 secretion, and IL-10 specifically inhibits TNF-α, as well as its own production, via an autoregulatory feedback loop [88]. Thus, IL-10 is produced late in the cytokine cascade as part of an endogenous protective strategy which down-regulates earlier synthesis of proinflammatory cytokines [84]. These studies suggest a heomeostatic mechanism whereby chronic exposure to trypanosome-derived LPS in the late-stages of HAT, triggers excessive TNF-α secretion, which in turn stimulates IL-10 release from activated monocyte-macrophages.

Overall, these data indicate that as in several other acute and chronic inflammatory conditions, IL-10 may have a therapeutic role in HAT. In fact, systemic intravenous IL-10 administration in humans drastically reduces inflammatory cytokine production [89] and is protective against lethal endotoxemia in animals [90]. By contrast, IL-10 knock-out mice and passive immunization against this cytokine enhances LPS-induced mortality and susceptibility to major inflammatory disorders [91, 92].

TGF-β is the prototype of a family of at least five peptides that regulate cell growth and differentiation, having both stimulatory and inhibitory effects on different cell types [93]. TGF-β is upregulated at the sites of tissue injury, inflammation, and repair. It is produced by numerous cell types including both $CD4^+$ and $CD8^+$ T cells. Similar to IL-10, expression of TGF-β by PBMC is delayed (2 – 3 days) following mitogen stimulation, suggesting that secreted TGF-β functions in an autocrine or paracrine manner to limit the extent of T-cell clonal expansion and cytokine release. TGF-β and TNF-α exert opposing activities, and TGF-β is a strong inhibitor of iNOS activity [67]. Th2 cells are preferentially inhibited by TGF-β through down-regulation of IL-4 production. Hence, both pre-B cells and mature B cell proliferation are inhibited by TGF-β [94]. Studies of TGF-β in African trypanosomiasis models are scarce. Bakhiet et al. [95] reported that TLTF induces TGF-β mRNA expression in mice and suggested that it may contribute to the immunosuppression characteristic of *T. b. brucei* infections. Both IL-10 and TGF-β expression have been detected in the CNS during inflammatory or infectious diseases, where they are thought to have a primarily neuroprotective role by inhibiting excessive microglial production of inflammatory monokines and NO [45, 96]. Initial results also suggest that IL-10 may contribute to central regulation of sleep and chronic fatigue [97].

The evidence accumulated so far suggests that the immune reactions associated with the progression to chronic disease in HAT infections provoke a shift toward expression of Th2 cytokines with concomitant suppression of the Th1 effector responses. The outcome of immune polarization is dependent upon the host's ability to generate the most effective type of immune response necessary to eradicate the particular parasite. Although experimental evidence points to a potentially protective role of IL-10 in late-stage HAT by countering the excessive production of TNF-α, one must keep in mind the beneficial roles of proinflammatory monokine synthesis in antiparasitic defences. Therefore, excessive IL-10, resulting in extreme downregulation of TNF-α, might be deleterious [51]. Exaggerated and uncontrolled production of IL-10 is likely to exacerbate disease through its ability to mediate suppression of monocyte-macrophage accessory and effector functions. Furthermore, excessive IL-10 production may contribute to the generalized immunosuppression characteristic of chronic HAT infections [5] either through its inhibition of NO synthesis [71], and/or via down-regulation of T cell-mediated immune responses [4, 7]. Thus, whether increased IL-10 is a cause or effect of increased disease susceptibility in HAT infections remains to be determined, but for now a definite association has been made.

Conclusion and prospects

The relationship between cytokines and altered host resistance in African trypanosomiasis is complex and only partially understood. It is evident from these and other studies discussed in accompanying chapters that cytokines are important mediators involved in the coordination of host defence mechanisms against African trypanosomes. A key finding is that the normal physiologic equilibrium between TNF-α and IL-10 is disrupted by chronic exposure to parasites. The initial release of TNF-α by IFN-γ-activated macrophages during acute trypanosomal infection may act as a protective response to limit disease severity. However, overproduction of TNF-α can become damaging to the host, and promotes the progression of immunopathologic symptoms of advanced disease. In this context, it is conceivable that the induction of anti-inflammatory or immunosuppressive cytokines, including IL-10 or TGF-β, is an attempt of the host to counteract the inflammatory cascade precipitated primarily by TNF-α. Excessive IL-10 can also have detrimental effects on host resistance through down-regulation of essential macrophage and T-cell activation and effector functions. Comprehension of the pathogenesis of HAT has been greatly facilitated by our recognition of the pivotal role that cytokines play in this disease, and novel immunotherapeutic approaches may result from this knowledge.

References

1. Ekwanzala M, Pépin J, Khonde N, Molisho S, Bruneel H, DeWals P (1996) In the heart of darkness: sleeping sickness in Zaire. Lancet 348:1427-1430
2. Gray AR (1967) Some principles of the immunology of trypanosomiasis. Bull WHO 37:177-193
3. Diggs CL (1982) Immunological research on African trypanosomiasis. Prog Allergy 31:268-300
4. Mansfield JM (1990) Immunology of African trypanosomiasis. In: Wyler DJ (eds) Modern Parasite Biology. WH Freeman, New York, pp 222-246
5. Sileghem M, Flynn JN, Darji A, De Baetselier P, Naessens J (1994) African trypanosomiasis. In: Kierszenbaum F (eds) Parasitic Infections and the Immune System. Academic Press, London, pp 1-51
6. Kotwal GJ (1996) The great escape. Immune evasion by pathogens. The Immunologist 4/5:157-164
7. De Baetselier P (1996) Mechanisms underlying trypanosome-induced T-cell immunosuppression. In: Mustafa AS, Al-Attiyah RJ, Nath I, Chugh TD (eds) T-cell subsets and cytokines interplay in infectious diseases. Karger, Basel, pp 124-139
8. Aggerwal BB, Purish RJ (1995) Human cytokines: their roles in disease and therapy. Blackwell Science, Oxford, UK
9. Barcinski MA, Costa-Moreira ME (1994) Cellular responses of protozoan parasites to host-derived cytokines. Parasitol Today 10:352
10. Cox FE, Liew FY (1992) T-cell subsets and cytokines in parasitic infections. Immunol Today 13:445-448
11. Abbas AK, Murphy KM, Sher A (1996) Functional diversity of helper T lymphocytes. Nature 383:787-793
12. Jankovic D, Sher A (1996) Initiation and regulation of $CD4^+$ T-cell function in host-parasite models. In: Romagnani S (eds) Th1 and Th2 cells in health and disease. Karger, Basel, pp 51-65
13. Kemp M, Theander TG, Kharazmi A (1996) The contrasting roles of $CD4^+$ T cells in intracellular infections in humans: leishmaniasis as an example. Immunol Today 17:13-16
14. Reiner SL (1994) Parasites and T helper cell development. Parasitol Today 10:485-488
15. Askonas BA (1985) Macrophages as mediators of immunosuppression in murine African trypanosomiasis. Curr Topics Microbiol Immunol 117:119-127
16. Urquhart GM (1980) The pathogenesis and immunology of African trypanosomiasis in domestic animals. Trans R Soc Trop Med Hyg 74:726-729
17. Lorenz P, Betschart B, Owen JS (1995) *Trypanosoma brucei brucei* and high-density lipoproteins: old and new thoughts on the identity and mechanism of the tryapanocidal factor in human serum. Parasitol Today 11:348-352
18. Rifkin MR (1978) Identification of the trypanocidal factor in normal human serum: High-density lipoprotein. Proc Natl Acad Sci USA 75:3540
19. Smith AB, Esko JD, Hajduk SL (1995) Killing of trypanosomes by the human haptoglobin-related protein. Science 268:284
20. Apted FIC (1970) Clinical manifestations and diagnosis of sleeping sickness. In: Mulligan HW (eds) The African Trypanosomiases. George Allen & Unwin, London, p 661
21. Pentreath VW (1995) Trypanosomiasis and the nervous system: pathology and immunology. Trans R Soc Trop Med Hyg 89:9-15
22. Buguet A, Bert J, Tapie P, Tabaraud F, Doua F, Lonsdorfer J, Bogui P, Dumas M (1993) Sleep-wake cycle in human African trypanosomiasis. J Clin Neurophysiol 10:190-195
23. Wakelin D (1996) Parasites and the immune system. BioScience 47:32-40
24. Borst P, Rudenko G (1994) Antigenic variation in African trypanosomes. Science 264:1872-1873
25. Mansfield JM (1994) T-cell responses to the trypanosome variant surface glycoprotein: A new paradigm? Parasitol Today 10:267-270
26. Jokiranta TS, Jokipii L, Meri S (1995) Complement resistance of parasites. Scand J Immunol 42:9-20
27. Stevens DR, Moulton JE (1978) Ultrastructural and immunological aspects of the phagocytosis of *Trypanosoma brucei* by mouse peritoneal macrophages. Infect Immun 19:972-982
28. Cross AGM (1996) Antigenic variation in trypanosomes: secrets surface slowly. BioAssays 18:283-291
29. Seyfang A, Mecke D, Duszenko M (1990) Degradation, recycling and shedding of *Trypanosoma brucei* variant surface glycoprotein. J Protozool 37:546
30. Darji A, Sileghem M, Heremans H, Brys L, DeBaetselier P (1993) Inhibition of T-cell responsiveness during experimental infections with *Trypanosoma brucei*: active involvement of endogenous gamma interferon. Infect Immun 61:3068-3102
31. Darji A, Beschin A, Sileghem M, Heremans H, Brys L, DeBaetselier P (1996) *In vitro* simulation of immunosuppression caused by *Trypanosoma brucei* active involvement of gamma interferon and tumor necrosis factor in the pathway of suppression. Infect Immun 64:1937-1943

32. Kierszenbaum F, Muthukkumar S, Beltz LA, Sztein MB (1991) Suppression by *Trypanosoma brucei rhodesiense* of the capacities of human lymphocytes to express interleukin-2 receptors and proliferate after mitogenic stimulation. Infect Immun 59:3518-3522
33. Mabbott NA, Sutherland IA, Sternberg JM (1995) Suppressor macrophages in *Trypanosoma brucei* infection: nitric oxide is related to both suppressive activity in vivo. Parasit Immunol 17:143-150
34. Sileghem M, Flynn JN (1992) Suppression of interleukin-2 secretion and interleukin-2 receptor expression during tsetse-transmitted trypanosomiasis in cattle. Eur J Immunol 22:767-773
35. Sternberg JM, McGuigan F (1992) Nitric oxide mediates suppression of T-cell responses in murine *Trypanosoma brucei* infection. Eur J Immunol 22:2741-2744
36. Sternberg JM, Mabbott NA (1996) Nitric oxide-mediated suppression of T cell responses during *Trypanosoma brucei* infection: soluble trypanosome products and interferon-γ are synergistic inducers of nitric oxide synthase. Eur J Immunol 26:539-543
37. Bakhiet M, Olsson T, Edlund C, Höjeberg B, Holmberg K, Lorentzen J, Kristensson K (1993) *Trypanosoma brucei brucei*-derived factor that triggers CD8$^+$ lymphocytes to interferon-γ secretion: purification, characterization and protective effects *in vivo* by treatment with a monoclonal antibody against the factor. Scand J Immunol 37:165
38. Belley AB, Chandee K (1995) Eicosanoid production by parasites: from pathogenesis to immunomodulation. Parasitol Today 11:327-334
39. Olsson T, Bakhiet M, Höjeberg B, Ljungdahl Å, Edlund C, Andersson G, Ekre H-P, Fung-Leung W-P, Mak T, Kristensson K (1993) CD8 is critically involved in lymphocyte activation by a *T. brucei brucei*-released molecule. Cell 72:715-727
40. Titus RG, Sherry B, Cerami A (1991) The involvement of TNF, IL-1 and IL-6 in the immune response to protozoan parasites. Immunoparasitol Today 12/7:A13-A16
41. Sweet MJ, Hume DA (1996) Endotoxin signal transduction in macrophages. J Leuk Biol 60:8-26
42. Bemelmans MHA, van Tits LJH, Buurman WA (1996) Tumor necrosis factor: production, release and clearance. Crit Rev Immunol 16:1-11
43. Carswell EA, Old LJ, Kassel RL, Green S, Fiore N, Williamson B (1975) An endotoxin-induced serum factor that causes necrosis of tumors. Proc Natl Acad Sci USA 72:3666-3670
44. Beutler B, Cerami A (1989) The biology of cachectin/TNF-α primary mediator of the host response. Ann Rev Immunol 7:625-656
45. Chao C, Hu S, Petersen PK (1995) Glia, cytokines, and neurotoxicity. Crit Rev Neurobiol 9:189-205
46. Pober JS, Cotran RS (1990) Cytokines and endothelial cell biology. Physiol Rev 70:427-451
47. Rink L, Kirchner H (1996) Recent progress in the tumor necrosis factor-α field. Int Arch Allergy Immunol 111:199-209
48. Cerami A (1992) Inflammatory cytokines. Clin Immunol Immunopathol 62:S3-S10
49. van der Poll T, Lowry SF (1995) Tumor necrosis factor in sepsis: mediator of multiple organ failure or essential part of host defence? Shock 3:1-9
50. Clark IA, Rockett KA (1995) TNF, malaria and sepsis. Lancet 345:75-76
51. Blackwell TS, Christman JW (1996) Sepsis and cytokines: current status. Brit J Anaesth 77:110-117
52. Barbara JAJ, Ostade XV, Lopez AF (1996) Tumor necrosis factor-alpha (TNF-α): the good, the bad and potentially very effective. Immunol Cell Biol 74:434-443
53. Lucas R, Magez S, Songa B, Darji A, Hamers R, deBaetselier P (1993) A role for TNF during African trypanosomiasis: involvement in parasite control, immunosuppression and pathology. Res Immunol 144:370
54. Flynn JN, Sileghem M (1991) The role of the macrophage in induction of immunosuppression in *Trypanosoma congolense*-infected cattle. Immunol 74:310-316
55. Mwangi SM, Odimba F, Logan-Henfrey L (1995) The effect of *Trypanosoma brucei brucei* infection on rabbit plasma iron and zinc concentrations. Acta Trop 59:283-291
56. Stadnyk AW, Gauldie J (1991) The acute phase response during parasitic infection. Immunoparasitol 12/7:A7-A12
57. Kongshavin PA, Ghadirian E (1988) Enhancing and suppressive effects of tumor necrosis factor/cachectin on growth of *Trypanosoma musculi*. Parasite Immunol 10:581-588
58. Magez S, Lucas R, Darji A, Songa B, Hamers R, De Baetselier P (1993) Murine tumor necrosis factor plays a protective role during the phase of the experimental infection with *Trypanosoma brucei brucei*. Parasit Immunol 15:635-641
59. Michie HR, Manogue KR, Spriggs DR, Revhaug A, O'Dwyer S, Dinarello CA, Cerami A, Wolfe SA, Wilmore DW (1988) Detection of circulating tumor necrosis factor after endotoxin administration. N Engl J Med 318:1481-1485
60. Radomski MW, Buguet A, Bogui P, Doua F, Lonsdorfer A, Tapie P, Dumas M (1994) Disruptions in the secretion of cortisol, prolactin, and certain cytokines in human African trypanosomiasis patients. Bull Soc Path Ex 87:376-379
61. Reincke M, Heppner C, Petzke F, Allolio B, Arlt W, Mbulamberi D, Siekmann L, Vollmer D, Winkelmann W, Chrousos GP (1994) Impairment of adrenocortical function associated with increased plasma tumor necrosis factor-alpha and interleukin-6 concentrations in African trypanosomiasis. Neuroimmunomodulation 1:14-22

62. Okomo-Assoumou M, Daulouède S, Lemesre J-L, N'Zila-Mouanda A, Vincendeau P (1995) Correlation of high serum levels of tumor necrosis factor-α with disease severity in human african trypanosomiasis. Am J Trop Med Hyg 53:539-543
63. Rhind SG, Sabiston BH, Shek PN, Buguet A, Muanga G, Stanghellini A, Dumas M, Radomski MW (1997) Effect of melarsoprol treatment on circulating IL-10 and TNF-α levels in human African trypanosomiasis. Clin Immunol Immunopathol 83:185-189
64. Green SJ, Nacy CA (1994) L-Arginine-derived nitric oxide is an antimicrobial effector molecule. ASM News 60:83-88
65. Moncada S, Higgs EA (1995) Molecular mechanisms and therapeutic strategies related to nitric oxide. FASEB 9:1319-1330
66. Kremsner PG, Winkler S, Brandts C, Wildling E, Jennie L, Graninge W, Prada J, Bienzle U, Juillard P, Grau GE (1995) Prediction of accelerated cure in *Plasmodium falciparum* malaria by the elevated capacity of tumor necrosis factor production. Am J Trop Med Hyg 53:532-538
67. Liew FY (1995) Regulation of lymphocyte function by nitric oxide. Curr Opin Immunol 7:396-399
68. Mendis KN, Carter R (1995) Clinical disease and pathogenesis in malaris. Parasitol Today 11:1-16
69. Anstey NM, Weinberg JB, Hassanali MY, Mwaikambo ED, Manyenga D, Misukonis MA, Arnelle DR, Hollis D, McDonald MI, Granger DL (1996) Nitric oxide in Tanzanian children with malaria: inverse relationship between malaria severity and nitric oxide production/nitric oxide synthase type 2 expression. J Exp Med 184:557-567
70. Vincendeau P, Daulouède S, Veyret B, Dardé ML, Bouteille B, Lemesre JL (1992) Nitric oxide-mediated cytostatic activity on *Trypanosoma brucei gambiense* and *T. brucei brucei*. Exp Parasitol 75:353-360
71. Buguet A, Burlet S, Auzelle F, Montmayeur A, Jouvet M, Cespuglio R (1996) Dual intervention of NO in experimental African trypanosomiasis. C R Acad Sci Paris, Sciences de la vie/Life sciences 319:201-207
72. Taylor K, Lutje V, Mertens B (1996) Nitric oxide synthesis is depressed in *Bos indicus* cattle infected with *Trypanosoma congolense* and *Trypanosoma vivax* and does not mediate T-Cell depression. Infect Immun 64:4115-4122
73. Szczepanik AM, Fishikin RJ, Rush DK, Wilmot CA (1996) Effects of chronic intrahippocampal infusion of lipopolysaccharide in the rat. Neuroscience 70:57-65
74. Rothwell NJ, Luheshi, G, Toulmond S (1996) Cytokines and their receptors in the central nervous system: physiology, pharmacology, pathology. Pharmacol Ther 69:85-95
75. Hunter CA, Jennings FW, Kennedy PGE, Murray M (1992) Astrocyte activation correlates with cytokine production in central nervous system pathology in experimental african trypanosomiasis. Lab Invest 67:635-642
76. Bruce AJ, Boling W, Kindly MS, Peschon J, Kraeme PJ, Carpenter MK, Holtsberg FW, Mattson MP (1996) Altered neuronal and microglial responses to excitotoxic and ischemic brain injury in mice lacking TNF receptors. Nature Med 2:788-795
77. Alafiatayo RA, Crawley B, Oppenheim BA, Pentreath VW (1993) Endotoxins and the pathogenesis of *Trypanosoma brucei brucei* infection in mice. Parasitology 107:49-53
78. Pentreath VW, Alafiatayo RA, Crawley B, Doua F, Oppenheim BA (1996) Endotoxins in the blood and cerebrospinal fluid of patients with African sleeping sickness. Parasitology 112:67-73
79. Krueger JM, Takahashi S, Kapás L, Bredow S, Roky R, Fang J, Floyd R, Renegar KB, Guha-Thakurta N, Novitsky, Obál F (1995) Cytokines in sleep regulation. Adv Neuroimmunol 5:171-188
80. Takahashi S, Kapás L, Krueger JM (1996) A tumor necrosis factor (TNF) receptor fragment attenuates TNF-α- and muramyl dipeptide-induced sleep and fever in rabbits. J Sleep Res 5:106-114
81. Kettlehut IC, Fiers W, Goldberg AL (1987) The toxic effects of tumor necrosis factor in vivo and their prevention by cyclooxygenase inhibitors. Proc Natl Acad Sci USA 84:4273-4277
82. Satomi N, Sakurai A, Haranaka R, Haranaka K (1988) Preventive effects of several chemicals on lethality of recombinant human tumor necrosis factor. J Biol Resp Modif 7:54-64
83. Moore KW, O'Garra A, Malefyt RD, Vieira P, Mosmann TR (1993) Interleukin-10. Ann Rev Immunol 11:165-179
84. Daftarian PM, Kumar A, Kryworuchko M, Diaz-Mitoma F (1996) IL-10 production is enhanced in human T cells by IL-12 and IL-6 in monocytes by tumor necrosis factor-α. J Immunol 157:12-20
85. Gazzienelli RT, Oswald IP, James SL, Sher A (1992) IL-10 inhibits parasite killing and nitrogen oxide production by IFN-γ-activated macrophages. J Immunol 148:1792-1796
86. Peyron F, Burdin N, Ringwald P, Vuillez JP, Rousset F, Banchereau J (1994) High levels of IL-10 in human malaria. Clin Exp Immunol 95:300-303
87. Brown WC, Woods VM, Chitko-McKown CG, Hash SM, Rice-Ficht AC (1994) Interleukin-10 is expressed by bovine type 1 helper, type 2 helper, and unrestricted parasite-specific T-cell clones and inhibits proliferation of all three subsets in an accessory-cell-dependent manner. Infect Immun 62:4697-4708
88. De Waal Malefyt R, Bennett B, Fidgor C, Vries JED (1991) Interleukin-10 (IL-10) inhibits cytokine synthesis by human monocytes: an autoregulatory role of IL-10 produced by monocytes. J Exp Med 174:1209-1220

89. Fuchs AC, Granowitz EV, Shapiro L, Vannier E, Dinarello CA (1996) Clinical, hematologic, and immunologic effects of interleukin-10 in humans. J Clin Immunol 16:291-303
90. Howard M, Muchamuel T, Anrade S, Menon S (1993) Interleukin-10 protects mice from leathal endotoxemia. J Exp Med 177:1205-1208
91. Kuhn R, Lohler J, Remmick D, Rajewsky K, Muller W (1993) Interleukin-10-deficient mice develop chronic enterocolitis. Cell 75:263-274
92. Marchant A, Vincent JL, Goldmam M (1996) Interleukin-10 as a protective cytokine produced during sepsis. In: Morrison DC, Ryan JL (eds) Novel Therapeutic Strategies in the Treatment of Sepsis. Marcel Dekker, New York, pp 301-311
93. Sporn MB, Roberts AB (1993) Transforming growth factor-β: recent progress, new challenges. J Cell Biol 119:1017-1021
94. McCartney-Francis NL, Wahl NL (1994) Transforming growth factor β: a matter of life and death. J Leuk Biol 55:401-409
95. Bakhiet M, Olsson T, Ljungdahl Å, Höjeberg B, van der Meide P, Kristensson K (1996) Induction of interferon-γ, transforming growth factor-β, and interleukin-4 in mouse strains with different susceptibilities to *Trypanosoma brucei brucei*. J Interfer Cytok Res 16:427-433
96. Lodge PA, Sriram S (1996) Regulation of microglial activation by TGF-beta, IL-10, and CSF-1. J Leuk Biol 60:502-508
97. Opp MR, Smith EM, Hughes TK (1995) Interleukin-10 (cytokine synthesis inhibitory factor) acts in the central nervous system of rats to reduce sleep. J Neuroimmunol 60:165-168
98. Reincke M, Arlt W, Heppner C, Petzke F, Chrousos GP, Allolio B (1998) Neuroendocrine dysfunction in African trypanosomiasis. The role of cytokines. Ann NY Acad Sci 840:809-821
99. Sternberg JM (1998) Immunobiology of African trypanosomiasis. Chem Immunol 70:186-190
100. Beschin A, Brys L, Magez S, Radwanska M, De Baetselier P (1998) *Trypanosoma brucei* infection elicits nitric oxide-dependent and nitric oxide-independent suppressive mechanisms. J Leukoc Biol 63:429-439
101. Uzonna JE, Kaushik RS, Gordon JR, Tabel H (1998) Immunoregulation in experimental murine *Trypanosoma congolense* infection: anti-IL-10 antibodies reverse trypanosome-mediated suppression of lymphocyte proliferation *in vitro* and moderately prolong the lifespan of genetically susceptible BALB/c mice. Parasite Immunol 20:293-302

CHAPTER 8

Immunology of African trypanosomiasis

P Vincendeau, MO Jauberteau-Marchan, S Daulouède, Z Ayed

Introduction

The immune response to parasitic organisms constitutes an important factor of the complex, multi-facetted host-parasite relationship. The inoculation of trypanosomes into their mammalian hosts triggers a series of events involving, at first, innate immunity and, secondarily, specific immunity. The latter requires an efficient presentation of parasitic antigens, activation of T and B cells implying specific antigen receptor recognition, and the development of effector cells and molecules. These mechanisms are highly regulated by multiple signals delivered through a large number of receptors transduced across the plasma membrane and processed.

During co-evolution with their hosts, trypanosomes have learnt to cope with host immune systems, by penetrating, diverting and altering the numerous steps leading to the generation of an effective immune response. Major modifications of immune systems have been observed in trypanosomiasis: lymphadenopathy, splenomegaly (up to thirty times the normal size) with destruction of lymphatic tissue architecture and hypergammaglobulinaemia. However, their effectiveness is limited as, most of the time, parasites cannot be eliminated and immunopathological phenomena appear (tissular alterations by autoimmune reactions and high incidence of infections, the sign of a severe immunosuppression) [1].

One of the major characteristic of trypanosomes is the presence of the Variant Surface Glycoprotein (VSG) which covers nearly all the membrane of trypanosomes in mammals and is the predominant surface antigen of African trypanosomes [2]. VSG constitutes an important molecular interface between trypanosomes and the host immune system. VSG enables them to avoid the specific immune response by antigenic variations (trypanosomes sequentially express antigenically distinct VSG) and also have several effects on immune elements such as induction of cytokines and autoantibodies [3-5]. Other trypanosome components and soluble factors are also involved in modulation of the immune system by acting on the synthesis of immune factors [7].

Understanding of the immune response was recently advanced by the discovery of the T and B cell subpopulations and, especially, of the T helper (Th) subsets, as well as the cytokines synthesized by Th1 and Th2 subsets. These factors control different aspects of the immune response, in particular the synthesis of nitric oxide by macrophages, which is probably involved in several steps in the immune mechanisms. The role of γδ T cells should also be taken into account as they have been implicated in other parasitic diseases such as malaria and leishmaniasis.

Most of the data concerning African trypanosomiasis have been obtained in animal diseases or experimental animal models. Few studies have concentrated only on the immunology of human African trypanosomiasis (HAT). Results obtained from animal diseases or experimental models can be investigated in HAT using adapted means. Genetic analysis of resistance and suceptibility to infection in inbred and congenic animal strains form the basis for research into equivalent genes in humans.

Genetic control of trypanosomiasis

The study of inbred and congenic mouse strains has contributed greatly to our understanding of the genetic regulation of infectious diseases. A number of genes controls infections of various pathogens by acting at the level of innate susceptibility, such as the N ramp gene, or at the level of acquired immunity. They may or may not be linked to the major histocompatibility locus H2.

Inbred strains of mice differ in their susceptibility to infection with *T. congolense*, as judged by duration of survival following infection. Balb/c and A/J mice were the most susceptible and C57BL/6 the most resistant [8]. The existence of inbred susceptible and resistant strains of mice has allowed to study the inheritance and mechanisms of host resistance. Resistance of mice to African trypanosomes is genetically determined. The control of resistance has been considered as dominant [9] or recessive [10]. The use of different inbred mouse strains and trypanosome clones may explain this result.

In *T. b. rhodesiense*-infected mice, survival was not correlated with the height of the first peak of parasitemia, but a strong negative correlation between the second peak of parasitemia and survival time was noted [11]. However, in *T. congolense*-infected mice, the efficiency of clearance of the first peak of parasitemia was correlated with the survival period [12]. Three loci influencing resistance of mice to *T. congolense* infection have been recently reported on chromosomes 5, 17 and 1 [13].

African cattle trypanosomiasis, mainly due to *T. congolense* and *T. vivax*, causes anemia and weight loss, leading to death. Some African cattle breeds (N'Dama) are, however, able to live and be productive in endemic areas and are considered to be trypanotolerant [14]. Trypanotolerance is genetically controlled and is an innate characteristic, but can be increased by repeated infections. Trypanotolerant cattle are not refractory to trypanosome infections but limit proliferation of trypanosomes. Parasite counts are lower than in trypanosensitive cattle (*Bos taurus*, *Bos indi-*

cus) [15]. This resistance depends on the nutritional, physiological and stress conditions of the animal. Besides, studies of other factors in relation to host defense and survival, and especially cytokine production, have revealed that the ability to produce IL-4 plays a role in the susceptibility to *T. brucei* infection [16].

Little is known about the effect of genetic polymorphism on infectious diseases in humans. Identification of human homologues for the murine genes controlling resistance and susceptibility to pathogens is in progress (recent identification of the human N ramp gene). Family studies should also be performed. Our knowledge of human trypanotolerance is based on reported cases [17, 18] and results from immunological screening [19]. Subjects with a positive CATT test were asymptomatic whereas the presence of blood parasites was observed. Moreover, although many Bantu people from Mbomo foci in the Congo were infected with *T. b. gambiense*, none of the pygmy population was infected. Serum from pygmies possesses a trypanolytic effect. This effect, present before infection and unrelated to antibody production, is dependent on innate immunity factors.

Innate immunity

Natural immunity was described by Laveran and Mesnil in 1900 [20]. They observed that normal human sera injected into *T. b. brucei*- infected mice caused a dramatic reduction in parasitemia. This phenomenon was not reproduced with the human trypanosome strains *T. b. gambiense* and *T. b. rhodesiense*. Trypanolytic factors (TLF) contained in normal human serum were identified as high-density lipoproteins [21]. Recently, two TLFs have been characterized in human serum. The first one (TLF1) belongs to a subclass of high-density lipoproteins and is inhibited by haptoglobin. In contrast, the second factor, TLF2, has a much higher molecular weight and does not appear to be a lipoprotein. Probably, the main trypanolytic effect is due to TLF2 which is not inhibited by haptoglobin [22] ([22], see review by Tomlinson and Raper [132]). The trypanocidal effect of cape buffalo serum has been attributed to xanthine oxidase [23].

Chancre

The local response in the skin corresponds to the first protection developped by the host. Following inoculation of *T. brucei* into mammalian hosts, by the tsetse fly, a local skin reaction is induced by trypanosome proliferation and appears a few days after inoculation. In efferent lymphatic vessels, trypanosomes have been detected in lymph 1-2 days before the chancre. Their number declined during development of the chancre (6 days) and later increased. They are detected in the blood 5 days after inoculation. In *T. congolense*-infected sheep, neutrophils predominate in the early days and then T and B lymphocytes infiltrate the chancre. Later, T lymphocytes predominate, especially CD8 T cells [24]. An early response due to an increase in CD4 and CD8 T cells was revealed by flow cytometry in the afferent lymph draining the chancre. As the chancre regressed, there was an increase in lymphoblasts and surface immunoglobulin-bearing cells [25]. During this first stage, trypanosomes expressed Variable Antigen Types (VATs) found characteristically in the tsetse fly, which changed after few days. An antibody response specific to these VATs appeared in the lymph and then in the plasma [26].

Complement

Both in humans and animals, complement activation by two pathways is detected in trypanosomiasis. The alternative pathway, independent on specific antibodies, was studied by the induction of trypanosome lysis (*T. congolense* and *T. b. brucei*) observed after the addition of fresh serum. Serum could induce trypanosome lysis only on uncoated VSG trypanosomes, as observed during the cycle of this parasite (procyclic forms). However, the appearance of VSG on parasites prevents trypanosome lysis by this alternative pathway [27, 28]. For another strain of *T. b. gambiense*, it was demonstrated that the alternative pathway was incompletely activated without generation of the terminal complex (C5-C9) able to induce membrane lysis [29]. The classical pathway, mediated by specific antibodies against trypanosomes, was also described and could be involved in parasite clearance by antibody-mediated lysis and/or opsonisation. The coated stages of *T. b. brucei* are lysed by antibodies and complement activation by the classical pathway. GP 63-like genes have been recently identified on trypanosomes [30]. The presence of GP 63 on *Leishmania* surface has been correlated with resistance to complement. Nevertheless, during these complement activations, the appearance of soluble fragments, including C3a and C5a anaphylatoxins and the C567 complex, could induce, on the one hand, the chemotaxis of neutrophils and monocytes and, on the other hand, the release of amines involved in vasoactivity and an increase in vascular permeability participating in the initial inflammatory response in the chancre. Immune complexes observed in HAT [31] can also activate the complement. These immune complexes are constituted by antibodies specific to trypanosomes (anti-VSG antibodies) leading to a rapid elimination of complement-fixing immune complexes [32] or by autoantibodies (see below), such as rheumatoid factor or anti-nucleic acid antibodies. These immune complexes with complement activation are also involved in some adverse effects, especially in tissue damage mediated by immune complex deposits [33], such as thrombosis and glomerular involvement [34, 35].

Natural killer cells

Natural killer (NK) cells are non-T-cell lymphocyte populations spontaneously mediating cytotoxicity. These cells have been shown to contribute to host defense against tumor cells and intracellular pathogens, especially viruses. They have functions in both innate and acquired immune responses. NK cells lyse extracellular parasites [126]. NK cells from *T. cruzi* -infected mice have been shown to exhibit significant lytic activity against extracellular trypomastigotes of *T. cruzi* [36].

NK cells secrete cytokines, especially IFN-γ and TNF-α, which play major roles in trypanosomiasis (see below). These cells are regulated by cytokines which can activate (TNF-α, IL-1β, IL-2, IL-7, IL-12 and IL-15) or inhibit (IL-10, TGF-β) NK cell functions. NK cells also participate in the initiation of the inflammatory response, through the synthesis of chemokines (IL-8, GM-CSF, M-CSF and IL-3) [37].

In *T. brucei*-infected mice, NK activity was not modified in the early stages of infection, but was dramatically reduced from day 9 onwards [38]. In contrast, NK cells were activated in mice infected with *T. musculi*, another natural trypanosome, and their critical role was demonstrated by the effects of NK cell depletion (by treatment with anti-asialo-GM1 antiserum) or NK cell activation by polyinosinic-polycytidylic acid copolymer (poly I:C) suggesting that NK cell cytotoxicity is an early innate response controlling the first stage of *T. musculi* infection [123].

T cells

Initial studies have evidenced alterations in T cell functions in trypanosomiasis, both *in vivo* and *in vitro*. Histological examination revealed a massive B cell expansion in the lymph nodes and spleen which replaced the thymus-dependent area in *T. b. brucei*-infected mice. These changes were detected on day 7 of infection and persisted for at least 70 days [39]. Moreover, the role of T cells in controlling infection was not clear, as the course of infection was similar in nude and normal mice [40] and in T cell-deprived mice [41].

Trypanosome antigen-specific T cell response was difficult to identify. In several studies, a transient proliferative T cell response to trypanosome antigens was noted in the first days of the infection, followed by an unresponsiveness [42].

In *T. b. brucei*-infected mice, an increased proliferation of T cells was observed in the first days of infection in spleen and bone marrow and disappeared later. In *T. congolense*-infected cattle, antigen-specific proliferation of T cells was obtained with more or less difficulty according to the antigen, the T cell population and the time used. However, a strong trypanosome-specific T cell proliferation occurred in infected cattle following treatment [43, 44]. Moreover, the kinetoplastid membrane protein 11, a lipophosphoglycan-associated protein, is a potent stimulator of T lymphocyte proliferation [45].

Most T cells in humans and mice bear $\alpha\beta$ T cell receptors (TCR) for the antigen. These cells possess surface markers which allow the discrimination of $CD4^+$ T cells (T helper cells) and $CD8^+$ T cells (cytotoxic T cells). The knowledge of T cell subsets has been deeply modified by the discovery of two subsets of T helper cells, Th1 and Th2 cells based on cytokine productions and, therefore, functional properties [127]. Th1 cells, secreting IL-2 and IFN-γ, activate T cytotoxic cells, contrasting with Th2 cells, secreting IL-4, IL-5, IL-6, IL-10, IL-13, which induce B cell response to T-dependent antigens. In *T. b. rhodesiense*-infected mice, VSG specific T cells were found predominantly in the peritoneum [46]. These $CD4^+$ cells bearing $\alpha\beta$ TCR specific to VSG epitopes did not proliferate but produce IFN-γ and IL-2 cytokines. These data suggest that VSG antigens preferentially stimulate the Th1 cell subset.

The studies on T-cell mediated reponses in regional lymph nodes during trypanosome infection, were performed in cattles infected by *T. congolense*. Two breeds were infected, N'Dama (trypanotolerant) and Boran (trypanosusceptible), and analysis were performed by flow cytometry. In both breeds a significant decrease of $CD4^+$ T cells was observed, contrasting with an increase of $CD8^+$ T cells, B cells and $\gamma\delta$ T cells. *In vitro*, it was detected a proliferative response to VSG and two invariant antigens, a 33 kDa cysteine protease and a 70 kDa heat-shock protein (hsp 70) antigen which is homologous to immunoglobulin heavy chain binding protein. No significant difference was observed between these two breeds in proliferative response of lymph node cells to VSG, Concanavalin A (Con A) or hsp 70. Only IFN-γ production in response to Con A was higher in Boran at 35 days post-infection [44, 47, 48].

Human and mouse immune systems contain few $\gamma\delta$ T cells, in contrast to those of ruminants [49]. Functions of $\gamma\delta$ T cells remain largely unknown. Involvement of $\gamma\delta$ T cells in malaria and leishmaniasis has been observed [50-52]. A proliferative response of $CD8^+$ T cells and $\gamma\delta$ T cells from trypanotolerant N'Dama to an antigen complex containing immunodominant epitopes was observed contrasting with the unresponsiveness in Boran. The role of this $\gamma\delta$ T cell response in parasite resistance remains unclear. So, $\gamma\delta$ T cells, as $CD4^+$ or $CD8^+$, do not proliferate when stimulated with soluble VSG *in vitro* [53]. It would be interesting to determine the role of cytokines synthesized by $\gamma\delta$ T cells.

Memory T, and also B cell responses, are induced by trypanosome infections. This data was demonstrated in susceptible-cattle infected with *T. congolense* and treated by trypanocidal drugs, allowing the disappearance of parasitemia during 3 years. T cell response (proliferation and IFN-γ production) was detected in cultured cells from lymph nodes in the presence of VSG antigens. An antibody production specific to *T. congolense* was also detected in supernatants after stimulation (LPS and IL-2) [44].

Indeed, although specific T cells do not act on trypanosomes in the same way as the cytotoxic T cells in several infectious diseases such as viral infections, they markedly modify immune responses, especially by the secretion of cytokines. They greatly modify functions of B cells (antibody synthesis, isotype switch) and macrophages (antigen presentation, effector mechanisms).

B cells

In African trypanosomiasis, the main feature is a dramatic increase in immunoglobulin levels (especially IgM), including trypanosome-specific antibodies and non-specific Ig production induced by cytokine activation of B cells. Some of these antibodies are also raised against autoantigens, corresponding to non-specific polyclonal activation of B cells producing natural autoantibodies [54] and also to antigen-driven antibodies induced by molecular mimicry.

Antibodies specific to trypanosomes are induced by several parasite antigens, including variant and invariant VSG epitopes, as well as membrane, cytoplasmic and nuclear antigens, through T-dependent and T-independent pathways [55]. Antibodies directed against trypanosome VSG components appeared in sera and their binding to the surface coat of the trypanosomes was able to induce a decrease in parasitemia, both in the blood and extravascular spaces, specifically by immune lysis of parasites and their destruction by the Kupffer cells in the liver. Only heterologous antigenic variants ($< 0.1\%$) remain to repopulate the blood and tissues [56]. VSG-specific IgM, appearing at high levels 3-4 days after trypanosome infection, participate in the parasite elimination. In contrast, VSG-specific IgG do not seem to be involved in the destruction of trypanosomes, as they appear after the disappearance of this VAT population. Another induction of antibodies, linked to the new VSG epitopes, appeared in sera and also contributed to decrease the new VAT-specific population. The VAT-specific antibodies, therefore, decreased to low levels whereas antibodies, belonging predominantly to the IgM class specific to invariant epitopes, remained at high levels. During infection, B cell non-specific stimulation was enhanced as T-independent B cell responses to the VSG successive parasitemias. In contrast, specific trypanosome B cell response, depending on T cell regulation, was depressed. Several factors may contribute to this immunosuppression.

Macrophages may become unable to present antigens to T cells (by defects in antigen processing and their inability to link any epitopes with major histocompatibility class II) and produce immunosuppressive factors (nitric oxide, prostaglandins and cytokines) (see below). An increase in immunosuppressive cytokines, such as IFN-γ and TGF-β, was also detected during infection. However, TGF-β is known to inhibit the production of IL-4, IL-5, IL-6, the major cytokines implied in B cell proliferation and differentiation [57].

Several autoantibodies are detected during African trypanosomiasis. High levels of polyclonal Igs were a marked feature of HAT. The specificity of these Igs is frequently characterized against a large range of autoantigens. Autoantibodies were directed against red blood cells [58], smooth muscle cells [59], liver and cardiolipids [60], nucleic acids: DNA and RNA [58, 61], intermediate filaments [62] and rheumatoid factors [63]. Autoantibodies directed against components of CNS myelin have also been reported. They are specific for the major glycosphingolipids of myelin, the galactocerebrosides, and were detected in sera from both experimentally infected animals [64] and patients from the Ivory Coast [65]. Other autoantibodies, directed against not yet characterized proteins have been described in HAT patients [66] as well as antibodies directed at myelin basic protein in experimentally infected animals [61]. Other antibodies were raised against L-tryptophan, a precursor to the neurotransmetter serotonin (Fig. 1), [5], or recognized some neuronal components of the cytoskeleton, neurofilament proteins. In some cases, these autoantibodies (anti-galactocerebrosides and anti-neurofilaments) are associated with the neurological stage of the disease and their detection in sera and CSF could contribute towards defining the neurological involvement of HAT [67]. There are several hypotheses for the origin of these antibodies. They may be induced by a non-specific stimulation of B cells producing natural autoantibodies [68, 69]. In other cases, antigen-driven autoantibodies are specific to epitopes of the causative infecting agent with molecular mimicry to self antigens. A cross-reactivity to intermediate filaments [70, 71] has been demonstrated for anti-neurofilament antibodies which recognized a common epitope expressed by neurons and flagellar components of trypanosomes [72].

A subpopulation of B cells, identified by the expression of high levels of surface Igs and of CD5 in humans and Ly-1 in mice is responsible for most serum IgM [73]. These CD5 B-cells produce autoantibodies, and antibodies to thymus-independant antigens. In cattle infected with *T. congolense*, a dramatic increase in these cells (more than four times the control value in blood) was measured and correlated with increases in serum Igs and in the absolute number of B cells [74]. An induction of these CD5 B-cells (directly by parasite products or indirectly through the cytokine network) could account for the alteration in Igs synthesis and antibody production observed in trypanosomiasis.

Furthermore, splenic B cells from *T. b. brucei*-infected mice display an aberrant activation phenotype and an impaired responsiveness to mitogenic stimulation *in vitro* [75]. These cells are cell cycle-arrested in G0/G1A suggesting another trypanosome induced mechanism of immune system alteration.

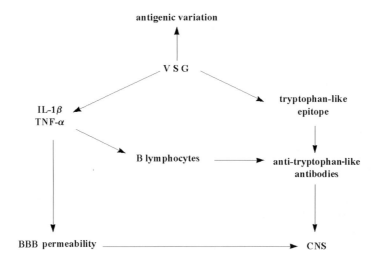

Fig. 1. VSG functions in the pathophysiology of HAT: protective mechanism of trypanosomes against immune system, induction of cytokines (TNF-α, IL1-β) and anti-tryptophan like autoantibodies which might display several pathophysiological effects which remain to be determined (BBB, blood brain barrier; CNS, central nervous system) [4, 5]

Macrophages

Mononuclear phagocytes play a key role in all steps of immune response in the inflammatory phase, as antigen presenting cells, in specific immunity, in synergy with antibodies and cytokines. They also can be involved in immunosuppressive and immunopathological phenomena. Quantitative, biochemical and functional changes of mononuclear phagocytes are observed in trypanosomiasis [76, 77]. In *T. brucei*-infected mice, histological examination showed a marked expansion in macrophages of the liver, spleen and bone marrow [78]. The Kupffer cells in the liver increased in number and were often found in mitosis [78]. The cells contained abundant phagolysosomes (vacuolated cytoplasm). An increased uptake of intravenously-injected sheep red blood cells was also detected.

Macrophages are highly sensitive to environmental factors, especially microorganisms, microorganism-derived products and cytokines. Macrophage modifications associated with the macrophage activation phenotype state have been reported in numerous infectious diseases. A clear reduction in mannose receptors, Fc receptors, C3bi receptors (Mac-1) and F4/80 occurs by day 4 post *T. b. brucei* infection. The expression of MHC Class II molecules is increased in *T. b. brucei*-infected mice and decreased in *T. b. rhodesiense*-infected mice [79, 80]. The antigen-presenting function was reported to be unmodified in *T. b. brucei*- infected mice and defective in *T. b. rhodesiense*-infected mice [78, 79].

Macrophages react to stimuli by adapted response. They secrete many factors with various functions, and synthesize cytokines and effector mediators. Macrophages may play an important role in protection against trypanosomes, particularly in the presence of homologous antiserum, although the role of activated macrophages alone has also been hypothesized [81, 82]. The immunological clearance of [^{75}Se]-methionine-labelled *T. brucei* in mice has been performed to investigate the respective roles of antibodies, macrophage activation and complement in the removal of circulating parasites. The clearance was largely supported by antibody-mediated phagocytosis in liver, which, at least in adoptive transferred-animals, is dependent on opsonization involving C3 component [83]. This data confortes the previous findings reporting that *in vitro* phagocytic function of macrophages was observed in the presence of immune serum [84]. As the existence of Fc receptors for IgM on macrophages is still controversial, the role of IgM antibodies on trypanosome phagocytosis in the absence of complement remains unlikely. Receptor-mediated phagocytosis is enhanced during infection [85]. It is possible that trypanosomes phagocytosed through receptors (C3b receptors, Fc receptors, etc.), or after destruction by complement-mediated lysis, trigger macrophage suppressor activity, although the participation of soluble factors or another cell types cannot be ruled out [1]. Furthermore, macrophages from *T. b. brucei*-infected mice are able to synthesize reactive oxygen intermediates (ROI) after triggering by phorbol myristate acetate [80]. Oxygen-derived species are among the most toxic products generated by macrophages. Trypanosomes are highly sensitive to these components, partularly to hydrogen peroxide and hypochlorous acid, synthesized during phagocytosis [86]. Macrophages from trypanosome-infected mice also synthesize reactive nitrogen intermediates (RNI). Trypanosomes are highly sensitive to the cytostatic/cytotoxic effects of these compounds [87, 88]. ROI and RNI, highly reactive radicals with short half-lives, can react together to form potent and more stable effector molecules able to act on distant targets such as extracellular parasites. We have recently shown that trypanosomes are highly sensitive to S nitroso-albumin, a new effector molecule synthesized by activated human macrophages *in vitro* [128]. Nitrosylated compounds (RSNO) could represent new effector molecules with a potent effect on targets distant from macrophages (Fig. 2).

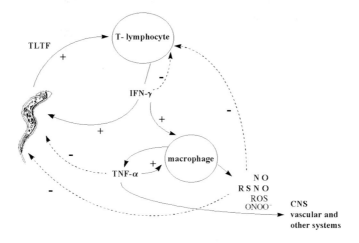

Fig. 2. Interactions between trypanosomes and immune system. TLTF produced by trypanosomes can induce the secretion of interferon-γ (IFN-γ) by T-lymphocytes. IFN-γ is a growth factor for trypanosomes and activates the macrophages. Macrophages synthetize TNF-α, reactive oxygen species (ROS), and reactive nitrogen intermediates [nitric oxide (NO)], nitrosylated compounds (RSNO) which display major inhibitor effects on trypanosomes and lymphocytes. Furthermore, these products act also on several host systems such as nervous and vascular systems. The presence of peroxynitrite (ONOO$^-$) has been demonstrate in the CNS of *T. b. brucei*-infected mice [129]

Macrophages are also active by secreting prostaglandins (PG) which modulate lymphocyte and macrophage functions. During a *T. b. brucei*-infection, the ratio of PGE2/PGF1α is reversed, with an overproduction of PGE2 [89].

Activated macrophages synthesize a large number of cytokines. The production of IL-1 is increased in *T. b. brucei*-infected mice, but this increase may be due to release rather than synthesis [90]. In murine macrophages, VSG induces IL-1 and TNF-α synthesis. Human monocytes can also be induced by trypanosomes (secreted factors from trypanosomes and VSG) and expressed IL-1 and TNF-α RNA transcripts, and secreted IL-1 and TNF-α in culture supernatants [4].

Cytokines

A profound dysregulation of the cytokine network is observed in trypanosomiasis. The first evidence of overproduction of TNF-α/cachectin was shown in *T. brucei*-infected rabbits [91]. TNF-α is known to induce fever, asthenia, cachexia and hypertriglyceridemia [92]. High levels of TNF-α are associated with the presence of patent inflammatory signs in the early phase of human trypanosomiasis and of major neurologic signs in the late phase [6]. A persistently-increased serum TNF-α level could contribute to the hypergammaglobulinemia observed in trypanosomiasis because the role of TNF-α on activation, proliferation and differentiation of B cells (Fig. 1) has already been shown [93]. Nevertheless, TNF-α participates in the mechanisms leading to trypanosome elimination: TNF-α acts indirectly in a cascade of events (Fig. 2) leading to cell activation or directly on parasites due to its cytotoxic properties [94]. Initial control of parasitemia in *T. b. brucei*-infected mice was diminished by the injection of anti-TNF-α antibodies [95]. VSG can trigger TNF-α production by macrophages, which are the cells producing the most of this molecule. Moreover, TNF-α production can be stimulated by IFN-γ. IFN-γ and TGF-β can be produced by CD8 T cells activated by a factor released by *T. b. brucei*, called TLTF (trypanosome derived lymphocyte-triggerring factor). TGF-β has immunosuppressive effects. An interesting fact is that IFN-γ stimulates parasite growth [7, 124]. The binding of Epidermal Growth Factor (EGF) on *T. b. brucei* receptors favoured parasite growth and was one of the first cytokine-parasite interactions noted [130, 131]. All these data show that by interfering with the cytokine network and by using cytokines as growth factors, trypanosomes can completely modify the effector functions of the immune system. The effects of cytokines could also be completely different according to the presence of co-stimulators and the time period during which they are produced in trypanosome-infected animals. Cytokines can also be involved in neurologic disorders. So, TNF-α has been reported to contribute to the pathogenesis of cerebral malaria. Mice chronically infected with *T. b. brucei* develop inflammatory lesions of the CNS after treatment with subcurative doses of a trypanocidal agent [96]. The presence of TNF-α RNA transcripts in the CNS of these mice suggests that TNF-α production could play a role in these lesions. Also, TNF-α and other cytokines contribute to the generation of somnogenic molecules such as IL-1 [97]. High levels of plasmatic IL-10 are also found in human trypanosomiasis [98]. A number of aspects deserve further investigation: the study of all the various cytokines and soluble cytokine receptors, the possible existence of membrane or soluble cytokine receptors synthesized by the parasite, and the interaction and modulation of all these elements. Cytokines have been shown to play an essential role in the synthesis of nitric oxide, whose effects on several features of immune response have been observed over the past few years.

Nitric oxide

Nitric oxide (NO) is a short-lived diatomic free radical synthesized from L-arginine by nitric oxide synthase (NOS) [99]. Calcium-dependent constitutive NOS release small amounts of NO (picomoles) within a short time, whereas calcium-independent inducible NOS (iNOS) release high levels of NO (nanomoles) for a long time. Expression of iNOS in macrophages, neutrophils, hepatocytes, endothelial cells and epithelial cells is regulated at transcription level by a number of agents, including microbial products and cytokines.

In vitro, murine cells produce large amounts of NO after exposure to a combination of stimuli (LPS, IFN-γ, IL-1β, TNF-α, etc.). Human monocytes treated by IL-4 express CD23 antigen [100]. The crosslink of CD23 induces iNOS expression, the release of NO and various other molecules (IL-6, TNF-α, oxygen radicals, lipid mediators). NO is involved in the inflammatory response mediated by endotoxin, cytokines or physiochemical stress. NO produced by cytokine-activated macrophages is important in host defense and plays a crucial role in controlling infections *in vivo*. The role of NO and cytokines has been studied in detail in mice infected by intracellular parasites [101]. In the murine *Leishmania* models, IFN-γ synthesized by Th1 cells leads to iNOS activity, whereas IL-4 and IL-10 synthesized by Th2 cells have a suppressive effect [102]. NO or other nitrogen intermediates can also react with the oxygen intermediates and form peroxynitrite and hydrogen radicals. Moreover, NO can form nitrosylated compounds which are able to transport and liberate NO on targets distant from NO producing cells. Nitrosylated compounds can not only act on extracellular parasites, but also modify host cell function [128]. These compounds may have various effects (parasite killing, alteration of tissue functions such as neurotransmission, etc.) according to their localization (spleen, liver, peritoneum, CNS, etc.). Elevated nitrate concentrations in serum from *T. b. rhodesiense*-infected patients provide evidence for the activation of NO synthesis [103]. By selective inhibition of Th1 cells, NO exerts a negative feedback effect. NO and PG synthesized by macrophages in experimental murine trypanosomiasis can act synergistically and mediate immunosuppression [104, 105]. The overproduction of NO, induced by dysregulation of the cytokine network, may lead to the alteration of immune response and may also be involved in the physiopathological mechanisms (Fig. 2).

Immunosuppression

The increased susceptibility of *T. gambiense* infected patients to secondary infections was pointed out in the initial observations and reports of the Sleeping Sickness Commission [106]. Cellular immunity (skin tests to PPD, *Candida*, streptococcal antigens and sensitization with DNCB) and humoral immunity (response to the H antigen of *Salmonella typhi*) were depressed in patients with HAT [107]. In a recent study, no statistical difference was found between the prevalence of HIV infection in HAT patients and controls [108].

Immunosuppression was also observed and investigated in experimental trypanosomiasis and trypanosome-infected cattle. In these models, immunosuppression was attributed to polyclonal B cell activation as well as the generation of suppressor T cells and suppressor macrophages [109, 125]. A polyclonal B cell activation was noted in trypanosomiasis (hypergammaglobulinemia and a large increase of B cells in the spleen, as well as the presence of numerous Mott cells in CSF and plasma cells in perivascular infiltrates), whereas specific antibody responses to trypanosome antigens were reduced [110].

A marked suppression of antibody response to *Brucella abortus* was reported in cattle infected with *T. congolense* [111]. *T. evansi* infection in sheep delayed and depressed the increase in total cell and lymphoblast output from a lymph node draining the site of a *Pasteurella haemolytica* vaccine administration. These reduced outputs may limit the dissemination of antigen specific cells [112].

Cells, cytokines and prostaglandins have been studied in order to know their contribution, alone or in synergy, and with or without parasitic components, to the immunosuppressive mechanisms. Trypanosome membrane fragments have been found to mimic the immunosuppressive effects of living parasites [113]. A deficient production of IL-2 and of IL-2 receptor expression has been shown in several models [90, 114]. The roles of macrophage-derived factors, especially prostaglandins and IFN-γ secreted by CD8 T cells in the suppression of IL-2 receptor expression on CD4 and CD8 T cells, were also shown [115]. Besides its action on the Th1 subset, rather than the Th2 subset, NO also acts on other elements, favouring immunosuppression.

Immunological therapies

The resistance of mice to African trypanosomes can be increased non-specifically by immunostimulants such as Bacille Calmette-Guérin and *Propionibacterium acnes* [116, 117]. These immunostimulants are considered to activate macrophages. *P. acnes*-treated macrophages inhibited *T. brucei* growth *in vitro* [117]. An acquired resistance has been observed in trypanocide-treated cattle. In a cohort study in Zaire during the 10 year-observation period of adults previously diagnosed and treated for HAT, the risk of a second episode of HAT was greatly reduced compared to the risk of a first episode in previously undiagnosed adults [118]. Induction of protective immunity by vaccination is an important goal to control infectious diseases. However, a vaccine must be very effective and not only delay development of the disease, but also include a large number of antigenic variants. A major aim is to identify invariant antigen(s) that elicit a protective immune response in trypanosomiasis. Invariant molecules and receptors of the flagellar pocket may represent potential candidates [119]. Identifying molecules inducing durable protection may lead to their production as recombinant antigens. Nucleic acid vaccines represent a new promising approach. They are able to induce all the elements of the specific immune response unlike killed microorganisms or defined protein. Studies using dead or living trypanosomes, soluble-released antigens, purified VSG and irradiated parasites have shown that protection is restricted to the VSG-specific epitopes.

Conclusion

The knowledge of host and parasite genomes and all immune response elements can help in understanding immune mechanisms (natural and acquired) developped in trypanosomiasis. These mechanisms are triggered starting with the first contact of hosts with trypanosomes and chancre formations. One of the initial questions is the influence of parasite inoculation, and especially of glossina saliva elements, on the modulation of both innate and specific immune responses of hosts. Studies have shown that tick saliva contains innate immunity inhibitors (complement, NK cells, etc.), reduces macrophage cytokine elaboration, and impairs the earliest stages of specific immunity [120]. Trypanosomes inhibit the proliferation of a human myeloid cell-line (HL-60) *in vitro* [121].

At the onset of specific response, IL-12 plays a major role in T cell shift towards a Th1 response. It would be interesting to investigate its role, as well as all the elements composing the cytokine network. Indeed, the presence of parasite molecules interacting with these elements, which indicates that a parasite can deeply penetrate the host immune system, might lead to paradoxal and adverse effects on immune response. Studies done on transgenic and knock-out mice have produced many major findings in the understanding of infectious diseases. It has been shown, for instance, that the disruption of IL-4 gene in *T. b. brucei*-infected B 10.Q mice alters the control of parasitemia and the production of anti-VSG antibodies, though shortening their life expectancy [16]. The use of different gene promoters/enhancers may contribute to define the role of a transgene in every type of tissue. The various immune effector mechanisms found to be efficient in controlling trypanosomiasis (antibodies, complement, phagocytosis, TNF-α, NO, etc.) might be effective on one organ, but ineffective elsewhere or target distinct parasite forms [122]. Further research on appropriate immunological means, associated with chemotherapeutic agents, might be useful in curing chemotherapy-resistant trypanosomiasis.

References

1. Vickerman K, Myler PJ, Stuart KD (1993) African trypanosomiasis. In: Warren KS (ed) Immunology and molecular biology of parasitic infections. Blackwell Scientific Publications, Boston, p 170
2. Pays E (1995) La variation antigénique et le problème du vaccin contre les trypanosomes africains. Bull Mém Acad Roy Méd Belgique 150:123-135
3. Tachado SD, Schofield L (1994) Glycosylphosphatidylinositol toxin of *Trypanosoma brucei* regulates Il-1α and TNF-α expression in macrophages by protein tyrosine kinase mediated signal transduction. Biochem Biophys Res Comm 205:984-991
4. Okomo-Assoumou MC (1995a) Mécanismes autoimmuns et perturbations du réseau des cytokines dans la physiopathologie de la trypanosomose humaine africaine. Thèse Sciences. Université de Bordeaux II, France
5. Okomo-Assoumou MC, Geffard M, Daulouède S, Chaugier C, Lemesre JL, Vincendeau P (1995b) Circulating antibodies directed against tryptophan-like epitopes in sera of patients with human African trypanosomiasis. Am J Trop Med Hyg 52:461-467
6. Okomo-Assoumou MC, Daulouède S, Lemesre JL, N'Zila-Mouanda A, Vincendeau P (1995c) Correlation of high serum levels of tumor necrosis factor-α with disease severity in human African trypanosomiasis. Am J Trop Med Hyg 53:539-543
7. Olsson T, Bakhiet M, Edlund C, Hojeberg B, Van Der Meide PH, Kristensson K (1991) Bidirectional activating signals between *Trypanosoma brucei* and CD8[+] T cells: a trypanosome-released factor triggers interferon-γ production that stimulates parasite growth. Eur J Immunol 21:2447-2454
8. Morrison WI, Murray M, Bovell DL (1981) Response of the murine lymphoid system to a chronic infection with *Trypanosoma congolense*. I. The spleen. Lab Invest 45:547-557
9. Greenblatt HC, Diggs CL, Rosenstreich DL (1984) *Trypanosoma rhodesiense*: analysis of the genetic control of resistance among mice. Infect Immun 44:107-111
10. De Gee AL, Levine RF, Mansfield JM (1988) Genetics of resistance to the African trypanosomes. VI. Heredity of resistance and variant surface glycoprotein specific immune responses. J Immunol 140:283-288
11. Seed JR, Sechelski J (1995) The inheritance of factors controlling resistance in mice infected with *Trypanosoma brucei rhodesiense*. J Parasitol 81:653-657
12. Ogunremi O, Tabel H (1995) Genetics of resistance to *Trypanosoma congolense* in inbred mice: efficiency of apparent clearance of parasites correlates with long-term survival. J Parasitol 81:876-881
13. Kemp SJ, Iraqi F, Darvasi A, Soller M, Teale AJ (1997) Localization of genes controlling resistance to trypanosomiasis in mice. Nat Genet 16:194-196
14. Pierre C (1906) L'élevage dans l'Afrique Occidentale Française. Gouvernement Général de l'Afrique Occidentale Française, Paris
15. Murray M, Morrison WI, Whitelaw DD (1982) Host susceptibility to African trypanosomiasis; trypanotolerance. Adv Parasitol 21:1-68
16. Bakhiet M, Jansson L, Büscher P, Holmdahl R, Kristensson K, Olsson T (1996) Control of parasitemia and survival during *Trypanosoma brucei brucei* infection is related to strain-dependent ability to produce IL-4. J Immunol 157:3518-3526
17. Marding RD, Hutchinson MP (1956) Sleeping sickness of an unusual type in Sierra Leone and its attempted control. Trans R Soc Trop Med Hyg 41:481-512
18. Lapeyssonie L (1960) Seconde note sur un cas exceptionnel de trypanosomiase. Présence de parasites pendant 21 ans, sans signes cliniques appréciables, chez une femme traitée sans succès pendant 10 ans. Bull Soc Path Ex 53:20-32
19. Lemesre JL, Noireau F, Makoundou ML, Louembet MT, Frézil JL (1988) Contribution of serologic technics to the analysis of the cerebrospinal fluid in Congolese patients with sleeping sickness. Bull Soc Path Ex 81:506-510
20. Laveran A, Mesnil F (1900) Sur l'agglutination des trypanosomes de rat par divers sérums. C R Soc Biol (Paris) 52:939-942
21. Rifkin MR (1978) Identification of the trypanocidal factor in normal human serum/high-density lipoprotein. Proc Natl Acad Sci USA 75:3450-3454
22. Rapper J, Nussenzweig V, Tomlinson S (1996) The main lytic factor of *Trypanosoma brucei brucei* in normal human serum is not high density lipoprotein. J Exp Med 183:1023-1029
23. Muranjan M, Wang Q, Li YL, Hamilton E, Otieno-Omondi FP, Wang J, Van Praagh A, Groothus JG, Black SJ (1997) The trypanocidal cape buffalo serum protein is xanthine oxidase. Infect Immun 65:3806-3814
24. Mwangi DM, Hopkins J, Luckins AG (1990) Cellular phenotypes in *Trypanosoma congolense* infected sheep: the local skin reaction. Parasit Immunol 12:647-658
25. Mwangi DM, Hopkins J, Luckins AG (1996) *Trypanosoma congolense* infection in sheep: cellular phenotypes in lymph and lymph nodes associated with skin reactions. J Comp Path 114:51-61
26. Barry JD, Emergy DL (1984) Parasite development and host responses during the establishment of *Trypanosoma brucei* infection transmitted by tsetse fly. Parasitology 88:67-84

27. Balber AE, Bangs JD, Jones SM, Proia RL (1979) Inactivation or elimination of potentially trypanolytic, complement-activating immune complexes by pathogenic trypanosomes. Infect Immun 24:617-627
28. Ferrante A, Allison AC (1983) Alternative pathway activation of complement by African trypanosomes lacking a glycoprotein coat. Parasite Immunol 5:491-498
29. Devine DV, Falk RJ, Balber AE (1986) Restriction of the alternative pathway of human complement by intact *Trypanosoma brucei* ssp. *gambiense*. Infect Immun 52:223-229
30. Elsayed NMA, Donelson JE (1997) African trypanosomes have differentially expressed genes encoding Homologues of the *Leishmania* GP63 surface protease. J Biol Chem 272:26742-26748
31. Lambert PH, Berney M, Kazyumba G (1981) Immune complexes in serum and cerebrospinal fluid in African trypanosomiasis. J Clin Invest 67:77-85
32. Russo DC, Williams DJ, Grab DJ (1994) Mechanisms for the elimination of potentially lytic complement-fixing variable surface glycoprotein antibody-complexes in *Trypanosoma brucei*. Parasitol Res 80:487-492
33. Nielsen KH (1985) Complement in trypanosomiasis. In : Tizard I (ed), Immunology and pathogenesis of trypanosomiasis. CRC Press, Boca Raton, Florida, USA, pp 133-144
34. Bruijn JA, Oemar BS, Ehrich JH, Fleuren GJ (1988) Immune complex formation in the kidney: recent observations in experimental trypanosomiasis. Ann Soc Belge Méd Trop 68:11-14
35. Van Velthuysen ML, Mayen AE, Prins FA, De Heer E, Bruijn JA, Fleuren GJ (1994) Phagocytosis by glomerular endothelial cells in infection-related glomerulopathy. Nephrol Dial Transplant 9:1077-1083
36. Hatcher FM, Kuhn RE (1982) Destruction of *T. cruzi* by natural killer cells. Science 218:295-296
37. Sharton-Kersten TM, Sher A (1997) Role of natural killer cells in innate resistance to protozoan infections. Curr Opin Immunol 9:44-51
38. Askonas BA, Bancroft GJ (1984) Interaction of African trypanosomes with the immune system. Philos Trans R Soc Lond [Biol] 307:41-50
39. Murray PK, Jennings W, Murray M, Urquhart GM (1974b) The nature of immunosuppression in *Trypanosoma brucei* infections in mice. II. The role of the T and B lymphocytes. Immunology 27:825-840
40. Jayawardena AN, Waksman BH (1977) Suppressor cells in experimentally trypanosomiasis. Nature 265:539-541
41. Askonas BA, Corsini AC, Clayton CE, Ogilvie BM (1979) Functional depletion of T- and B-memory cells and other lymphoid cell subpopulations during trypanosomiasis. Immunology 36:313-321
42. Gasbarre LC, Hug K, Louis JA (1980) Murine T lymphocyte specificity for African trypanosomes. I. Induction of a T lymphocyte dependent proliferative response to *Trypanosoma brucei*. Clin Exp Immunol 41:97-106
43. Emery DL, Wells PW, Tenywa T (1980) *Trypanosoma congolense*: specific transformation *in vitro* of leukocytes from infected or immunised cattle. Exp Parasitol 50:358
44. Lutje V, Mertens B, Boulange A, Williams DJ, Authie E (1995) *Trypanosoma congolense*: proliferative responses and interleukin production in lymph node cells of infected cattle. Exp Parasitol 81:154-164
45. Tolson DL, Jardin A, Schnur LF, Stebeck C, Tuckez C, Beecraft RP, Teb HS, Olafson RW, Pearson TW (1994) The kinetoplastid membrane protein 11 of *Leishmania donovani* and African trypanosomes is a potent stimulator of T lymphocytes proliferation. Infect Immun 62:4893-4899
46. Schleifer KW, Filutowicz LR, Schopf LR, Mansfield JM (1993) Characterization of T helper cell responses to the trypanosome variant surface glycoprotein. J Immunol 150:2910-2919
47. Authié E, Muteti DK, Mbawa ZR, Lonsdale-Eccles JD, Webster P, Wells CW (1992) Idenfication of a 33-kilodalton immunodominant antigen of *Trypanosoma congolense* as a cysteine protease. Mol Biochem Parasitol 56:103-116
48. Boulangé A, Authié E (1994) A 69 kDa immunodominant antigen of *Trypanosoma (Nannomonas) congolense* is homologous to immunoglobulin heavy chain binding protein (BiP). Parasitology 109:163-173
49. Hein WR, Mac Kay (1991) Prominence of $\gamma\delta$ T cells in the ruminant immune system. Immunol Today 12:30-31
50. Rosat JP, Conceicao-Silva F, Waanders GA, Beermann F, Wilson A, Owen MJ, Hayday AC, Huang S, Aguet M, MacDonald HR (1995) Expansion of gamma delta$^+$ T cells in BALB/c mice infected with *Leishmania major* is dependent upon Th2-type CD4$^+$ T cells. Infect Immun 63:3000-3004
51. Rzepczyk CM, Anderson K, Stamatiou S, Townsend E, Allworth A, McCormack J, Whitby M (1997) Gamma delta T cells: their immunobiology and role in malaria infections. Int J Parasit 27:191-200
52. Behr C (1996) Immunopathologie du paludisme: émergence d'une nouvelle réactivité lymphocytaire T. C R Séances Soc Biol Fil. 190:357-368
53. Flynn JN, Sileghem M (1994) Involvement of $\gamma\delta$ T cells in immunity to trypanosomiasis. Immunology 83:86-92

54. Guilbert B, Dighiero G, Avrameas S (1982) Naturally occurring antibodies against nine common antigens in human sera. I. Detection, isolation and characterization. J Immunol 128:2779-2787
55. Reinitz DM, Mansfield JM (1990) T-cell-independent and T-cell-dependent B-cell responses to exposed variant surface glycoprotein epitopes in trypanosome-infected mice. Infect Immun 58:2337-2342
56. Molyneux DH, Pentreath V, Doua F (1996) African Trypanosomiasis in man. In: Cook GC (ed) Manson's Tropical Diseases. WB Saunders Company Ltd, London, UK, pp 1183-1196
57. Fargeas C, Wu CY, Nakajima T, Cox D, Nutman T, Delespesse G (1992) Differential effects of transforming growth factor β on the synthesis of Th1 and Th2-like lymphokines by human T lymphocytes. Eur J Immunol 22:2173-2176
58. Kobayakawa T, Louis J, Shozo I, Lambert PH (1979) Autoimmune response to DNA, red cells, and thymocyte antigens in association with polyclonal antibody synthesis during experimental African trypanosomiasis. J Immunol 122:296-301
59. Wolga JI, Ribeiro CD, Gaillat JP, Stahl JP, Micoud M, Gentilini M (1981) Autoanticorps dans les trypanosomiases humaines africaines (anticorps anti-muscle lisse au cours d'une maladie à *Trypanosoma gambiense*). Bull Soc Path Ex 74:676-684
60. Mackenzie AR, Boreham PLF (1974) Autoimmunity in trypanosome infections. I. Tissue autoantibodies in *Trypanosoma (Trypanozoon) brucei* infections in the rabbit. Immunology 26:1225-1228
61. Hunter CA, Jennings FW, Tierney JF, Murray M, Kennedy PGE (1992) Correlation of autoantibody titres with central nervous system pathology in experimental African trypanosomiasis. J Neuroimmunol 41:143-148
62. Anthoons JA, Van Marck EA, Gigase PL (1986) Autoantibodies to intermediate filaments in experimental infections with *Trypanosoma brucei gambiense*. Z Parasitenkd 72:443-452
63. Kazyumba G, Berney M, Brighouse G, Cruchaud A, Lambert PH (1986) Expression of B cell repertoire and autoantibodies in human African trypanosomiasis. Clin Exp Immunol 65:10-18
64. Jauberteau MO, Ben Younes-Chenoufi A, Amevigbe M, Bouteille B, Dumas M, Breton JC, Baumann N (1991) Galactocerebrosides are antigens for immunoglobulins in sera of an experimental model of trypanosomiasis in sheep. J Neurol Sci 101:82-86
65. Amevigbe M, Jauberteau-Marchan MO, Bouteille B, Doua F, Breton JC, Nicolas JA, Dumas M (1992) Human African trypanosomiasis: presence of antibodies to galactocerebrosides. Am J Trop Med Hyg 45:652-662
66. Asonganyi T, Lando G, Ngu JL (1989) Serum antibodies against human brain myelin proteins in Gambian trypanosomiasis. Ann Soc Belge Méd Trop 69:213-221
67. Pentreath VW (1994) Trypanosomiasis and the nervous system. Pathology and immunology. Trans R Soc Trop Med Hyg 89:9-15
68. Arneborn P, Biberfeld G, Forsgren M, Von Stedingk LV (1983) Specific and non specific B-cell activation in measles and varicella. Clin Exp Immunol 51:165-172
69. Mortazavi-Milani SM, Facer CA, Holborow EJ (1984) Induction of anti-intermediate filament antibody in rabbit experimentally infected with *Trypanosoma brucei brucei*. Immunology 52:423-426
70. Dales S, Fujinami RS, Oldstone MBA (1983) Infection with vaccinia favors the selection of hybridomas synthesizing autoantibodies against intermediate filaments, one of them cross-reacting with the virus hemagglutinin. J Immunol 131:1546-1553
71. Fujinami RS, Oldstone MBA, Wroblewska Z, Frankel ME, Koprowski H (1983) Molecular mimicry in virus infection: cross-reaction of measles virus phosphoprotein or herpes simplex virus protein with human intermediate filaments. Proc Natl Acad Sci USA 80:2346-2350
72. Ayed Z, Brindel I, Bouteille B, Van Meirvenne N, Doua F, Houinato D, Dumas M, Jauberteau MO (1997) Detection and characterization of autoantibodies directed against neurofilament proteins in human African trypanosomiasis. Am J Trop Med Hyg 57:1-6
73. Kipps TJ (1990) The CD5 B cell. Adv Immunol 47:117-185
74. Naessens J, Williams DJL (1992) Characterization and measurement of CD5[+] B cells in normal and *Trypanosoma congolense*-infected cattle. Eur J Immunol 22: 1713-1718
75. Sacco RE, Hagen M, Donelson JE, Lynch RG (1994) B lymphocytes of mice display an aberrant activation phenotype and are cell cycle arrested in G0/G1A during acute infection with *Trypanosoma brucei*. J Immunol 153:1714-1723
76. Vincendeau P, Caristan A, Pautrizel R (1981) Macrophage function during *Trypanosoma musculi* infection in mice. Infect Immun 34:376-381
77. Anosa VO, Kaneko JJ (1983) Pathogenesis of *Trypanosoma brucei* infection in deer mice (*Peromyscus maniculatus*). V. Macrophage ultrastructure and function. Vet Pathol 20:617-631
78. Murray PK, Jennings W, Murray M, Urquhart GM (1974a) The nature of immunosuppression in *Trypanosoma brucei* infections in mice. I. The role of macrophage. Immunology 27:815-824
79. Bagasra O, Schell RF, LeFrock JL (1981) Evidence for depletion of Ia[+] macrophages and associated immunosuppression in African trypanosomiasis. Infect Immun 32: 188
80. Grosskinsky CM, Ezekowitz RA, Berton G, Gordon S, Askonas BA (1983) Macrophage activation in murine African trypanosomiasis. Infect Immun 39:1080-1086
81. Lumsden WH (1967) Trends in research on the immunology of trypanosomiasis. Bull WHO 37:165-175

82. Stevens DR, Moulton JE (1978) Ultrastructural and immunological aspects of the phagocytosis of *Trypanosoma brucei* by mouse peritoneal macrophages. Infect Immun 19:972-982
83. Macaskill JA, Holmes PH, Whitelaw DD, McConnel JM, Jennings EW, Urquhart GM (1980) Immunological clearance of 75Se-labelled *Trypanosoma brucei* in mice. II. Mechanisms in immune animals. Immunology 40:629-635
84. Takayanagi T, Kawaguchi H, Yabu Y, Itoh M, Yano K (1992) Inhibition of IgM antibody-mediated aggregation of *Trypanosoma gambiense* in the presence of complement. Experientia 48:1002-1006
85. Fierer J, Askonas BA (1982) *Trypanosoma brucei* infection stimulates receptor-mediated phagocytosis by murine peritoneal macrophages. Infect Immun 37:1282-1284
86. Vincendeau P, Daulouède S, Veyret B (1989) Role of hypochlorous acid in *Trypanosoma musculi* killing by phagocytes. Parasitology 98:253-257
87. Vincendeau P, Daulouède S (1991) Macrophage cytostatic effect on *Trypanosoma musculi* involves an L-arginine-dependent mechanism. J Immunol 146:4338-4343
88. Vincendeau P, Daulouède S, Veyret B, Dardé ML, Bouteille B, Lemesre JL (1992) Nitric oxide-mediated cytostatic activity on *Trypanosoma brucei gambiense* and *Trypanosoma brucei brucei*. Exp Parasitol 75:353-360
89. Askonas BA (1985) Macrophages as mediators of immunosuppression in murine African trypanosomiasis. Curr Top Microbiol Immunol 117:119-127
90. Sileghem M, Darji A, Remels L, Hamers R, De Baetselier P (1989) Different mechanisms account for the suppression of interleukin-2 production and suppression of interleukin-2 receptor expression in *Trypanosoma brucei*-infection mice. Eur J Immunol 19:119-124
91. Rouzer CA, Cerami A (1980) Hypertriglyceridemia associated with *Trypanosoma brucei brucei* infection in rabbits: role of defective triglyceride removal. Mol Biochem Parasitol 2:31-38
92. Cerami A, Beutler B (1988) The role of cachectin/TNF in endotoxic shock and cachexia. Immunol Today 9:28-31
93. Roldan E, Rodriguez C, Navas G, Parra C, Brieva JE (1992) Cytokine network regulating terminal maturation of human bone narrow B cells capable of spontaneous and high rate of Ig secretion *in vitro*. J Immunol 149:2367-2371
94. Lucas R, Magez S, De Leys R, Fransen L, Scheerlinck JP, Rampelberg M, Sablon E, De Baetselier P (1994) Mapping the lectin-like activity of tumor necrosis factor. Science 263:814-817
95. Lucas R, Magez S, Songa B, Darji A, Hamers R, De Baetselier P (1993) A role for TNF during African trypanosomiasis: involvement in parasite control, immunosuppression and pathology. Res Immunol 144:370-375
96. Hunter CA, Cow JW, Kennedy PGE, Jennings FW, Murray M (1991) Immunopathology of experimental African sleeping sickness: detection of cytokine mRNA in the brains of *Trypanosoma brucei brucei*-infected mice. Infect Immun 59:4636-4640
97. Pentreath VW (1989) Neurobiology of sleeping sickness. Parasitol Today 7:215-218
98. Rhind SG, Sabiston BH, Shek PN, Buguet A, Muanga G, Stanghellini A, Dumas M, Radomski MW (1997) Effect of melarsoprol treatment on circulating IL-10 and TNF-α in human African trypanosomiasis. Clin Immunol Immunopathol 83:185-189
99. Moncada S, Higgs, EA (1993) The L-arginine-nitric oxide pathway. N Engl J Med 329:2002-2012
100. Dugas B, Mossalayi MD, Damais C, Kolb JP (1995) Nitric oxide production by human monocytes: evidence for a role of CD 23. Immunol Today 16:574-580
101. Adams LB, Hibbs JB Jr, Taintor RR, Krahenbuhl J (1990) Microbiostatic effect of murine-activated macrophages for *Toxoplasma gondii*. Role for synthesis of inorganic nitrogen oxides from L-arginine. J Immunol 144:2725-2729
102. James SL, Nacy CA (1995) Effector functions of activated macrophages against parasites. Curr Opin Immunol 5:518-523
103. Sternberg JM (1996) Elevated serum nitrate in *Trypanosoma brucei rhodesiense* infections: evidence for inducible nitric oxide synthesis in trypanosomiasis. Trans R Soc Trop Med Hyg 90:395
104. Schleifer KW, Mansfield JM (1993) Suppressor macrophages in African trypanosomiasis inhibit T cell proliferative responses by nitric oxide and prostaglandins. J Immunol 151:5492-5503
105. Mabbot NA, Sutherland IA, Stenberg JM (1995) Suppressor macrophages in *Trypanosoma brucei* infection: nitric oxide is related to both suppressive activity and lifespan *in vivo*. Parasite Immunol 17:143-150
106. Low GC, Castellani A (1903) Report of the Sleeping Sickness Commission. Proc R Soc London 2:14
107. Greenwood BM, Whittle HC, Molyneux DH (1973) Immunosuppression in Gambian trypanosomiasis. Trans R Soc Trop Med Hyg 67:846-850
108. Meda HA, Doua F, Laveissière C, Miezan TW, Gaens F, Brattegaard K, De Muynck A, De Cock KM (1995) Human immunodeficiency virus infection and human African trypanosomiasis: a case control study in Côte d'Ivoire. Trans R Soc Trop Med Hyg 89:639-643
109. Oka M, Ito Y, Furuya M, Osaki H (1984) *Trypanosoma gambiense*: immunosuppression and polyclonal B-cell activation in mice. Exp Parasitol 58:209-214
110. Sacks DL, Askonas BA (1980) Trypanosome-induced suppression of antiparasite responses during experimental African trypanosomiasis. Eur J Immunol 10:971-974

111. Rurangirwa FR, Musoke AJ, Nantulya VM, Tabel H (1983) Immune depression in bovine trypanosomiasis effects of acute and chronic *Trypanosoma congolense* and chronic *Trypanosoma vivax* infections on antibody response to *Brucella abortus* vaccine. Parasite Immunol 5:207-276
112. Onah DN, Hopkins J, Luckins AG (1997) Effects of *Trypanosoma evansi* on the output of cells from a lymph node draining the site of *Pasteurella haemolytica* vaccine administration. J Compar Path 117:73-82
113. Clayton CE, Sacks DL, Ogilvie BM, Askonas BA (1979) Membrane fractions of trypanosomes mimic the immunosuppressive and mitogenic effects of living parasites on the host. Parasite Immunol 1:241-249
114. Alcino O, Fresno M (1985) Suppressor factor of T-cell activation and decreased interleukin-2 activity in experimental African trypanosomiasis. Infect Immun 50:382-387
115. Darji A, Sileghem M, Heremans H, Brys L, De Baetselier P (1993) Inhibition of T-cell responsiveness during experimental infections with *Trypanosoma brucei*: active involvement of endogenous IFN-γ. Infect Immun 61:3098-3102
116. Murray M, Morrison WI (1979) Non-specific induction of increased resistance in mice to *Trypanosoma congolense* and *T. brucei* by immunostimulants. Parasitology 79:349-366
117. Black SJ, Murray M, Shapiro SZ, Kaminsky R, Borowy NK, Musanga R, Otieno-Omondi F (1989) Analysis of *Propionibacterium acnes*-induced non-specific immunity to *Trypanosoma brucei* in mice. Parasite Immunol 11:371-383
118. Khonde N, Pépin J, Niyonsenga T, Milord F, De Wals P (1995) Epidemiological evidence for immunity following *Trypanosoma brucei gambiense* sleeping sickness. Trans R Soc Trop Med Hyg 89:607-611
119. Olecnick JG, Wolfe R, Nayman RK, McLaughlin J (1988) A flagellar pocket membrane fraction from *Trypanosoma brucei rhodesiense*. Immunogold localization and nonvariant immunoprotection. Infect Immun 56:92-98
120. Wikel SK (1996) Tick modulation of host cytokines. Exp Parasitol 84:304-309
121. Keku TO, Seed JR, Sechelski JB, Balber A (1993) *Trypanosoma brucei rhodesiense*: the inhibition of HL-60 cell growth by the African trypanosomes *in vitro*. Exp Parasitol 77:306-314
122. Mc Lintock LM, Turner CM, Vickerman K (1993) Comparison of the effects of immune killing mechanisms on *Trypanosoma brucei* parasites of slender and stumpy morphology. Parasite Immunol 15:475-480
123. Albright JW, Jiang D, Albright JF (1997) Innate Control of the Early Course of Infection in Mice Inoculated with *Trypanosoma musculi*. Cellular Immunol 176:146-152
124. Bakhiet M, Olsson T, Van Der Meide P, Kristensson L (1990) Depletion of $CD8^+$ T cells suppresses growth of *Trypanosoma brucei brucei* and interferon-γ production in infected rats. Clin Exp Immunol 81:195-199
125. Flynn JN, Sileghem M (1991) The role of the macrophage in induction of immunosuppression in *Trypanosoma congolense*-infected cattle. Immunology 74:310-316
126. Hunter CA (1996) How are NK cell responses regulated during infection? Exp Parasitol 84:444-448
127. Lichtman AH, Abbas AK (1997) T-cell subsets: recruiting the right kind of help. Curr Biol 7:R242-R244
128. Mnaimneh S, Geffard M, Veyret B, Vincendeau P (1997) Albumin nitrosylated by activated macrophages possesses antiparasitic effects, neutralized by anti-NO-acetylated-cysteine antibodies. J Immunol 158:308-314
129. Keita M (1998) Études histologiques et immunohistologiques de l'évolution de la pathologie du système nerveux central au cours de la trypanosome humaine Africaine. Thèse Sciences. Université de Limoges, France
130. Hide G, Gray A, Harrison CM, Tait A (1989) Identification of an epidermal growth factor receptor homologue in trypanosomes. Mol Biochem Parasitol 36:51-59
131. Sternberg JM, Mc Guigan F (1994) *Trypanosoma brucei*: mammalian epidermal growth factor promotes the growth of the African trypanosome bloodstream form. Exp Parasitol 78:422-424
132. Tomlinson S, Raper J (1998) Natural human immunity to trypanosomes. Parasitol Today 14:354-359

CHAPTER 9

Pathology of African trypanosomiasis

K Kristensson, M Bentivoglio

Introduction

African sleeping sickness is characterized by a number of distinct neurological symptoms, which include dysaesthesia, extrapyramidal motor disturbances, disruption of sleep, as well as neuropsychiatric changes including mood disturbances which could represent the initial symptoms (see other chapters in this volume). Although the pathological reactions in tissues of both humans and experimental animals during trypanosome infection have been extensively investigated, the correlation between such pathological-anatomical changes and the clinical symptoms of the disease remains to be clarified. Furthermore, the pathogenetic mechanisms behind the clinical and pathological alterations are still not understood. In this review of the pathology of human African sleeping sickness, the initial pathological reactions to the invading parasite will be first described, and then those in the organism at the different stages of disease. A focus will be given to alterations in the nervous system, since neurological and neuropsychiatric changes are the most prominent features of the disease in humans. In order to understand the pathogenic mechanisms behind these disturbances, different experimental models have been used and these will be summarized. Finally, the potential influence of molecules released from the trypanosome or the immune system during the infection on different regions of the nervous system will be discussed.

Early pathological reactions to the invading parasites

As discussed in other chapters of this volume, the trypanosomes are usually transmitted by the bite of a tsetse fly. In both humans and cattle a chancre may develop at the site of the bite from the infected fly. A few histological examinations of biopsies from such chancres have revealed a marked edema and an infiltration of lymphocytes and macrophages; a proliferation of endothelial cells and fibroblasts has also been reported. Trypanosomes are present in these chancres, where they can be detected by the use of fine-needle aspirations before they appear in the circulation. From these chancres the parasites can reach the circulation either through the lymphatic drainage or by invading dermal venules.

A reaction in the local lymph nodes often occurs simultaneously with the development of the chancre, but these reactions have rarely been defined at the histological level in biopsy material. The host antibody response to the parasite, and parasites with antigenic variations, develop at this stage. Intact parasites that have thus escaped the humoral immune response may finally reach the thoracic duct and enter the blood circulation to become disseminated in the organism. The parasites will then circulate in successive waves of proliferation in the blood, and inflammatory changes in the various tissues and organs have been described as the parasites escape the circulation. Multiplying parasites occur both in the lymph and the blood, as well as in extravascular tissues. The tissues most prominently involved at this stage of disease are the skin, heart, skeletal muscles and serous membranes in the pericardium, pleurae and peritoneum. In these tissues an interstitial infiltration of mononuclear inflammatory cells has been observed and in experimental studies the occurrence of vasculitis has been reported. These early pathological reactions have been reviewed by Poltera [1].

The end of this first stage of parasite invasion has by convention been set by the appearance of parasites or mononuclear inflammatory cells in the cerebrospinal fluid. For natural reasons most pathological reports on the human disease derive from this later stage with an involvement of the nervous system.

Alterations in the nervous system in human African trypanosomiasis

Brain pathology

The first autopsies of African sleeping sickness were reported in 1840 by Clarke (quoted by Bertrand et al. [2]), who observed signs of cerebrospinal meningitis. A number of brief examinations of the brains from patients deceased from African sleeping sickness were then reported at the end of the 19th century and these have been reviewed by Bertrand et al. [2]. The first detailed description of the histopathological changes in the brain was made by Mott [3], who examined the brains of two patients from Congo who died at Charing Cross Hospital in London. In these brains he observed an infiltration of mononuclear inflammatory cells in the leptomeninges and around vessels in the brain parenchyma. He emphasized that the nerve cells were structurally intact, and this feature could be used to differentiate neuropathologically the disease from general paralysis of the insane in syphilis. In 1905, at a time when Dutton and Todd (quoted by Spielmeyer [4]) and Castellani [5], (reviewed in Bentivoglio et al. [6]) had observed trypanosomes in the cerebrospinal fluid in sleeping sickness patients, Mott had extended his study to 24 cases of African trypanosomiasis and in these he had confirmed the main findings from his first patients. By the use of silver impregnation a marked increase in numbers of astrocytes was also revealed in the brain [7]. In addition, Mott [8] described the occurrence in the infected brains, as well as in other organs, of peculiar morular-shaped cells that were later given his name. The Mott's cells were originally considered pathognomonic for trypanosome infections, but they can also be found occasionally in other chronic inflammatory diseases of the brain, such as neurosyphilis.

Following these reports, Spielmeyer [4, 9] published reports on the changes in the brain in four sleeping sickness patients. He emphasized that plasma cells dominated among the perivascular inflammatory cells, that there was a marked glial cell proliferation and that alterations were more pronounced in the white matter than in the cerebral cortex. He also pointed out that the changes in African sleeping sickness could be distinguished from those of neurosyphilis due to the more widespread plasma cell infiltration, and less superficial gliosis and nerve cell degeneration in the cortex in the former condition.

In a detailed study of two cases of sleeping sickness from West Africa, Bertrand et al. [2] clearly showed that the perivascular inflammation was diffusely distributed mainly in the white matter and that, as a consequence, the disease should be classified as a leucoencephalitis. In spite of this, the myelin was generally well preserved. These authors also pointed out that, in addition to astrocytes, the microglial cells showed a marked reaction.

In an overview of the pathology of African sleeping sickness, Serra [10] described that macroscopically the brains showed signs of pachymeningitis and granular thickening of the periventricular walls. The choroid plexus had been found to be often congested and thickened even in cases who had died soon after diagnosis. Some cases had also shown dilatation of the ventricles, probably as a consequence of stenosis of the aqueduct. Microscopically, diffuse perivascular infiltration of plasma cells was mainly limited to the white matter, in which there was also a pronounced astrocytic reaction and proliferation of microglial cells. In spite of the marked inflammatory cell infiltration in the white matter, the myelin was well preserved. The Mott's morular cells seemed to be most numerous in the chronic cases.

In a later review, Radermecker [11] also emphasized that in sleeping sickness the inflammatory changes in the brain affected especially the white matter in the cerebral hemispheres, that plasma cells dominated among the inflammatory cells, and that Mott's morular cells were often numerous. In contrast to the observations by Serra [10], Radermecker [11] mentioned that a diffuse pallor of the myelin staining could be noted in several of the cases.

A systematic clinical-neuropathological analysis of 16 patients from Congo affected by sleeping sickness was performed by van Bogaert and

Janssen [12]. They divided their patients into four major clinical groups, i.e. patients with: 1. predominance of mental disturbances; 2. extrapyramidal changes; 3. ataxia, and 4. signs of radiculo-polyneuritis. Like in previous studies, these authors observed pachymeningitis, and perivascular cell infiltrations and glial reactions that were most pronounced in the periventricular white matter. In the first group of patients, changes were particularly prominent in the thalamus and hypothalamus, while the patients in the second group with extrapyramidal signs of disease showed inflammatory changes with a prevalence in the globus pallidus, caudate nucleus and in the hypothalamic paraventricular and supraoptic nuclei. In the ataxic form, inflammatory changes seemed to prevail at the bulbo-cerebellar level, and in the last group of patients inflammation occurred in the spinal cord as well as in the spinal roots and peripheral nerves (described below). Like in the report by Radermecker [11], a number of the cases described by van Bogaert and Janssen [12] showed some pallor of the myelin staining in the brain white matter. Similarly to Bertrand et al. [2], these authors concluded that African trypanosomiasis should be regarded as a leucoencephalitis, since changes in the white matter were more marked than those in the gray matter.

Collomb [13] summarized the anatomical-clinical findings in human African trypanosomiasis by dividing the disease into an initial phase characterized by a generalized perivascular infiltration of lymphocytes and plasma cells in the organism, followed by a more advanced and a terminal stage. The advanced stage was described as characterized by inflammation in the leptomeninges and, in the brain, the lesions predominated in the meso-diencephalic region with periadventitial inflammatory cell infiltrations followed by perivascular lymphocyte plasma cell infiltration in the white matter. Mott's morular cells, on the other hand, were reported to occur throughout the brain parenchyma. In the terminal stage, demyelination in the white matter and in the basal ganglia were described, whereas the cerebral cortex was well preserved. In summary, the inflammatory process was reported to affect mainly subcortical regions, with a marked involvement of the meso-diencephalic structures. Thus, the distribution of the inflammatory lesions in sleeping sickness relatively spared the cerebral cortex, in contrast to the cortical lesions observed in general paralysis in neurosyphilis.

These early descriptions seem to have dealt with the Gambian form of African trypanosomiasis in West Africa. The first larger study on the pathology of the brain in Rhodesian trypanosomiasis, which is the East African form of sleeping sickness, was given by Calwell [14], who examined 17 brains obtained from autopsies which he had performed under what he described as 'African bush conditions'. Most of these brains displayed an abundant perivascular infiltration of lymphocytes and plasma cells. Mott's morular cells were found in all the brains, and these were abundant in 11 of the cases and scanty in the other 6. In an analysis of 252 histological sections, Calwell noted that the morular cells occurred in every part of the brain with the following descending order of frequency: medulla, cerebellum, parietal cortex, frontal cortex, occipital cortex, meninges, corpus callosum and basal ganglia. Mott's cells were instead notably absent in the choroid plexus, which showed the occurrence of inflammatory cells, fibrinous exudate and edema in most of the cases. There were no neuronal changes, and areas of demyelination, when present, were very restricted and limited to perivascular sites. Trypanosomes could not be detected in any part of the brain. Because of treatment, the Rhodesian form of trypanosomiasis was considered to have become more similar to the most chronic Gambian form of the disease and the neuropathological changes of these two conditions were similar. The findings in the Rhodesian form of sleeping sickness were corroborated by Manuelidis et al. [15] and Poltera et al. [16] in a study of 6 and 14 cases, respectively, of this disease. The former authors described that although the perivascular cell infiltration was most abundant in the white matter, one untreated case of short duration showed a remarkable astrocytic response and microglial activation also in the cerebral and cerebellar cortices. Inflammatory cell infiltrates and an intense microglia activation were present also in the caudate nucleus, putamen and thalamus; in the brainstem the dorsal part seemed to be more affected than the ventral tegmentum. The number of infiltrates in the hypothalamus varied. In an acute untreated case a single trypanosome was seen by one of the observers in the brain.

Based on these earlier studies and others, the pathological changes in the human brain in African trypanosomiasis can be summarized as follows:
- *macroscopically*, signs of edema with widening of the gyri and flattening of the sulci of the cerebral hemisphere may occur, as well as signs of a leptomeningitis and granular ependymitis. Occasionally, a moderate internal hydrocephalus is observed, probably as a result of stenosis of the aqueduct secondary to the granular ependymitis. Signs of cerebral or cerebellar herniation are rarely described.
- *microscopically*, the dominating change is a non-specific inflammatory cell infiltration in the leptomeninges and perivascularly, mainly in the white matter (Figs. 1, 2c). The perivascular cuffs consist to a large extent of lymphocytes and plasma cells. Hemosiderin-loaded macrophages can occur around blood vessels, indicating the occurrence of small hemorrhages. In addition, there is an infiltration in the white matter of Mott's morular cells, which can occur in association with small groups of inflammatory cells, but also isolated in the brain parenchyma (Fig. 2). These morular cells have been shown to contain immunoglobulins of the IgM type in large intracellular vacuoles [17] and they are probably derived from plasma cells. Their number may vary in the different cases, and they are considered to be most numerous in the brains

from patients with a longer history of the disease. Morular cells can occur also in other chronic inflammatory diseases such as syphilis, tuberculosis and blastomycosis, but not in such large numbers as in trypanosomiasis.

In the brains of sleeping sickness patients there is consistently a very marked activation of microglial cells, and a diffuse astrocytosis or astrogliosis. The distribution of the inflammatory changes is diffuse and prevails in the periventricular areas and in the white matter of the cerebral hemispheres, as well as in the basal ganglia and in the brain stem and white matter of the cerebellum. In periventricular zones, signs of tissue necrosis may occur, while neurons seem to be well–preserved in general, and in particular in the cerebral cortex. In spite of the perivascular inflammation in the white matter, the myelin is in many cases also well–preserved (Fig. 2d) and does not seem, therefore, to be a primary target of the pathological process. Only in some cases a demyelination limited to the perivascular regions has been described and occasionally a more diffuse pallor of the white matter occurs. In none of these studies have trypanosomes been detected in the brain parenchyma, although they have been documented in other organs. The changes in the brain seem to be qualitatively similar in the Rhodesian and Gambian forms of the disease, and eventual differences seem to be related more to the slower progress in the latter.

Thus, the histopathological changes in the brain in African sleeping sickness are those of a leucoencephalitis with an accentuation in periventricular regions, which may partly undergo necrosis. The inflammatory cells are dominated by plasma cells, among which the morular cells are

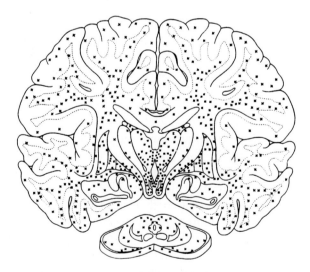

Fig. 1. Schematic drawing of the distribution of the inflammatory changes in the human brain in the Gambian form of African trypanosomiasis (reproduced from [11])

Pathology of African trypanosomiasis

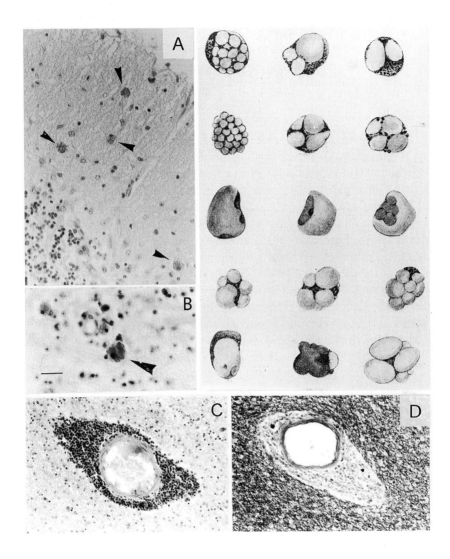

Fig. 2. Histological alterations in the brain from human patients with African sleeping sickness. A: scattered morular cells (*arrowheads*) in the molecular layer of the cerebellar cortex; B: morular cell in the cerebral white matter (*arrowhead*); C: perivascular infiltration of lymphocytes and plasma cells in the cerebral white matter; D: well–preserved myelin fibers in the vicinity of a large perivascular inflammatory cell infiltration in the cerebral white matter (A-C: hematoxylin-eosin; D: Luxol fast blue-cresyl violet; bars: A, 35 µm; B, 20 µm; C-D, 100 µm; these pictures are taken from sections of a brain coming from the Born-Bunge Foundation, Antwerpen, Belgium). TOP RIGHT: different appearance of morular cells drawn from human cases of chronic sleeping sickness (reproduced from [10])

prominent. In the absence of trypanosomes in the brain parenchyma, it has to be explained why this inflammation with immunoglobulin-secreting cells appears in the white matter and why the periventricular zone can be a predilection site for destructive lesions. Furthermore, these uncharacteristic inflammatory changes in the brain do not obviously correlate to the rather distinct clinical symptoms from the central nervous system.

Changes in the spinal cord and peripheral nervous system

In their study of three patients of African sleeping sickness of the Gambian type, who showed clinical signs of polyradiculitis and polyneuritis, Janssen et al. [18] and van Bogaert and Janssen [12] described infiltration of inflammatory cells and the occurrence of Mott's cells in the leptomeninges of the spinal cord. In one case, they also noted perivascular inflammatory cell infiltration in the posterior columns of the spinal cord and around the dorsal horns. The spinal cord parenchyma seemed, however, spared by detectable changes. Perivascular inflammation also occurred in cranial and spinal nerve roots as well as in peripheral nerves. The dorsal roots seemed to be particularly afflicted by the inflammation. Mott's cells were present also in the peripheral nervous system. In one case, signs of demyelination with a loss of myelin and intact axons were observed at the level of the spinal roots. In the peripheral nerves of this case, nerve fibers undergoing Wallerian degeneration were also noted. This indicated an additive effect of pathogenetic stimuli in a proximo-distal direction in the nerve.

In a study of the Rhodesian form of the disease, Poltera et al. [16] described chronic cellular infiltration in the roots of the cranial nerves in two cases. In one of these, samples from the brachial plexus and sciatic nerve were also examined and these appeared histologically normal, while small intramuscular nerve branches showed a patchy inflammatory cell infiltration; such infiltratates were also noted in epicardial nerves and around the epicardial ganglion.

Arsenical encephalopathy

During the late stage of sleeping sickness with clinical involvement of the nervous system, organic arsenicals are still the most powerful trypanocidic drugs available. However, these drugs may have serious side-effects and in about 10% of the cases the treatment is complicated by an acute arsenical encephalopathy, which has a high mortality rate. In a neuropathological study of 16 fatal cases of African sleeping sickness (Gambian form) from Daloa, Ivory Coast, an acute reactive arsenical encephalopathy seemed to have been the principal cause of death in at least 10 cases [19].

Neuropathologically, two types of changes are considered to be related to this reaction [20]. Some patients develop clinically an acute convulsive state associated with high intracerebral pressure. Brain edema, sometimes with signs of herniation, and a severe hypoxic brain damage in the Ammon's horn occur in these patients. In some cases the damage extends throughout the cerebral hemispheres. Interestingly, the Purkinje cells in the cerebellum seem to be spared, which may indicate that hypoglycemia, that results in a similar pattern of neuronal degeneration, may contribute to this condition [19]. In other patients, who rapidly develop a coma without convulsions, the neuropathological picture is that of an acute hemorrhagic leucoencephalopathy. Macroscopically these brains show petechial hemorrhages in the pons and midbrain, and microscopically fibrinoid necrosis is present in small vessels and ring or ball hemorrhages. The mechanism behind acute hemorrhagic leucoencephalopathy is not clear, but it has been generally proposed to represent an intense form of delayed hypersensitivity response to a pathogen. Acute reactive arsenical encephalopathy of both types of the above-described pathology may develop 3 to 19 days after the first injection of the arsenical compound and the condition, thus, seems related more to a delayed hypersensitivity reaction than to direct toxicity to the brain from disrupted trypanosomes or from the drug [19].

Skeletal muscle changes

A profound hyperaesthesia with muscle pain is common in African trypanosomiasis. Histopathological examinations have revealed a patchy infiltration of inflammatory cells in the peri- and endomysium. However, only in a few cases the skeletal muscles have been subjected to systematic analysis. For instance, Poltera et al. [16] described the findings in muscles of the lower and upper limbs, the recti abdominis and the diaphragm of a patient with the Rhodesian form of the disease. Groups of atrophied muscle fibers were observed, but overall the inflammation was found to be mild. Surprisingly, the changes were most pronounced in the diaphragm.

Alterations in visceral organs in human African trypanosomiasis

Almost all visceral organs can be the site of alterations or inflammatory cell infiltrations. A most conspicuous alteration involves the adipose tissue that is markedly lost. The spleen and lymph nodes are increased in size and histologically they display an increase in the number of plasma cells and macrophages, while the number of lymphocytes is reduced. In the myocardium, large groups of parasites and a heavy inflammatory cell infiltration are seen in the interstitial tissue. Myocardial muscle fibers may be atrophic. Mild cell infiltration and sclerosis have been observed in the thyroid and in the medulla of the adrenal glands. No detailed study of the kidneys has been reported.

Neuropathology of experimental infections

In order to decipher the pathogenetic mechanisms of the disease, a number of different animal species, ranging from different strains of mice and rats to mini-pigs, dogs and monkeys, have been subjected to experimental infections with trypanosomes. A review on experimental findings in monkeys, cats, goats and dogs was given by van Bogaert [21]. Losos and Ikede [22] reviewed in detail the pathological changes in both naturally occurring and experimental infections. The present short overview will be limited to infections with *Trypanosoma brucei*, which include the human pathogenic strains *T.b. rhodesiense* and *T. b. gambiense*. Furthermore, since *T. congolense* and *T. vivax* are confined to the plasma in the blood vessels and may interact with endothelial cells, their pathogenic mechanisms differ markedly from *T. brucei* sp which can invade the interstitial tissue of various organs. The immunological responses, as well as the factors that regulate growth of the parasite, most likely also differ between trypanosomes that are restricted to the blood circulation and those which can interact with the cells in the tissues of the host animal [22, 23].

Both the human pathogenic strains *T. b. rhodesiense* and *T. b. gambiense*, as well as the rodent pathogenic strain *T. b. brucei*, have been used in experimental studies. In general, after infection the animals have developed inflammatory diseases often with a relatively rapid course compared to the human diseases. Most of these studies have been devoted to analyze the immunological reaction and the immune response of an animal to the invading parasites [22] and are discussed elsewhere in this volume. The present chapter will be restricted to experimental studies which have employed more chronic models of the disease and have focused on the pathogenesis of the nervous system alterations.

Early experimental studies

Rodents

In mice, *T. b. rhodesiense* and *T. b. brucei* usually give rise to an acute disease, while it has been more difficult to infect these animals with *T. b. gambiense*. However, one strain of *T. b. gambiense* was found to cause a chronic infection in mice and rats [24, 25], and the brains displayed leptomeningeal and perivascular cell infiltration; Mott's morular cells could also be identified in the infected rats [26]. Poltera et al. [27] described the cerebral immunopathology caused by another strain of *T. b. gambiense*, which was known to produce low levels of parasitemia and a chronic infection in a fraction of the injected mice. Such mice could live up to 22 months after infection, and showed histologically a meningoencephalitis with a mild deposition of immune complexes in the choroid plexus and infiltration of plasma cells. By immunofluorescence, trypanosomes could be detected in the choroid plexus, and also diffusely in the brain. The question whether or not the parasites occurred extravascularly in the brain was addressed in an ultrastructural study by Van Marck et al. [28] in mice and rats infected with *T. b. gambiense*. Such extravascular parasites were abundant in the choroid plexus, while in the cerebral parenchyma only a few degenerating trypanosomes were seen adjacent to the lateral ventricles. Some of these periventricular trypanosomes may be taken up by degenerating ependymal cells [29].

Although *T. b. rhodesiense* usually cause acute diseases in rodents, one isolated strain of this subspecies could produce a relatively chronic disease with a survival of 6 to 9 weeks in most of the infected mice. These mice showed a meningoencephalitis with infiltration of lymphocytes, plasma cells and macrophages. There was no destruction of the parenchyma, and the nerve cells were intact. Masses of trypanosomes were found in the choroid plexus and in some mice parasites also occurred in the brain tissue.

Deer mice (*Peromyscus maniculatus*) have been reported to develop a chronic disease after infection with *T. b. brucei* [30]. Such mice, examined histologically up to 222 days after infection, showed an infiltration of lymphocytes in the leptomeninges and perivascularly in the brain parenchyma; plasma cells were few and Mott's cells rarely encountered, whereas astrocytic proliferation and microglial reaction were very pronounced. In spite of this, neuronal or myelin lesions were not apparent. Parasites were present in the choroid plexus and near the ependyma in the periventricular neuropil.

A subacute-chronic infection in mountain voles (*Microtus montanus*) has been observed following infection with *T. b. rhodesiense* [31]. These animals survived 5-8 weeks after the infection, but showed only a moderate meningitis and therefore were not considered to be a suitable model for chronic African sleeping sickness.

Dogs

They develop an acute disease after infection with *T. b. brucei* and *T. b. rhodesiense,* while after infection with *T. b. gambiense* the disease is more chronic. Examination of the brain after *T. b. brucei* infection has revealed an infiltration of lymphocytes and plasma cells particularly in the choroid plexus and the pituitary gland, with a massive invasion of trypanosomes in the former [32].

Pigs

Mini-pigs recovered from infections with *T. b. gambiense*, but developed a chronic infection with *T. b. brucei* and survived up to 172 days postinfection. These pigs showed a moderate meningoencephalitis with very few trypanosomes [33].

Monkeys

They have been extensively used in several studies and the first detailed description of *T. b. brucei*, *T. b. rhodesiense* and *T. b. gambiense* was given in 1928 by Peruzzi who examined 43 infected animals and 10 controls [82]. A marked meningoencephalitis occurred in prolonged cases, particularly after infections with *T. b. gambiense*. Peruzzi suggested that the choroid plexus was the site of entry of the parasites to the brain and from this site they invaded the cerebrospinal fluid. A correlation between the clinical alterations and the neuropathological changes following infection by *T. b. gambiense* in monkeys was made by van Bogaert [21]. On the basis of his systematic observations on the topography and nature of the lesions, van Bogaert suggested a hematogenous spread of the infection via the Virchow-Robin spaces and through the spinal fluid into the brain. The encephalitis in the monkeys resembled the one in humans, although the occurrence of morular cells was rare. A meningoencephalitis with marked perivascular cuffings and microglial reaction was also reported in brains from monkeys subjected to chemotherapeutical trials after infections with *T. b. rhodesiense* or *T. b. brucei*. Interestingly, trypanosomes were not seen in the choroid plexus of such treated monkeys, whereas they were observed to be scattered in the neuropil [34].

Rabbits

They usually develop a response to the infection different from that of other animals. The disease may last 1 to 2 months or longer after *T. b. brucei* infections [22], and is hallmarked by anemia and emaciation. From such emaciated rabbits cachectin was isolated, which is identical to the later defined tumor necrosis factor, i.e. TNF-α. These animals have been the subject of various metabolic studies [35], while neuropathological examinations seem to have been rare.

Sheep and goats

Infected with *T. b. brucei*, they may survive 2 to 3 months [22, 36] and have been the subject for analyses of immunoreactivity to cerebral autoantigens, i.e. galactocerebrosides [37], and therapeutical trials [38] as described elsewhere in this volume. In van Bogaert's study [21] one of four goats infected with *T. b. gambiense* developed a severe subacute encephalitis and showed neuropathologically a marked inflammatory cell infiltration in the white matter of the cerebral hemispheres and cerebellum. There were also severe inflammatory changes in the globus pallidus, thalamus, substantia nigra and ventral pontine nuclei.

Recent experimental studies

Our series of experimental studies on the pathogenesis of the nervous system dysfunction in African trypanosomiasis started by an immunohistochemical mapping of the invasion of the parasites in the rodent nervous system using a primary antibody raised in rabbits against whole trypanosomes [39]. Thirteen days after infection, the parasites could be seen to invade dorsal root ganglia and cranial ganglia. In the brain, the parasites did not seem to cross the so-called blood-brain barrier even at 33 days p.i., but invaded the circumventricular organs, in which a barrier to the diffusion of proteins from the vessels is lacking. Thus, trypanosomes were detected in the choroid plexus, the subfornical organ, the organum vasculosum of the lamina terminalis, the median eminence, the neural lobe of the pituitary gland, the subcommissural organ, the pineal gland and the area postrema in the brain stem. Also the vessels of the cranial and dorsal root ganglia lack a diffusion barrier to macromolecules, and in these ganglia too the parasites invaded the interstitial tissue and surrounded the nerve cell bodies that were structurally intact [39] (Fig. 3). A similar distribution of parasites was seen in deer mice and no signs of invasion into the brain parenchyma was seen at 68 days p.i. In the circumventricular organs and the cranial and dorsal root ganglia an infiltration of lymphocytes, which were dominated by the $CD8^+$ phenotype, was observed in rats. Inflammatory cell infiltrates were also found in the leptomeninges and were sparsely distributed around intracerebral vessels. Macrophage-like cells showing immunopositivity for major histocompatibility complex (MHC) antigen class II were numerous throughout the brain, but especially around the circumventricular organs [40]. Hunter et al. [41] described in detail astrocytic reaction with a similar distribution in experimentally infected mice and correlated this finding to the expression of cytokines from activated astrocytes. There were no signs of neuronal degeneration, and immunohistochemical staining for various neurotransmitters did not reveal any selective loss of neurons; also the myelin was intact [39, 42].

The only alteration in neurons observed in the rats in our experiments involved the supraoptic and paraventricular nuclei of the hypothalamus, which expressed MHC class I antigens during the infection [40]. In normal brains, the expression of MHC class I antigens is limited to

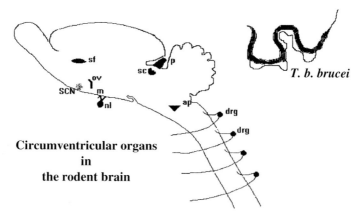

Fig. 3. Schematic drawing a mid-sagittal section of a rodent brain showing the distribution of the circumventricular organs: sf, subfornical organ; ov, organum vasculosum of the lamina terminalis; m, median eminence; nl, neural lobe of the pituitary gland; sc, subcommissural organ; p, pineal gland; ap, area postrema. These organs as well as the choroid plexus and the dorsal root ganglia (drg) are the primary targets of trypanosome invasion in the rodent brain. The approximate position of the suprachiasmatic nuclei (SCN) is given

leptomeninges, choroid plexus and luminal surface of intracerebral vessels. Thirteen days after infection with *T. b. brucei*, a marked immunolabeling for MHC class I antigens appeared in neurons of the supraoptic and paraventricular hypothalamic nuclei. Since both nuclei project directly to the median eminence, which is a site for infiltration of both trypanosomes and inflammatory cells, factors released in this region could have interacted with the axon terminals of the affected neurons to send retrograde signals to the hypothalamic nuclei for the MHC antigen induction. Whether or not this antigen activation also signifies functional alterations in the supraoptic and paraventricular hypothalamic nuclei remains to be clarified.

In summary, these experimental neuropathological studies have disclosed that there are no overt structural changes in the brain of either nerve cell bodies or myelin sheaths, which implies that dysfunction of the nervous system in such experimental infections should be related to an influence on neuronal function at the molecular level.

Pathogenetic considerations

A number of pathogenetic mechanisms have been discussed to underlie the pathological changes in the nervous system function during trypanosome infection. Since a number of these mechanisms have been examined in other contexts, as described in this volume, we will highlight only the mechanisms relevant to an understanding of the brain dysfunction and then describe in greater detail our experimental studies aimed at understanding how signaling molecules between the trypanosomes and the immune system may interact with the nervous system to cause disease.

Toxins

The idea that trypanosomes, like bacteria, can produce toxins was originally put forth by Laveran and Mesnil [43] and Laveran and Pettit [44] who showed that dried *T. b. brucei* injected into mice resulted in weakness, convulsions and loss of reflexes. The mice usually recovered and did not show any histopathological alteration. Other investigators, however, failed to demonstrate a trypanotoxin and only later evidence accumulated that biologically active factors can be released by living, dying and disintegrating trypanosomes (reviewed in Tizard et al. [45]).

Trypanosomes may generate a number of metabolic compounds that may have toxic properties. As regards brain function, particular attention has been paid to trypanosome tryptophan metabolism [45]. *T. b. gambiense* can metabolize tryptophan to indole-3-ethanol (tryptophol) and this indole injected intraperitoneally into experimental animals can produce a sleep-like behavior state [46, 47]. Thus, trypanosomes are capable to synthesize indoles that can affect sleep, but a quantitative study that correlates the levels of such released metabolites to the behavior and sleep of the infected individual is still lacking (as discussed by Tizard et al. [45]). Dying or dead trypanosomes can also release bioactive substances that may potentially influence the brain function. These include enzymes such as proteases and phospholipases, as well as lipopolysaccharides and free fatty acids, as reviewed by Tizard et al. [45].

Interest has also been paid to the possibility that living parasites may release molecules that affect the host. These include the variant surface antigens that can be shed during antigenic switches and molecules released from the flagellar pocket.

The variant surface antigen of the trypanosomes is inserted into the membrane by a glycosyl phosphatidylinositol (GPI) anchor. After release, such GPI-anchored proteins may be inserted into heterologous cell membranes, and recently a transfer of GPI-anchored proteins between different cells has been observed also *in vivo* [48]. It has also been reported that the variant surface protein of different *T. brucei* sp strains can be transferred to sheep erythrocytes. Such a transfer of the membrane, but not of the soluble form of the protein, could sensitize the erythrocytes to complement-mediated immunolysis [49]. A potential transfer of variant surface glycoproteins from trypanosomes to neurons opens up an interesting possibility of interference with neuronal function, but as yet no trypanosome-derived surface proteins have been detected in the nervous parenchyma.

Stumpy forms of *T. b. brucei* may produce secretory filaments that are surrounded by a glycoprotein coat and secreted from the flagellar pocket.

The significance of these filaments is unknown [45]. Other proteins that may affect the host are also secreted from the flagellar pocket. One of such proteins, trypanosome-derived lymphocyte triggering factor (TLTF) that we have recently defined, is briefly described below.

Auto-antibodies

During trypanosome infections in both humans and experimental animals, the induction of antibodies against a number of autoantigens has been described [50]. A fraction of the high levels of immunoglobulins (Ig) M induced during trypanosome infections are anti-trypanosome, but others are directed against autoantigens. Interestingly, the Mott's morular cells that are widely dispersed in the brain during trypanosome infections contain IgM [17]. Some of these auto–antibodies are directed against nervous tissue elements, such as myelin basic proteins and cerebrosides and gangliosides [37, 51]. The pathogenetic significance of these auto–antibodies remains to be evaluated.

Immune complexes

Such complexes are detected in the cerebrospinal fluid in a majority of patients with the Gambian form of African trypanosomiasis, and their occurrence is related to the clinical evidence of involvement of the central nervous system [52]. In a chronic mouse model of *T. b. gambiense* infection, deposits of immunoglobulins were seen in the choroid plexus in the brain [27]. However, immune complexes are not a consistent finding in human [53] or experimental trypanosome infections [51], and their role in the pathogenesis of the nervous system dysfunction is not clear.

Cytokines

Infections with *T. b. brucei* are associated with a marked induction and release of cytokines. Release of TNF-α (which, as mentioned above, was originally called cachectin in trypanosome-infected rabbits) may cause the pronounced loss of weight and cachexia that often follows trypanosome infections. Interleukin (IL)-1 is released from activated macrophages [54] and interferon (IFN)-α/β as well as IFN-γ are released into the serum early during infection [55]. Such cytokines may also be produced from activated glial cells during the infection. RNA transcripts for TNF-α, IL-1 and IL-4, but not IL-6, have been observed in astrocytes during the infection and the role of glia-derived cytokines in the pathogenesis of the brain disease has been evaluated [51] (see also other chapters of this book).

Kinins

In a series of studies, Boreham [56] and Goodwin [57] described the release of kinins into the blood during experimental *T. b. brucei* infections in rodents and rabbits, and correlated the level of kinins to that of the parasitaemia. It has been proposed in these studies that kinins may be involved in the vascular permeability changes observed in chronic infections. Increased amounts of histamine have also been described in trypanosome-infected mice [58], but the relation between histamine release and the intense pruritus described in human trypanosome infection has not been examined.

Nitric oxide

During infections with *T. b. brucei* in mice, both peritoneal and splenic macrophages release high levels of nitric oxide (NO) [59]. It has been suggested that this free radical favors growth of the parasites in the host, since NO may mediate immunosuppression and does not seem to exert any lytic effect on the extracellular parasites [60]. Since NO plays a role of intercellular messenger in the nervous system, it is of interest to evaluate whether its release by activated microglia/macrophages in the brain during trypanosome infection can be of pathogenetic relevance (see other chapters in this book).

Nervous system dysfunction in experimental animals

In order to reveal molecular mechanisms of pathogenetic importance for the nervous system dysfunction in African trypanosomiasis, we have pursued a series of investigations on peripheral and central neural structures. Since human African trypanosomiasis is characterized by somatosensory symptoms, we have focused our studies on dorsal root ganglia and primary afferents. In addition, in the search for alterations in the brain in the course of the disease, immediate early gene expression was studied during 24 hours. These latter studies brought our attention to the circadian rhythm generating system in the course of experimental trypanosome infection.

Disturbances of the somatosensory system

The majority of patients with African trypanosomiasis complain of pruritus and cutaneous hyperaesthesia and deep hyperpathia [61, 62]. Our observation of an early invasion of trypanosomes into cranial and dorsal root ganglia in experimental animals prompted us to study whether rats develop sensory signs of disease infection with *T. b. brucei*. Using a hot plate analgesimeter, signs of hyperalgesia were observed already 7 days p.i. Although the general conditions of the animals were good until day 17 p.i., a further significant decrease in the response latency in the hot plate test was observed between days 14 and 17 p.i. [42]. Thus, hyperalgesia is an early event also in experimental infections and the hot plate analgesimeter provides a simple and reliable method to follow the disease progression. Nerve fibers implicated in pain propagation, i.e. those containing the neuropeptides substance P and calcitonin gene-related peptide, did not show structural changes when analyzed immunohistochemically in the skin and spinal cord of *T. b. brucei* infected rats [42], which indicates a disturbance of neurotransmission at the functional level. As described in the previous section, kinin concentration may increase in the blood [56] and the secretion of IL-1 from macrophages may be enhanced during trypanosome infections [54]. Both molecules may be involved in the observed hyperalgesia since bradykinin can sensitize nociceptors [63] and IL-1 is a potent hyperalgesic agent [64].

On the basis of our observation that IFN-γ is induced already within one day after infection with *T. b. brucei* in rats, and that small neurons in dorsal root ganglia express a molecule that immunologically cross-reacts with IFN-γ [65, 66], we have undertaken a series of investigations on the interactions between *T. b. brucei* and sensory neurons, as well as between IFN-γ and pain-related behavior.

In a series of experiments we have shown that *T. b. brucei* releases TLTF that triggers lymphocytes to produce IFN-γ and that IFN-γ can promote growth of the parasites [67, 68, 83]. In cultures of dorsal root ganglia neurons we observed that the survival of the trypanosomes was promoted by the sensory neurons and that such a growth-promoting effect could be blocked by antibodies to TLTF and to IFN-γ [69]. This indicates that TLTF released from trypanosomes can induce the release of the IFN-γ-like immunoreactive molecule from neurons; this molecule has been isolated from sensory ganglia and has been shown to stimulate parasite growth, similarly to lymphocyte-derived IFN-γ [70].

We then addressed the question as to whether IFN-γ can be involved in pain-related behavior. Recombinant rat IFN-γ was injected intrathecally in decerebrated, spinalized rats. This injection elicited an initial intensive, brief facilitation of the flexor reflex followed by sustained reflex facilitation that lasted about 40 min. Pretreatment with an inhibitor of NO synthase (NOS), nitro-L-arginine-ester, could totally block the initial response and partially the prolonged response. This study showed a possible hyperalgesic action of IFN-γ and indicated that this action is mediated, at least in part, through an activation of the L-arginine-NO pathway [71].

In order to pursue the investigation of the potential role of IFN-γ in pain-related behavior, further series of experiments have been performed using gene-manipulated mice with disruption of the functional part of the IFN-γ receptor. In wild-type mice, intrathecal injections of recombinant mouse IFN-γ evoked biting behavior, which was not observed in the mice with disrupted gene for the receptor. Both types of mice had similar withdrawal thresholds to mechanical stimulation and reacted similarly to foot-pad carrageenan injections. However, in contrast to wild-type mice,

IFN-γ receptor knock-out mice did not show autotomy after sciatic nerve section, which indicates that functional IFN-γ receptors in the spinal nociceptive pathways may be related to neuropathic pain [72].

By the use of immunohistochemistry, IFN-γ receptors were localized to the superficial dorsal horn and the lateral spinal nucleus in both the rat and mouse spinal cord. The immunoreactivity to this receptor overlapped that to neuronal NOS-1, the synthetic enzyme of NO in neurons [73], and was not reduced after rhizotomy indicating a postsynaptic localization in the spinal cord [72]. In view of the tentative interaction between IFN-γ receptor activation and production of NO, it is interesting to note that inhibitors of NOS have been found to suppress neuropathic pain-related behavior [74]. A working hypothesis may therefore be put forth that the deep pain sensations frequently encountered in African trypanosomiasis are in part related to release of either IFN-γ from activated lymphocytes in the vicinity of IFN-γ receptors in the nervous tissue, or of the neuronal IFN-γ-like molecule in sensory neurons, which project to the areas in the spinal cord where the receptor is expressed [66], thus funding a basis for a specific molecular interaction between the extracellular parasites and subsets of neurons.

Disturbances in the circadian rhythm generating system

In view of the clinical evidence of alterations of the sleep/wake cycle during the human disease, and of changes in the sleep pattern in experimental trypanosome infection in rats, we started an investigation aimed at the study of gene expression during sleep and wake. This study was based on previous evidence that the expression of the immediate early gene c-*fos*, a transcription factor which is rapidly and transiently induced in activated neurons, oscillates in the brain during 24 hours [75]. In our initial investigations on the basal c-*fos* expression in the brain of trypanosome-infected rats, we noticed that, in an advanced stage of the infection, the induction of its protein product Fos was dysregulated in the suprachiasmatic nuclei (SCN) during the light hours, which correspond in nocturnal animals to the period of sleep [76].

The SCN are represented by paired densely aggregated cell groups lying dorsal to the optic chiasm in the anterior hypothalamus. A wealth of studies have indicated that the SCN are the principal pacemaker responsible in the mammalian brain for the generation and synchronization of biological rhythms, entraining endogenous rhythms of behavior and hormone secretion to environmental cues and, in particular, to the light-dark cycle (reviewed in Klein et al. [77]). Light is conveyed to the SCN through the retinohypothalamic tract. Retinal fibers, which terminate in the rat in the ventrolateral portion of the SCN, are glutamatergic, and the downstream activation involves glutamate receptors (reviewed by Ebling [78]). In the cascade of molecular events involved in the SCN function, a potential regulatory role for immediate early genes has been suggested (reviewed by Takahashi [79]). In particular, the expression of the c-*fos* gene may be involved in the signal transduction pathway conveying environmental information for entrainment of the biological clock. c-*fos* is induced in the retinorecipient sector of the SCN in response to photic stimulation, and light-induced c-*fos* expression in the SCN is circadian time-dependent (reviewed by Takahashi [79]). On this basis, we verified the occurrence of Fos protein induction in the SCN after photic stimulation of trypanosome-infected animals during the early subjective night. The number of Fos-immunoreactive neurons turned out to be significantly reduced in the SCN under these experimental conditions [80]. These findings indicated that the events at the transcriptional level which are part of the mechanism for photic entrainment of the circadian pacemaker and/or their regulation could be severely disrupted during *T. b. brucei* infection. In particular, the lack of genomic response to light observed in the SCN pointed out an alteration of the photic entrainment of the circadian clock in trypanosome-infected rats.

Together with the evidence of hypothalamic changes during experimental trypanosome infection in the rat [40], which has been reviewed in the preceding sections, these data strongly suggested a functional involvement of the hypothalamus during the infection.

On this basis, the intrinsic activity of SCN neurons was examined in slice preparations from trypanosome-infected rats [81, 84]. These experiments derived from previous evidence that the spontaneous activity of SCN neurons undergoes *in vitro* a circadian oscillation with a characteristic peak of activity at a defined time during 24 hours (see Klein et al. [77]). In our investigation, the rhythm of spontaneous activity recorded from the SCN of trypanosome-infected rats was found to be markedly altered, displaying a phase advance of its peak during 24 hours and a reduced firing rate. Thus, such *in vitro* experiments provided a striking demonstration of severe disturbances of the rhythmic activity of the SCN during *T. b. brucei* infection.

We then investigated *in vivo*, by means of an anterograde tracer injected into the eye, whether the organization of the retinal fibers terminating in the SCN was modified in trypanosome-infected rats [84]. However, these experiments did not point out alterations in the distribution and density of retinohypothalamic fibers during the infection. Thus, the presynaptic input to SCN neurons did not seem to be affected in experimental trypanosomiasis.

On the other hand, remarkable changes were detected in the SCN at the postsynaptic level when the expression of ionotropic glutamate receptor subunits was investigated in rats in an advanced stage of the infection [84]. Under these experimental conditions, the expression of the glutamate AMPA Glu R2/3 and NMDA R1 receptors was found to be significantly decreased, pointing out an impairment of excitatory neurotransmission in the circadian pacemaker in experimental trypanosomiasis. In addition, since blockade of glutamatergic neurotransmission disrupts the cellular and behavioral responses of the SCN to light, and NMDA receptor antagonists can block c-*fos* induction in the SCN in response to a light pulse (see Ebling [78]), the alterations in glutamate receptor expression disclosed during *T. b. brucei* infection could account for the above-mentioned impairment of light-induced Fos expression.

Altogether, these latter data pointed out that *T. b. brucei* infection causes selective alterations in the circadian timing system, with alterations in molecular mechanisms of light transduction and their postsynaptic regulation and with a decrease of the SCN oscillatory activity and a disruption of its circadian rhythmicity. Such alterations could potentially play a role in endogenous rhythm dysregulation in African sleeping sickness.

Concluding remarks

The neuropathology of African sleeping sickness points essentially to a periventricular inflammatory process with strong involvement of the white matter. This topographical pattern of the changes may be related to a localization of the extracellular trypanosomes to circumventricular organs with seeding of parasites into the cerebrospinal fluid. Factors released from living or disintegrating trypanosomes may then diffuse into the periventricular brain parenchyma, and such a diffusion is facilitated in the white matter, which contains wide extracellular spaces. Different trypanosomal factors may then elicit the inflammatory cell response. Mott's morular cells are conspicuous, but it is still not clear why these cells are widely spread in the parenchyma, often isolated from other inflammatory cells, and whether the IgM produced in these cells is directed against trypanosomal proteins or auto–antigens. The different neurological symptoms of the disease may be related to interactions between molecules released from the trypanosomes, or the host animals defense system, and specific regions of the nervous system. We have outlined how such molecular interactions may occur in systems for pain-related behavior and for disturbances in circadian rhythms, which are major clinical features of the disease.

Acknowledgements

The Authors are indebted to Prof. Dr. J.J. Martin, Universitaire Instelling Antwerpen (Wilrijk), Belgium, for making materials of human African trypanosomiasis from the Born-Bunge Foundation available for our study. Karolina Kristensson is acknowledged for figure drawing. This study received financial support from the UNDP/World Bank/WHO Special Programme for Research and Training in Tropical Diseases and the Italian National Research Council (CNR).

References

1. Poltera AA (1985) Pathology of human African trypanosomiasis with reference to experimental African trypanosomiasis and infections of the central nervous system. Brit Med Bull 41:169-174
2. Bertrand I, Bablet J, Sicé A (1935) Lésions histologiques des centres nerveux dans la trypanosomiase humaine. Ann Inst Pasteur 54:91-144
3. Mott FW (1899) The changes in the central nervous system of two cases of negro lethargy: sequel to Dr. Manson's clinical report. Brit Med J 1:1666-1669
4. Spielmeyer W (1908) Die Trypanosomenkrankheiten und ihre Beziehungen zu den syphilogenen Nervenkrankheiten. Fischer, Jena
5. Castellani A (1903) On the discovery of a species of *Trypanosoma* in the cerebrospinal fluid of cases of sleeping sickness. Proc Roy Soc Lond 71:501-508
6. Bentivoglio M, Grassi-Zucconi G, Kristensson K (1994a) From trypanosomes to the nervous system, from molecules to behavior: a survey, on the occasion of the 90th anniversary of Castellani's discovery of the parasites in sleeping sickness. Ital J Neurol Sci 15:75-87
7. Mott FW (1906) The microscopic changes in the nervous system in a case of chronic Dourine or Mal de Coit, and comparison of the same with those found in sleeping sickness. Proc Roy Soc Biol 78:1-12
8. Mott FW (1905) Observations on the brains of men and animals infected with various forms of trypanosomes. Preliminary notes. Proc Roy Soc Biol 76:235-242
9. Spielmeyer W (1907) Schlafkrankheit und progressive Paralyse. Münch Med Wochenschr 54:1065-1068
10. Serra G (1940) La malattia del sonno o Castellanosi. Soc Acc Stamperia Zanetti, Venezia
11. Radermecker J (1956) Leucoencéphalite à parasites connus: la trypanosomiase. Systématique et électroencéphalographie des encéphalites et encéphalopathies. Electroenceph Clin Neurophysiol 117-124
12. van Bogaert L, Janssen P (1957) Contribution à l'étude de la neurologie et neuropathologie de la trypanosomiase humaine. Ann Soc Belge Méd Trop 37:379-426
13. Collomb H (1957) Encéphalite de la trypanosomiase humaine africaine. Gaz Méd France 64:1064-1078
14. Calwell HG (1937) The pathology of the brain in Rhodesian trypanosomiasis. Trans R Soc Trop Med Hyg 30:611-624
15. Manuelidis EE, Robertson DHH, Amberson JM, Polak M, Haymaker W (1965) *Trypanosoma rhodesiense* encephalitis. Clinicopathological study of five cases of encephalitis and one of Mel B hemorrhagic encephalopathy. Acta Neuropathol 5:176-204
16. Poltera AA, Owor R, Cox JN (1977) Pathological aspects of human African trypanosomiasis (HAT) in Uganda. Virch Arch A Path Anat Histol 373:249-265
17. Greenwood BM, Whittle HC (1980) The pathogenesis of sleeping sickness. Trans R Soc Trop Med Hyg 74:716-723
18. Janssen P, van Bogaert L, Haymaker W (1956) Pathology of the peripheral nervous system in African trypanosomiasis. A study of seven cases. J Neuropath Exp Neurol 15:269-287
19. Adams JH, Haller L, Boa FY, Doua F, Dago A, Konian K (1986) Human African trypanosomiasis (*T. b. gambiense*): a study of 16 fatal cases of sleeping sickness with some observations on acute reactive arsenical encephalopathy. Neuropath Appl Neurobiol 12:81-94
20. Haller L, Adams H, Merouze F, Dago A (1986) Clinical and pathological aspects of human African trypanosomiasis (*T. b. gambiense*) with particular reference to reactive arsenical encephalopathy. Am J Trop Med 35:94-99
21. van Bogaert L (1962) Experimental trypanosomal encephalitis. In: Innes JRM, Saunders LZ (eds) Comparative Neuropathology. Academic Press, New York and London, pp 465-481
22. Losos GJ, Ikede BO (1972) Review of pathology of diseases in domestic and laboratory animals caused by *Trypanosoma congolense*, *T. vivax*, *T. brucei*, *T. rhodesiense* and *T. gambiense*. Vet Pathol Suppl 9:1-71
23. Sileghem M, Naessens J (1995) Are CD8 T cells involved in control of African trypanosomiasis in a natural host environment? Eur J Immunol 25:1965-1971
24. Stefanopoulo G, Etévé (1943) Méningo-encéphalo-myélite de la souris blanche due à une souche "neurotrope" de *Trypanosoma gambiense*. Bull Soc Path Ex 36:43-46

25. Roubaud E, Stefanopoulo GJ, Duvolon S (1944) Etude chez le rat d'une souche neurotrope de *Trypanosoma gambiense*. Bull Soc Path Ex 37:292-296
26. Stefanopoulo G, Caubet P, Duvolon S (1943) Présence de cellules muriformes de Mott chez les rats infectés de *Trypanosoma gambiense*. Bull Soc Path Ex 37:296-302
27. Poltera AA, Hochmann PH, Lambert PH (1982) *Trypanosoma brucei gambiense*: cerebral immunopathology in mice. Acta Trop 39:205-218
28. Van Marck EAE, le Ray D, Beckers A, Jacob W, Wery M, Gigase PLJ (1981) Light and electron microscope studies on extravascular *Trypanosoma brucei gambiense* in the brain of chronically infected rodents. Ann Soc Belge Méd Trop 61:57-78
29. Ormerod WE, Hussein MS-A (1986) The ventricular ependyma of mice infected with *Trypanosoma brucei*. Trans R Soc Trop Med Hyg 80:626-633
30. Stevens DR, Moulton JE (1977) Experimental meningoencephalitis in *Trypanosoma brucei* infection of deer mice (*Peromyscus maniculatus*). Acta Neuropath 38:175-180
31. Bafort JM, Schmidt H (1983) Experimental chronic *Trypanosoma brucei rhodesiense* infections in *Microtus montanus*. Am J Trop Med Hyg 32:968-975
32. Morrison WI, Murray M, Sayer P, Preston JM (1981) The pathogenesis of experimentally induced *Trypanosoma brucei* infection in the dog. I Tissue and organ damage. Am J Pathol 102:168-181
33. Bungener W, Mehlitz D (1984) Histopathological findings in mini-pigs infected with different strains of *Trypanosoma brucei*. Tropenmed Parasit 35:109-114
34. Poltera AA, Sayer PD, Brighouse G, Bowell D, Rudin W (1985) Immunopathological aspects of trypanosomal meningoencephalitis in vervet monkeys after relapse following Berenil® treatment. Trans R Soc Trop Med Hyg 79:527-531
35. Rouzer CA, Cerami A (1980) Hypertriglyceridemia associated with *Trypanosoma brucei brucei* infection in rabbits: role of defective triglyceride removal. Mol Biochem Parasitol 2:31-38
36. Bouteille B, Dardé ML, Dumas M, Catanzano G, Pestre-Alexandre M, Breton JC, Nicolas A, N'Do DC (1988a) The sheep as an experimental model for African trypanosomiasis. I. Clinical study. Ann Trop Med Parasitol 82:141-148
37. Jauberteau MO, Younes-Chennoufi AB, Amevigbe M, Bouteille B, Dumas M, Breton JC, Baumann N (1991) Galactocerebrosides are antigens for immunoglobulins in sera of an experimental model of trypanosomiasis in sheep. J Neurol Sci 101:82-86
38. Bouteille B, Dardé ML, Pestre-Alexandre M, Dumas M, Breton JC, Nicolas JA, Catazano G, Munoz M (1988b) Traitement de la trypanosomiase expérimentale du mouton à *Trypanosoma brucei brucei*: efficacité du Ro 15-0216 (dérivé 2-nitroimidazolé). Bull Soc Path Ex 81:616-622
39. Schultzberg M, Ambatsis M, Samuelsson E-B, Kristensson K, van Meirvenne N (1988) Spread of *Trypanosoma brucei* to the nervous system: early attack on circumventricular organs and sensory ganglia. J Neurosci Res 21:56-61
40. Schultzberg M, Olsson T, Samuelsson E-B, Maehlen J, Kristensson K (1989) Early major histocompatibility complex (MHC) class I antigen induction in hypothalamic supraoptic and paraventricular nuclei in trypanosome-infected rats. J Neuroimmunol 24:105-112
41. Hunter CA, Jennings FW, Kennedy PG, Murray M (1992) Astrocyte activation correlates with cytokine production in central nervous system of *Trypanosoma brucei brucei*-infected mice. Lab Invest 67:635-642
42. Wiesenfeld-Hallin Z, Kristensson K, Samuelsson E-B, Schultzberg M (1991) Studies of hyperalgesia induced by *Trypanosoma brucei brucei* infection in rats. Acta Trop 48:215-222
43. Laveran A, Mesnil F (1902) Recherches morphologiques et expérimentales sur le Trypanosome du nagana ou maladie de la mouche tsetse. Ann Inst Pasteur 16:1-45
44. Laveran A, Pettit A (1911) Des trypanotoxines. Bull Soc Path Ex 4:42-45
45. Tizard I, Nielsen KH, Seed JR, Hall JE (1978) Biologically active products from African trypanosomes. Microbiol Rev 42:661-681
46. Morgane PJ, Stein WC (1973) Effects of serotonin metabolites on sleep-waking activity in cats. Brain Res 50:205-213
47. Seed JR, Sechelski J (1977) Tryptophol levels in mice injected with pharmacological doses of tryptophol and the effect of pyrazole and ethanol on these levels. Life Sci 21:1603-1610
48. Kooyman DL, Byrne GW, McClellan S, Nielsen D, Tone M, Waldmann H, Coffman TM, McCurry KR, Platt JL, Logan JS (1995) In vivo transfer of GPI-linked complement restriction factors from erythrocytes to the endothelium. Science 269:89-92
49. Rifkin MR, Landsberger FR (1990) Trypanosome variant surface glycoprotein transfer to target membranes: A model for the pathogenesis of trypanosomiasis. Proc Natl Acad Sci USA 87:801-805
50. MacKenzie AR, Boreman PFL (1974) Autoimmunity in trypanosome infections. I. Tissue autoantibodies in *Trypanosoma* (Trypanozoon) *brucei* infections of the rabbit. Immunology 26:1225-1238
51. Hunter CA, Kennedy PGE (1992) Immunopathology in central nervous system human African trypanosomiasis. J Neuroimmunol 36:91-95
52. Lambert PH, Berney M, Kazyumba G (1981) Immune complexes in serum and in cerebrospinal fluid in African trypanosomiasis. J Clin Invest 67:77-85

53. Whittle HC, Mohammed P, Greenwood BM (1980) Immune complexes in Gambian sleeping sickness. Trans R Soc Trop Med Hyg 74:833-834
54. Askonas BA, Bancroft GJ (1984) Interaction of African trypanosomes with the immune system. Phil Trans R Soc London 307:41-50
55. Bancroft GJ, Sutton CJ, Morris AG, Askonas BA (1983) Production of interferons during experimental African trypanosomiasis. Clin Exp Immunol 52:135-143
56. Boreham PFL (1970) Kinin release and the immune reaction in human trypanosomiasis caused by *Trypanosoma rhodesiense*. Trans R Soc Trop Med Hyg 64:394-400
57. Goodwin LG (1970) The pathology of African trypanosomiasis. Trans R Soc Trop Med Hyg 64:797-817
58. Richards WHG (1965) Pharmacologically active substances in the blood tissue and urine of mice infected with *Trypanosoma brucei*. Brit J Pharmac Chemother 24:124-131
59. Sternberg J, McGuigan F (1992) Nitric oxide mediates suppression of T cell responses in murine *Trypanosoma brucei* infection. Eur J Immunol 22:2741-2744
60. Sternberg J, Mabbott N, Sutherland I, Liew FY (1994) Inhibition of nitric oxide synthesis leads to reduced parasitemia in murine *Trypanosoma brucei* infection. Infect Immun 62:2135-2137
61. Giordano C (1973) Les signes neurologiques et électroencéphalographiques de la trypanosomiase humaine africaine. Méd Afr Noire 20:317-324
62. Boa YF, Traore MA, Doua F, Kouassi-Traore MT, Kouassi BE, Giordano C (1988) Les différents tableaux cliniques actuels de la trypanosomiase humaine africane à *T. b. gambiense*. Bull Soc Path Exot 81:427-444
63. Beck PW, Handwerker HO (1984) Bradykinin and serotonin effects on various types of cutaneous nerve fibers. Pflügl Arch Eur J Physiol (Berlin) 347:209-222
64. Ferreira SH, Lorengetti BB, Bristow AF, Poole S (1988) Interleukin-1β as a potent hyperalgesic agent antagonized by a tripeptide analogue. Nature 334:698-700
65. Ljungdahl Å, Olsson T, van der Meide PH, Holmdahl R, Klareskog L, Höjeberg B (1989) Interferon-gamma-like immunoreactivity in certain neurons of the central and peripheral nervous system. J Neurosci Res 24:451-456
66. Robertson B, Grant G, Kristensson K (1996) Characterization of neuronal interferon-γ immunoreactive dorsal root ganglion cells and their projections to the spinal cord of the rat. Prim Sens Neur 1:221-230
67. Olsson T, Bakhiet M, Höjeberg B, Ljungdahl Å, Edlund C, Andersson G, Ekre H-P, Fung-Leung W-P, Mak T, Wigzell H, Fiszer U, Kristensson K (1993) CD8 is critically involved in lymphocyte activation by a *T. brucei brucei*-released molecule. Cell 72:715-728
68. Bakhiet M, Olsson T, Mhlanga J, Büscher P, Lycke N, van der Meide PH, Kristensson K (1996) Human and rodent interferon-γ as a growth factor for *Trypanosoma brucei*. Eur J Immunol 26:1359-1364
69. Eneroth A, Bakhiet M, Olsson T, Kristensson K (1992) Bidirectional signals between *Trypanosoma brucei brucei* and dorsal root ganglion neurons. J Neurocytol 21:846-852
70. Olsson T, Kelic S, Edlund C, Bakhiet M, Höjeberg B, van der Meide PH, Ljungdahl Å, Kristensson K (1994) Neuronal interferon-gamma immunoreactive molecule, bioactivities and purification. Eur J Immunol 24:308-314
71. Xu X-J, Hao J-X, Olsson T, Kristensson K, van der Meide PH, Wiesenfeld-Hallin Z (1994) Intrathecal interferon-γ facilitates the spinal nociceptive flexor reflex in the rat. Neurosci Lett 182:263-266
72. Robertson B, Xu X-J, Hao J-X, Wiesenfeld-Hallin Z, Mhlanga M, Grant G, Kristensson K (1997) Interferon-γ receptors in nociceptive pathways: role in neuropathic pain-related behaviour. NeuroReport 8:1311-1316
73. Valtschanoff JG, Weinberg RJ, Rustioni A (1992) NADPH-diaphorase in the spinal cord of rats. J Comp Neurol 321:209-222
74. Meller ST, Pechman PS, Gebhart GF, Maves TJ (1992) Nitric oxide mediates the thermal hyperalgesia produced in a model of neuropathic pain in the rat. Neuroscience 50:7-10
75. Grassi-Zucconi G, Menegazzi M, Carcereri A, Bassetti A, Montagnese P, Cosi C, Bentivoglio M (1993) c-*fos* mRNA is spontaneously induced in the rat brain during the activity period of the circadian cycle. Eur J Neurosci 5:1071-1078
76. Bentivoglio M, Grassi-Zucconi G, Peng Z-C, Bassetti A, Edlund C, Kristensson K (1994b) Trypanosomes cause dysregulation of c-fos expression in the rat suprachiasmatic nucleus. NeuroReport 5:712-714
77. Klein DC, Moore RY, Reppert SM (1991) Suprachiasmatic nucleus. The mind's clock. Oxford University Press, New York Oxford
78. Ebling FJP (1996) The role of glutamate in the photic regulation of the suprachiasmatic nucleus. Pro Neurobiol 50:109-132
79. Takahashi JS (1993) Biological rhythms: From gene expression to behavior. In: Wetterberg L (ed) Light and biological rhythms in man. Pergamon Press, Oxford, pp 3-20
80. Peng Z-C, Kristensson K, Bentivoglio M (1994) Dysregulation of photic induction of Fos-related protein in the biological clock during experimental trypanosomiasis. Neurosci Lett 182:104-106

81. Christenson J, Lundkvist G, el Tayeb RAK, Peng Z-C, Bentivoglio M, Kristensson K (1996) *Trypanosoma brucei* dysregulates the circadian pacemaker in the rat suprachiasmatic nucleus *in vitro*. Soc Neurosci Abstr 22:2057
82. Peruzzi M (1928) Changes in the choroid plexuses and entry of trypanosomes into the cerebrospinal fluid. Final report of the league of Nations on human trypanosomiasis, pp 295-308
83. Vaidya T, Bakhiet M, Hill KL, Olsson T, Kristensson K, Donelson JE (1997) The gene for a T lymphocyte triggering factor from African trypanosomes. J Exp Med 186:433-438
84. Lundkvist GB, Christenson J, El Tayeb RAK, Peng ZC, Grillner P, Mhlanga JDM, Bentivoglio M, Kristensson K (1998) Altered neuronal activity rythm and glutamate receptor expression in the suprachiasmatic nuclei of *Trypanosoma brucei*-infected rats. J Neuropathol Exp Neurol 57:21-29

CHAPTER 10

Hormones in human African trypanosomiasis

MW Radomski, G Brandenberger

Introduction

Humans suffering from African trypanosomiasis or sleeping sickness demonstrate major disruptions in their circadian sleep-wake distribution, with the intensity of the sleep-wake disturbances increasing with the stage of advancement of the disease [1]. Buguet et al. [1] postulated that the disruptions in the circadian sleep-wake cycle in infected humans at the stage of meningoencephalitis may be due to a disease-induced disturbance of the circadian timing system.

We undertook a series of hormonal studies on African patients with varying degrees of severity of infection to assess whether the disruptions in the circadian sleep-wake cycle extended to hormones with well–known circadian rhythms in healthy humans [2, 3]. Several hormones exhibit circadian rhythms which may be controlled by a circadian timing system synchronized to the sleep-wake cycle, or are dependent upon both of these influences [4]. Three types of interactions have been described between the sleep-wake cycle and endocrine rhythms: hormones as cortisol that are relatively independent of sleep and are reflections of endogenous circadian rhythms independent of shifts in the sleep-wake cycle; those as prolactin which demonstrate sleep-dependent secretory patterns increasing during sleep even when sleep is displaced; and those as plasma renin activity (PRA) and growth hormone (GH) which are related to the internal sleep structure. To determine the extent of circadian rhythm disruptions in sleeping sickness patients, we examined the above three groups of hormones in healthy and in infected African subjects.

Two approaches were used in the examination of healthy and infected Africans from the Ivory Coast and the Congo. To establish whether circadian rhythms in certain hormones were disrupted as the disease progressed in severity, hourly blood samples were taken over a 24h period and analyzed for cortisol, prolactin, and GH [2, 3, 5]. To assess whether the relationship between certain hormones to the internal structure of sleep was maintained with disruptions in the sleep-wake cycle in infected humans, blood was sampled every 10 min over 24 h and assayed for PRA, prolactin and cortisol [6, 7].

Disruptions in cortisol and prolactin rhythms in sleeping sickness

Twenty-four hour patterns of plasma cortisol and prolactin were measured in 8 sleeping sickness patients (Daloa, Ivory Coast) at varying degrees of progression of the disease as well as in a control group of 6 healthy subjects of similar ethnic origin and from the same geographical area. Blood was sampled every hour over a 24h period and polysomnographic recordings were taken to monitor sleep and sleep composition in the subjects during the same period. The plasma values for cortisol, prolactin, and GH were subjected to a Z-statistical transformation and the Z-values over 24 h were tested for the presence of circadian rhythmicity by applying standard cosinor analyses.

Figure 1 shows the mean cortisol and prolactin circadian variations for the control group, the group of patients (n = 5) in the early stages of the disease, and the group (n = 3) in the late meningoencephalitic stage. The corresponding 24h variations in NREM and REM sleep are shown in the last row of the Figure. A significant periodicity in the 24h cortisol and prolactin variations were found in both the healthy group and the sick group in the early stages of the disease. These rhythms corresponded closely to the periodicity observed in the sleep parameters, cortisol increasing towards the latter half of the nocturnal and sleep period, and prolactin increasing with the onset of sleep. The advanced stage of the disease was characterized by a total disruption in cortisol, prolactin and sleep rhythms with sporadic increases distributed randomly throughout the 24h/day and independent of the nocturnal period.

Buguet et al. [1] had previously demonstrated the relationship between the disappearance of the circadian rhythm of the sleep-wake cycle and the severity of the clinical symptoms of the disease. Patients in the advanced stages of the infection in which the parasite has become established in the brain, have meninges that are infiltrated with lymphocytes, plasma cells, and morular cells, and exhibit no sleep rhythms. We have found above that as the disease progresses in severity, the extent of circadian disruption in infected patients also extends to other well-known hormonal circadian rhythms as cortisol and prolactin. Whether a close temporal relationship with the disrupted rhythms in cortisol, prolactin and the sleep episodes persists in the absence of any rhythmicity could not be determined because of the infrequent hourly sampling. This question is addressed in the next section of this paper. However, the above findings indicate that the pathogenesis of this disease is related to dysfunctions of circadian pacemakers in the brain.

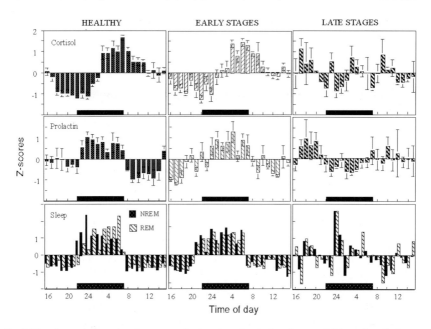

Fig. 1. Hourly variations in plasma cortisol and prolactin levels over 24 h and variations in sleep composition expressed as % NREM and REM for healthy subjects and sleeping sickness patients in the early stages and late advanced stages of the disease. The data are shown as Z-transformed scores. The dark horizontal bar on the x-axis represents the nocturnal lights-out period of the 24h/ day. The variations in the hormones and sleep stages for the healthy subjects and early-stage patients show a significant circadian rhythmicity

The well-known temporal relationship between SWS and GH persists in many conditions of disrupted sleep. We examined the hourly levels of GH in our subjects above and compared them to the occurrence of SWS episodes. We hoped that the sleeping sickness model of circadian disruption might shed further light on the association of plasma levels of GH and the occurrence of SWS. A strong correlation between plasma GH levels and the occurrence of SWS (applying a lag time of 16 min as in Holl et al. [8]) was found in all three groups of subjects indicating that this association persists in even the most severely ill patients, despite the absence of any circadian rhythmicity in GH or SWS [5]. Although dissociations between sleep and GH secretion have been described in the literature, sleep and stimulation of GH secretion are regarded as outputs of a common central nervous system mechanism [9, 10] with hypothalamic GHRH being the likely factor for synchronizing GH secretion and SWS [11]. The fact that the association between GH levels and the occurrence of SWS persisted even under conditions of severe disruptions of the sleep-wake cycle in our sleeping sickness patients suggests that the trypanosomiasis-induced disruption in the synchronization of endogenous rhythms does not impact on the hypothalamopituitary axis, but perhaps on mechanisms activating GHRH.

Pulsatile hormone release in relation to the internal sleep structure in sleeping sickness

Pulsatile hormone secretion appears to be a common characteristic of several endocrine systems and, through frequent blood sampling (every 10 min), some well-defined ultradian rhythms have been identified with periodicities of about 90-100 min similar to that of the REM-NREM sleep cycles, suggesting that both processes could be temporally linked [12]. Cortisol, little influenced by sleep and rather controlled by an internal oscillator, exhibits a temporal association with slow wave sleep which invariably occurs during the descending phases of the secretory pulse [13]. Prolactin is strongly influenced by sleep, increasing during sleep even when this is displaced, and REM sleep onset is associated with a low secretory rate of prolactin [14]. Renin, a key enzyme of the renin-angiotensin system, is related to a specific stage of sleep with nocturnal oscillations in PRA mirroring changes in the NREM-REM sleep cycle, such that PRA can be considered as a biological marker of sleep stage alternation [15, 16]. The sleeping sickness model provided an opportunity to further test this association between hormone pulses and sleep stages under conditions of disrupted REM-NREM sleep cycles in sick patients.

As well as examining the same subjects in the previous section, an additional 6 patients from the Congo were also included. Continuous withdrawal of small quantities of blood via a catheter occurred over 24 h and analysis of cortisol, prolactin and PRA every 10 min. These analyses were then compared to polysomnographic recordings of sleep and the temporal relationship of the pulses of these hormones to sleep stages examined [7].

In sleeping sickness patients, the temporal organisation of cortisol pulses within the 24h period differed widely. The 24h profiles were characterized by a succession of pulses with a reduced circadian amplitude and a shorter quiescent period, as compared to healthy subjects. However, despite these alterations, the previously described relationship between cortisol and SWS episodes was preserved. SWS remained associated with the declining phases of cortisol pulses even in the most severely ill patients (Fig. 2). Sleep-wake episodes in sleeping sickness patients were fragmented throughout the 24h period in association with attenuated nocturnal prolactin levels and higher daytime levels. Prolactin release remained pulsatile with the same number of pulses per 24 h in both groups. The relationship between prolactin pulses and sleep stages [14] was maintained with REM sleep onset beginning in the non-ascending

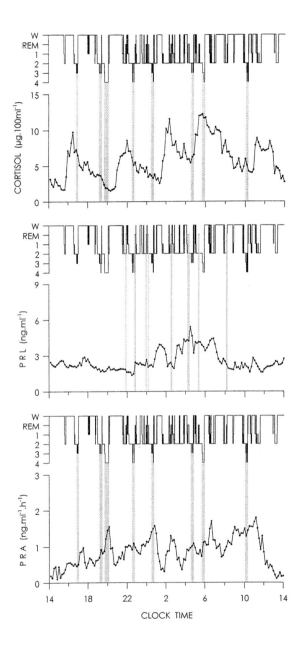

Fig. 2. Individual 24h cortisol, prolactin and plasma renin activity (PRA) profiles in one patient with sleeping sickness illustrating the relationship between hormone pulses and sleep stages. As in healthy subjects, slow wave sleep occurs in the descending phases of cortisol pulses (*top panel*); the onset of REM sleep occurs in the descending phases of prolactin pulses (*middle panel*); SWS occurs in the ascending phases of PRA oscillations (*bottom panel*)

portions of the pulses in both healthy subjects and patients, despite the disruptions in the rhythms of both sleep and prolactin in the latter group (Fig. 2).

A significant rhythm in PRA was observed in healthy subjects, with increases during nocturnal sleep, and it was not evident in the patients. The mean levels of PRA showed a general ascending trend over 24 h in sleeping sickness which was not related to any trend in sleep quality. Pulse analysis of the individual profiles revealed that in healthy subjects the number of oscillations was significantly greater when asleep rather than awake in contrast to the sleeping sickness patients where oscillations in PRA were distributed throughout the 24h period. The previously described relationship between oscillations in PRA and alternations in internal sleep structure [15, 16] was observed in both healthy subjects and patients, with increasing levels of PRA occurring during SWS episodes and decreasing levels during REM sleep. In the 6 patients studied, 32 of 33 SWS episodes recorded were linked to significant increases in PRA, and 40 of 42 REM sleep episodes occurred when PRA was declining [7]. Thus, the normal relationship between PRA and sleep structure is preserved in sleeping sickness (Fig. 2).

Conclusions

The two approaches described above demonstrate that there are profound disruptions in the circadian rhythms of cortisol, prolactin, GH and PRA in human African trypanosomiasis. These studies of the sleeping sickness disease model have confirmed that 24h variations in prolactin, GH and PRA are not circadian in nature but are principally related to sleep. Despite the disorganization of the 24h rhythms, the relationship between hormonal pulses and internal sleep structure found in healthy subjects was preserved in sleeping sickness. The SWS episodes remained associated with increases in GH and decreases in cortisol, and REM sleep onset with the non-ascending phases of prolactin pulses. Similarly, the association between PRA oscillations and NREM-REM sleep cycles, when long enough, persisted in all patients.

Buguet et al. [1] demonstrated that the disappearance of the circadian variation of the sleep-wake cycle was related to the severity of clinical progression of the disease. Our findings demonstrate that disruptions in the mechanisms controlling the sleep-wake cycle also extend to other circadian rhythms supporting the hypothesis that the networks of the biological clock are altered in the course of the disease. Such disappearances of circadian rhythms have been observed in animals after lesions of the suprachiasmatic nucleus [17] and dysregulation of gene expression in the suprachiasmatic nucleus has been shown in trypanosome-infected rat brains [18].

Thus, the pathogenesis of the disease, particularly in its advanced state, appears to be related to dysfunctions of the circadian pacemaker. However, the temporal relationship between pulsatile hormone release and specific sleep stages remains persistent, emphasizing the strength of the processes coupling hormonal release and the internal sleep structure and its independence of the circadian timing system.

Acknowledgements

This work was supported by the World Health Organization (grant n° TDR/ID 910048), the French Ministry of Cooperation, the Société Elis, and companies UTA and Air Afrique. Institutions which co-participated in this work were as follows: Defence and Civil Institute of Environmental Medicine, Toronto, Canada; Institut de Neurologie Tropicale, Limoges, France; Laboratoire de Physiologie, Faculté de Médecine, Abidjan, Côte d'Ivoire; Laboratoire de Physiologie et de Psychologie Environmentales, CNRS, Strasbourg, France; Projet de Recherches Cliniques sur la Trypanosomiase, Daloa, Côte d'Ivoire; Service des Grandes Endémies, Brazzaville, Congo; Service de Neurologie, CHU de Brazzaville, Congo.

References

1. Buguet A, Bert J, Tapie P, Tabaraud F, Doua F, Lonsdorfer J, Bogui P, Dumas M (1993) Sleep-wake cycle in human African trypanosomiasis. J Clin Neurophysiol 10:190-196
2. Radomski MW, Buguet A, Bogui P, Doua F, Lonsdorfer A, Tapie P, Dumas M (1994) Disruptions in the secretion of cortisol, prolactin, and certain cytokines in human African trypanosomiasis patients. Bull Soc Path Ex 87:376-379
3. Radomski MW, Buguet A, Montmayeur A, Bogui P, Bourdon L, Doua F, Lonsdorfer A, Tapie P, Dumas M (1995) Twenty-four hour plasma cortisol and prolactin in human African trypanosomiasis patients and healthy African controls. Am J Trop Med Hyg 52:281-286
4. Mullen PE (1983) Sleep and its interaction with endocrine rhythms. Br J Psychiatr 142:215-220
5. Radomski MW, Buguet A, Doua F, Bogui P, Tapie P (1996) Relationships of plasma growth hormone to slow-wave sleep in African sleeping sickness. Neuroendocrinol 63:393-396
6. Brandenberger G, Buguet A, Spiegel K, Stanghellini A, Muanga G, Bogui P, Montmayeur A, Dumas M (1994a) Maintien des relations entre la sécrétion pulsatile des hormones et la structure interne du sommeil dans la trypanosomiase humaine africaine. Bull Soc Path Ex 87:383-398
7. Brandenberger G, Buguet A, Spiegel K, Stanghellini A, Muanga G, Bogui P, Dumas M (1996) Disruption of endocrine rhythms in sleeping sickness with preserved relationship between hormonal pulsatility and the REM-NREM sleep cycles. J Biol Rhythms 11:258-267
8. Holl RW, Hartman ML, Velduis JD, Taylor WM, Thormer MO (1991) Thirty-second sampling of growth hormone in man: correlation with sleep stages. J Clin Endocrinol Metab 72:854-861
9. Obal F, Payne L, Opp M, Alfoldi P, Kapas L, Krueger JM (1992) Growth-hormone releasing hormone antibodies suppress sleep and prevent enhancement of sleep after sleep deprivation. Am J Physiol 263:R1078-R1085
10. Parker DC, Sassin JF, Mace JW, Gotline RW, Rossman LG (1969) Human growth hormone release during sleep: electroencephalographic correlation. J Clin Endocrinol Metab 29:871-874
11. Obal F, Alfoldi P, Cady AB, Johannsen L, Sary G, Krueger JM (1988) Growth hormone-releasing factor enhances sleep in rats and rabbits. Am J Physiol 255:R310-R316
12. Brandenberger G (1993) Episodic hormone release in relation to REM sleep. J Sleep Res 2:193-198
13. Follenius M, Brandenberger G, Bandesapt JJ, Libert JP, Ehrhart J (1992) Nocturnal cortisol release in relation to sleep structure. Sleep 15:21-27

14. Spiegel K, Follenius M, Simon C, Saini J, Ehrhart J, Brandenberger G (1994) Prolactin secretion and sleep. Sleep 17:20-27
15. Brandenberger G, Follenius M, Simon C, Ehrhart J, Libert JP (1988) Nocturnal oscillations in plasma renin activity and REM-NREM sleep cycles in humans: a common regulatory mechanism? Sleep 11:242-250
16. Brandenberger G, Follenius M, Goichot B, Saini J, Spiegel K, Ehrhart J, Simon C (1994b) Twenty-four hour profiles of plasma renin activity in relation to the sleep-wake cycle. J Hypert 12:277-283
17. Rusak B, Bina KG (1990) Neurotransmitters in the mammalian circadian system. Ann Rev Neurosci 13:387-401
18. Bentivoglio M, Grassi-Zucconi G, Peng ZC, Bassetti A, Edlund C, Kristensson K (1994b) Trypanosomes cause dysregulation of c-fos expression in the rat suprachiasmatic nucleus. NeuroReport 5:712-714

CHAPTER 11

Sleeping sickness: a disease of the clock with nitric oxide involvement

A Buguet, R Cespuglio

Introduction

Although it was practically eradicated by the 1960s, sleeping sickness or human African trypanosomiasis (HAT) is now spreading rapidly and affecting more than 10% of the population in certain endemic areas [1]. This disease, described more than a century ago, was thought to be specific of Melanoids [2]. It is anchored by the geographical distribution of the tse-tse fly, which infects humans with a protozoa, *Trypanosoma brucei gambiense* (Western and Central Africa) or *T. b. rhodesiense* (Eastern Africa). The hemolymphatic phase (stage I) represents the invasion of the blood and lymphatic tissues and is followed by a meningoencephalitis (stage II), with the presence of trypanosomes in the cerebrospinal fluid (CSF) and perivascular infiltrates of lymphoplasmocytic cells [3].

The disease bears its name from original descriptions of sleep-wake disorders in meningoencephalitic patients. The patients were thought to suffer from hypersomnia, being sleepy by day and restless by night [4]. However, in 1890, Mackensie [5] had observed that the patients slept often, but in short bouts evenly distributed throughout the nycthemeron, the total sleep duration not exceeding that of a healthy sleeper. This excellent clinical description could have evoked a major disturbance in the circadian (24-hour) rhythmicity of the sleep-wake cycle. However, chronobiology is a new science and the definition of circadian rhythms was proposed by Halberg in 1959 [6]. In the same year, the muscular atonia of paradoxical sleep (rapid eye movement or REM sleep in humans) was first described using polygraphic sleep recordings [7]. In fact, polysomnography (electro-encephalogram, EEG; electro-oculogram, EOG; electromyogram, EMG) represents the only objective means to distinguish the different stages of vigilance, i.e. wakefulness, non-REM sleep and its 4 stages, stages 3 and 4 representing slow-wave sleep (SWS) and REM sleep. This technique is difficult, however, to set up in developing countries, especially in the bush. This may be why around the clock sleep recordings were not made until the late 1980s.

Although such recordings had been made at night [8, 9] and/or during diurnal naps [10], the first 24-hour sleep recording in sleeping sickness was in fact performed in 1988 in Niamey [11]. We were then able to confirm that sleeping sickness is not a hypersomnia, but rather a disease of the circadian timing system.

The patient was a migrant worker who had contracted the disease on the Ivory Coast before returning to Niger and presented a severe neurological disorder with mainly extrapyramidal symptomatology. Although the EEG trace was loaded with slow waves, sleeping, waking and REM sleep were identifiable. However, stage 1 was hard to differentiate from stage 2, and stage 3 from stage 4. The 24-hour recording revealed the disappearance of the circadian distribution of sleeping and waking, differing from its biphasic distribution observed frequently in Africa, with a major night sleep episode and a daytime nap [12]. Sleep and wakefulness occurred indifferently throughout the nycthemeron.

Following this observation, our group also demonstrated that other biological rhythms were disturbed in sleeping sickness, such as body temperature [11, 13], cortisol and prolactin secretion [14] or growth hormone release [15]. However, the relationship between plasma hormone pulses to sleep-wake stages remains similar to that of healthy subjects. This is true for plasma renin activity, cortisol and prolactin pulses [16, 17], and growth hormone [15]. Curiously, the circadian rhythm of melatonin, the pineal hormone known to interfere with the circadian timing system, remains untouched [18]. Hormonal data will be exposed in a joint chapter of this volume (see Chapt 10).

We propose here to review the polysomnographic data obtained in investigations conducted in Western (Daloa, The Ivory Coast) and Central Africa (Brazzaville, Congo). Healthy West-Africans (Abidjan, The Ivory Coast) observed in similar light and bed conditions served as controls. Sixty-eight 24-hour recordings were collected from 20 patients and 6 healthy subjects. Another 46 recordings from 23 patients have not yet been analysed.

We shall also examine the hypothesis of the involvement of nitric oxide in the pathogenesis of the disease.

Methodological aspects of the recording of the sleep-wake cycle in sleeping sickness

Patients with meningoencephalitis were hospitalized after a medical prospection in exposed villages from known epidemiological foci. On the Ivory Coast, the 8 patients lived in two villages at a distance of 13 and 17 km from the town of Daloa [19]. In the Republic of Congo [20, 21], 10 patients were hospitalized after a field investigation in the area of

N'Gabé, along the Congo river, and from the Bouenza focus (Boko Songho). In both cases, the field investigation was conducted with a trained team. Human African trypanosomiasis was diagnosed on the basis of the presence of clinical signs such as fever, meagerness, somnolence and the presence of lymph nodes, and from laboratory investigations. In the sequence of laboratory tests, people who were revealed positive to the CATT (Card Agglutination Test for African Trypanosomiasis) had a new blood sample. Blood was examined directly at a light microscope and/or centrifuged for the ion exchange minicolumn test with further microscopic examination of the filtrate. When the diagnosis had been confirmed, the patients were proposed to be transferred and treated at the Research Center for Trypanosomiasis Clinic at Daloa or at the Neurology ward of the University Hospital of Brazzaville. Once hospitalized, the patients or their families gave their informed consent to participate in the study, and the experimental set up and protocol were approved by National Health authorities. The diagnosis of sleeping sickness was confirmed by the presence of *T. b. gambiense* in the blood or a lymph gland puncture and in the cerebrospinal fluid, and by a serologic immunofluorescence test.

At the hospital, the patients were examined by two neurologists. All patients were at the stage of meningoencephalitis and ranked according to the clinical and biological severity of the illness [22]. The principal clinical and laboratory criteria used to determine the severity of the disease included: presence of lymph nodes, daytime somnolence, intermittent fever, headaches, sensory disturbances with uncomfortable diffuse superficial or deep sensations (hyperpathia), presence of primitive reflexes (palmo-mental reflex, sucking reflex), exaggerated deep tendon reflexes, psychiatric disorders (confusion, mood swings, agitation, aggressive behavior, euphoria, absent gaze, mutism, indifference), pruritus, with or without skin lesions, and tremor (fine and diffuse without any myoclonic jerk at rest or during movement), besides an abnormal number of cells (lymphocytes at microscopic examination) in the cerebrospinal fluid. The patients did not present any Babinski sign, nor any alteration in muscle tone or numbness. They had no objective sensory deficiency. The duration of the illness prior to selection for the study could not be established with any degree of certainty. No cases were seropositive for HIV or HTLV-1. In both countries, the most severely ill patient was identified as P1 and the least severe one as P8 (Daloa) and P10 (Brazzaville).

The Daloa patients had two 24-hour recordings, one with an intravenous indwelling catheter, blood samples being taken every hour, and the other without any blood collection. The two situations were in random order. In both cases, the patients remained in bed for 24 hours and were allowed to sleep ad libitum, in an air conditioned room (dry bulb temperature: 24-25°C). Specific treatment with melarsoprol was undertaken immediately after each recording. Polysomnographic recordings were taken on an 8-channel polygraph following a standard procedure [23]. Two channels were devoted to the EEG, while the EMG and EOG were taken on separate channels. Three channels served to analyse the respiratory cycle, nasal and buccal airflow with Cu-Ct thermocouples and chest movements with a strain gauge, and electrocardiogram was also taken. Paper speed was 15 mm.s^{-1} and determined 20 s scoring periods. In 7 patients (P2-P8), classical scoring of the states of vigilance [24] could be done (wakefulness; rapid eye movement (REM) sleep; non-REM sleep or stages 1 to 4; slow-wave sleep, SWS or stages 3 and 4). In P1, however, the scoring technique had to be adapted, following Schwartz and Escande [10] and Buguet et al. [11], because of the ubiquity of EEG slow waves. Non-REM sleep was thus divided into light sleep and SWS. In all patients wakefulness and REM sleep were easily identified.

In the Brazzaville investigation, each patient had 3 treatment sessions during 3 consecutive weeks. Twenty-four hour polysomnographic recordings were taken before and one day after each treatment session. The latter consisted of corticotherapy (30 mg/day) during 2 days, followed by 3 days of melarsoprol (Arsobal®) intravenous injections (one injection per day at the dose 3.6 mg.kg^{-1}), with an antihistaminic associated on the first day of the arsenical treatment. Blood samples were collected over 10 min epochs during the first two recording sessions, using a constant withdrawal pump, to allow the analysis of hormonal secretion pulses.

Another investigation was conducted in similar conditions (with continuous blood withdrawal) in Abidjan on 6 healthy subjects recorded over 24 hours with a venous catheter. Again, the protocol was approved by Ivorian Health authorities and accepted by the volunteers [21].

In the three investigations, blood withdrawal was kept below 300 ml/24 hours. Blood sampling was terminated either after 24 hours or at the request of the patients or if the patient's hemoglobin dropped below 10 g.dl^{-1}.

The circadian disruption of the sleep-wake cycle in sleeping sickness

The healthy subjects had a major sleep episode at night and a small nap in the afternoon or late morning [20]. They did not show any circadian disturbance of their sleep-wake cycle. However, according to the observation of Adam [25] in 7 healthy subjects who slept with an indwelling catheter, there was a reduction in total sleep time and REM sleep, and sleep efficiency was impaired due to numerous awakenings. On the contrary, sleep of the Daloa patients did not change whether they had a catheter or not [19]. It seems that these hyperpathic patients had a lower sensitivity to environmental stimuli. The patients showed a major disturbance in the circadian distribution of sleeping and waking (Fig. 1). The sleep-wake disturbances observed were related to the severity of the disease. The most severely ill patients extended their sleep throughout the nycthemeron, showing shorter latencies to SWS and REM sleep. The average duration of wakefulness and sleep episodes was inversely related to the severity of the disease. This means that the sleep-wake alternation tended to occur in short cycles, with even shorter cycles being observed in the most severely sick, sleeping and waking occurring equally by day and by night. These data demonstrated that, despite its appellation, sleeping sickness is not a hypersomnia. Instead, there was a disappearance of the circadian rhythm of the sleep-wake cycle. Particularly, REM sleep occurred throughout the nycthemeron. Sleep onset REM sleep phases (SOREM) occurred often, meaning that REM sleep appeared during wakefulness episodes, without any intermediary non-REM sleep.

Data from the Brazzaville investigation [21] show that the circadian disturbance of the sleep-wake organization is reversible after treatment (Fig. 2). The rhythmicity of the sleep-wake alternation, evidenced by the number of sleep episodes, proved to be related to the severity of the disease, and decreased throughout the 4 successive recording sessions. The least severly ill patient resumed one sleep episode after the third treatment session. Conversely to the evolution of the number of sleep episodes, the mean duration of sleep episodes increased with the progression of the treatment. The number of SOREM and/or REM sleep phases occurring before SWS within a given sleep episode was high in all patients before treatment. It decreased in the least sick patients after two treatment sessions.

The treatment-related improvement of the 24-hour alternation of sleep and wakefulness was also observed recently in two Caucasians who were infected by *T. b. rhodesiense* in Rwanda, treated with melarsoprol and examined 4, 6 and 11 months later [27]. The number of sleep episodes was normalized in these patients after at least 6 months following treatment with melarsoprol.

Therefore, sleeping sickness is not only a sleep disorder, but more so a disturbance of the circadian rhythm of the sleep-wake cycle along with that of cortisol, prolactin and growth hormone secretions, which keep their links to the sleep/wake stages (see Chapt 10). However, the secretion of the endogenous synchronizer melatonin remains normal [18]. These findings could suggest a dysregulation at the level of the main pacemaker of the circadian timing system, the suprachiasmatic nuclei [28], which are entrained by the ambient light cycles through the retino-hypothalamic tract. In fact, such a disappearance of circadian rhythms has been observed in animals after a lesion of the suprachiasmatic nucleus [29-31]. However, although the function of the internal clock is impaired in sleeping sickness, the reversibility of the circadian rhythm disturbance after treatment implies that there are no neuronal lesions in the suprachiasmatic nuclei. Such reversible circadian disturbances of the sleep-wake cycle have also been shown in the rat after lesions of the nucleus raphe dorsalis [32]. Furthermore, the occurrence of SOREM episodes in sleeping sickness patients is also in favour of the involvement of the serotonergic nuclei of the raphe system in the dysregulation of the internal clock.

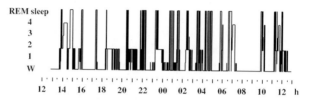

Fig. 1. Twenty-four-hour sleep (REM sleep and stages 1-4 of non-REM sleep)-wake (W) distribution (hypnogram) in a patient examined at Daloa [19], showing the almost homogeneous daytime and night–time occurrence of sleep and wake episodes (with permission from [20]). This patient, aged 17, showed REM sleep episodes occurring on a "narcoleptic" mode, immediately after a wakefulness episode. REM sleep episodes occurred with a periodicity of 90 min, realizing an example of the "basic rest-activity cycle" postulated by Kleitman [26]

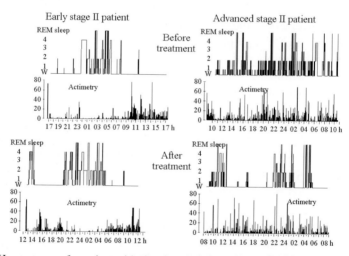

Fig. 2. Hypnograms of a patient with sleeping sickness at the beginning of meningoencephalitis (early stage II patient) and in a severely sick patient (advanced stage II patient) recorded at Brazzaville, before (*upper graph*) and after (*lower graph*) specific treatment. Before treatment, the severely sick patient had a very disrupted sleep (REM sleep and stages 1-4 of non-REM sleep) and wake (W) distribution throughout the 24-hour recording. The sleep-wake cycle began to be normalized after the 5-day treatment session (corticoids and a 3-day melarsoprol cure). However, although sleep onset REM sleep phases had decreased, REM sleep still initiated most sleep episodes. In both patients, the actigraphic activity was recorded with a Gehwiller wrist actigraph. The alternation of sleep and wakefulness is evident in the early stage II patient. Contrarily, the sleep wake cycle cannot be determined precisely in the severely sick patient who is almost permanently restless

This nucleus intervenes in the triggering of REM sleep [33], which occurs concomitantly with the inhibition of the serotonergic neurones discharge [34, 35] and the concomitant drop in the axonal nerve endings serotonin release [36, 37]. It provides also a heavy serotonergic projection to the vasoactive intestinal peptide-containing neurons of the suprachiasmatic nuclei [38, 39]. The implication of raphe nuclei in African trypanosomiasis is further endorsed by the recent finding of the circulating anti-tryptophan-like antibodies present in patients with sleeping sickness [40], which lead to a degeneration of serotonergic fibers when injected intracerebroventricularly in rats [41].

It therefore seems that in sleeping sickness the raphe-suprachiasmatic nuclei liaison is involved in the reversible disturbance of the circadian rhythms of the sleep-wake cycle and of hormonal secretions, while the suprachiasmatic nuclei-pineal gland pathway remains undisturbed.

Nevertheless, as in the 1980s [42], the pathogenetic mechanisms of sleeping sickness have still not been established. In any case, it is linked to the impossibility of the immune system to get rid of the trypanosome [43], which is approached in this volume by Vincendeau et al. (Chapter 8) and by Rhind and Shek (Chapter 7). The alterations in the immune function and nervous system in African trypanosomiasis led us to investigate the involvement of nitric oxide (NO), a key molecule in immune and neurophysiological mechanisms, in experimental trypanosomiasis. Our attention was focused on the possible involvement of NO after Vincendeau et al. [44] had demonstrated that interferon-γ-activated macrophages are not cytostatic on cultured *T. b. brucei* when inhibitors of NO synthase are used. The cytostatic effect is resumed when NO is added to the medium. More specifically, the disturbances in the functioning of the circadian timing system may also be related to the increase in the permeability of the blood-brain barrier, which is suggested by the occurrence of the well-documented perivascular infiltrates of lymphoplasmocytic cells [45, 46]. In this chapter, we shall concentrate on the involvement of cerebral NO in the development of the disease.

Trypanosomiasis and nitric oxide involvement in the brain

Nitric oxide is an ubiquitous messenger, which can be massively produced during brain lesions induced, for example, by ischemia [47, 48]. Recently, NO has been proposed as a messenger between the trypanosome, the immune system and the brain [49]. It can therefore be postulated that an increased NO production may be involved in the development of trypanosomiasis. To study the NO involvement in the development of brain damages in trypanosomiasis, we used an animal model developed in the rat. In fact, the progressive disruption of the circadian rhythmicity of the sleep-wake cycle was also found in rats infected with *T. b. brucei* [50]. This is not surprising as the classical perivascularitis and meningovascularitis with plasmocytic infiltrates and gliosis [45, 46] are predominant in the periventricular area around the third ventricle, maximum in the infundibular region, where lays the major structure of the circadian timing system, the suprachiasmatic nuclei [51]. It is known that the brain influences immune function through hormones and that in turn, immune messengers affect the brain [52]. In fact, a soluble factor is involved in the interaction between trypanosomes and the brain via the immune system [53]. Furthermore, we have shown that the L-arginine-NO pathway is also involved in the development of experimental trypanosomiasis due to the infestation of rats by *T. b. brucei* [54].

NO was measured in the sensorimotor cerebral cortex with a porphyrine-Nafion®-coated probe by differential normal pulse voltammetry [55]. The working electrode is made of a carbon fiber (Ø 30 μm, Textron C_{ny}) inserted into a pulled glass micropipette, the end of the fiber emerging beyond the tip of the glass by about 500 μm. Watertightness between glass and carbon fiber is insured with a resin. The pipette is filled up with a conductive resin and a silver wire inserted into the non-pulled tip of the pipette. The sensor is maintained at 40°C for 12 hours. Before use, the carbon fiber is coated with porphyrin-nickel and Nafion®. A reference (Ag/AgCl) and an auxiliary electrode (tungsten) complete the potentiostat. The potential applied between the reference working electrodes crosses from 400 to 1350 mV in 2 min. It allows the detection of an oxidation NO peak at 650 mV. Values of the NO peak were expressed in percent of control values obtained in healthy rats after electrode calibration done with solutions of varied concentrations of NO.

In rats infected since 2 weeks, NO was higher in the cortex and the lateral ventricle than in healthy animals. After 3 weeks of infection, the increase in cortical NO was further enhanced. Similar data were recently obtained in mice, with a progressive recruitment of brain NO parallel to the development of the disease.

This overproduction of NO could be responsible for the circadian rhythm disturbance in the sleep-wake cycle. The increased constitutive NO synthase activity could affect the blood-brain barrier, allowing the occurrence of the inflammatory process caused by the infiltration of lymphoplasmocytes. These cells infiltrate preferentially areas of the brain close to feeble blood–brain barrier zones. As a matter of fact, some of these areas, such as the suprachiasmatic nuclei, are also implicated in the determination of major clinical signs of sleeping sickness, such as the disruption of sleeping and waking and other circadian rhythms.

The implication of NO in trypanosomiasis can be examined from recent experiments on heatstroke [56]. In heat-exposed rats, an increase in NO production is observed at the end of heat stress, before the development of heatstroke and death. As the glutamatergic N-methyl-D-aspartate (NMDA) receptors are involved in the activation of NO synthase [57], the blockade of the NMDA receptors should reduce NO production. However, a non competitive antagonist of the glutamatergic NMDA receptors, MK 801 [58], did not modify the time course of NO production during heat stress. This could mean that NO synthase could be activated by a voltage-dependent Ca^{2+} channel rather than that of the NMDA receptor [59]. NO may also have inhibited the NMDA receptor and masked the pharmacological blockade of the NMDA receptor by MK 801 [60]. Alternatively, NO synthase activation may have been produced by triggering agents other than glutamate. In this case, interleukin-1 and TNF-α (tumor necrosis factor), which synergistically mediate neurotoxicity through an increase in NO production, could be good candidates [61].

Acknowledgements

This work was supported by the World Health Organization (grant TDR/ID 910048), the French Ministry of Cooperation, the Société Elis, and companies UTA and Air Afrique. Institutions which co-participated in this work were as follows: Defence and Civil Institute of Environmental Medicine, Toronto, Canada; Département de Médecine Expérimentale, INSERM-U480, Lyon, France; Institut de Neurologie Tropicale, Limoges, France; Laboratoire de Physiologie, Faculté de Médecine, Abidjan, Côte d'Ivoire; Laboratoire de Physiologie et de Psychologie Environnementales, C.N.R.S., Strasbourg, France; Nestle Ltd Research Center, Lausanne, Switzerland; Projet de Recherches Cliniques sur la Trypanosomiase, Daloa, Côte d'Ivoire; Service des Grandes Endémies, Brazzaville, Congo; Service de Neurologie, C.HU. de Brazzaville, Congo.

References

1. De Raadt P (1994) Trente années de recherche sur la trypanosomose. Second Congress of Tropical Neurology, Limoges, September 21-23
2. Nicolas A (1861) De la maladie du sommeil. Gaz Hebdo Méd Chir 21:670-673
3. Dumas M, Boa FY (1988) Human African trypanosomiasis. In: Vinken PJ, Bruyin GW, Klawans HL (eds) Handbook of Clinical Neurology, revised Series. , Elsevier Science Publishers, Amsterdam, pp 339-344
4. Manson-Bahr P (1942) Manson's tropical diseases. A manual of diseases of warm climates. Cassel, London
5. Mackensie S (1890) La maladie du sommeil en Afrique. Mercredi Méd 47:597-598
6. Halberg F (1959) Physiologic 24-hour periodicity: general and procedural considerations with reference to the adrenal cycle. Z Vitamin Hormon Fermentforsch 10:225-296
7. Jouvet M, Michel F, Courjon J (1959) Sur un stade d'activité électrique cérébrale rapide au cours du sommeil physiologique. C R Soc Biol 152:1024-1028

8. Bert J, Collomb H, Fressy J, Gastaud H (1965) Etude électroencéphalographique du sommeil nocturne au cours de la trypanosomiase humaine africaine. In: Fischgold H (ed) Le sommeil de nuit normal et pathologique. Etudes électroencéphalographiques. Masson, Paris, pp 334-352
9. Giordano C, Dumas M, Kouassi B, Boa F, Piquemal M (1984) Les aspects électroencéphalographiques de la maladie du sommeil (trypanosomiase humaine africaine). XIth International Congress of Tropical Medicine and Malaria, Calgary, September 16-22
10. Schwartz BA, Escande C (1970) Sleeping sickness: sleep study of a case. Electroenceph Clin Neurophysiol 29:83-87
11. Buguet A, Gati R, Sèvre JP, Develoux M, Bogui P, Lonsdorfer J (1989) 24 hour polysomnographic evaluation in a patient with sleeping sickness. Electroenceph Clin Neurophysiol 72:471-178
12. Buguet A, Hankourao O, Gati R (1990) Self-estimates of sleep in African students in a dry tropical climate. J Environ Psychol 10:363-369
13. Gati R, Tabaraud F, Buguet A, Bert J, Tapie P, Bittel J, Sparkes B, Breton JC, Doua F, Bogui P, Lonsdorfer A, Lonsdorfer J, Moulin J, Chameaud J, Dumas M (1990) Analyse circadienne du sommeil, de la température rectale et de variables immunologiques et endocrinologiques dans la maladie du sommeil: étude préliminaire. Bull Soc Path Ex 83:275-282
14. Radomski MW, Buguet A, Montmayeur A, Bogui P, Bourdon L, Doua F, Lonsdorfer A, Tapie P, Dumas M (1995) 24-hour plasma cortisol and prolactin in human African trypanosomiasis patients and healthy african controls. Am J Trop Med Hyg 52:281-286
15. Radomski MW, Buguet A, Doua F, Bogui P, Tapie P (1996) Relationship of plasma growth hormone to slow-wave sleep in African sleeping sickness. Neuroendocrinology 63:393-396
16. Brandenberger G, Buguet A, Spiegel K, Stanghellini A, Mouanga G, Bogui P, Montmayeur A, Dumas M (1994) Maintien des relations entre la sécrétion pulsatile des hormones et la structure interne du sommeil dans la trypanosomiase humaine africaine. Bull Soc Path Ex 87:383-389
17. Brandenberger G, Buguet A, Spiegel K, Stanghellini A, Muanga G, Bogui P, Dumas M (1996) Disruption of endocrine rhythms in sleeping sickness with preserved relationship between hormonal pulsatility and the REM-NREM sleep cycles. J Biol Rhythms 11:258-267
18. Claustrat B, Buguet A, Geoffriau M, Montmayeur A, Bogui P, Muanga G, Stanghellini A, Dumas M (1994) Le rythme nycthéméral de la mélatonine (MLT) est conservé dans la trypanosomose humaine africaine. Bull Soc Path Ex 87:380-382
19. Buguet A, Bert J, Tapie P, Tabaraud F, Doua F, Lonsdorfer J, Bogui P, Dumas M (1993) Sleep-wake cycle in human african trypanosomiasis. J Clin Neurophysiol 10:190-196
20. Buguet A, Bert J, Tapie P, Bogui P, Doua F, Mouanga G, Stanghellini A, Sarda J, Tabaraud F, Gati R, Montmayeur A, Chauffard F, Lonsdorfer J, Dumas M (1994) Distribution du sommeil et de la veille dans la trypanosomose humaine africaine. Bull Soc Path Ex 87:362-367
21. Buguet A, Montmayeur A, Bourdon L, Auzelle F, Tapie P, Bert J, Tabaraud F, Dumas M, Bogui P, Doua F, Stanghellini A, Sarda J, Muanga G, Chauffard F, Lonsdorfer J, Brandenberger G, Radomski MW, Claustrat B, Cespuglio R (1995) La maladie du sommeil: trouble majeur des rythmes circadiens. Rev Sci Technique Défense 29:107-117
22. Tapie P, Buguet A, Tabaraud F, Bogui P, Doua F, Bert J (1996) Electroencephalographic and polygraphic features in 24-hour recordings in sleeping sickness and healthy African subjects. J Clin Neurophysiol 13:339-344
23. Buguet A, Rivolier J, Jouvet M (1987) Human sleep patterns in Antarctica. Sleep 10:374-382
24. Rechtschaffen A, Kales A (1968) A manual of standardized terminology, techniques and scoring system for sleep stages of human subjects. NIH publications n° 204, Public Health Service, US Govt Printing Office, Washington DC
25. Adam K (1982) Sleep is changed by blood sampling through an indwelling venous catheter. Sleep 5:154-158
26. Kleitman N (1969) Basic rest-activity cycle in relation to sleep and wakefulness. In: Kleitman N (ed) Sleep physiology and pathology. Lippincott, Philadelphia, pp 39-52
27. Montmayeur A, Brosset C, Imbert P, Buguet A (1994) Cycle veille-sommeil au décours d'une trypanosomose humaine africaine à *Trypanosoma brucei rhodesiense* chez deux parachutistes français. Bull Soc Path Ex 87:368-371
28. Moore RY (1983) Organisation and function of a central nervous system oscillator: the suprachiasmatic hypothalamic nucleus. Fed Proc 42:2783-2789
29. Coindet J, Chouvet G, Mouret J (1975) Effects of lesions of the suprachiasmatic nuclei on paradoxical sleep and slow wave sleep circadian rhythms in the rat. Neurosci Lett 1:243-247
30. Mouret J, Coindet J, Debilly G, Chouvet G (1978) Suprachiasmatic nuclei lesions in the rat: alterations in sleep circadian rhythms. Electroenceph Clin Neurophysiol 45:402-408
31. Rusak B, Bina KG (1990) Neurotransmitters in the mammalian circadian system. Annu Rev Neurosci 13:387-401
32. Coindet J, Debilly G, Mouret J (1979) Role of the raphe nuclei: II. On sleep circadian organization in the rat. Third European Neuroscience Meeting, Rome
33. El Kafi B, Cespuglio R, Léger L, Marinesco S, Jouvet M (1994) Is the nucleus raphe dorsalis a target for the peptides possessing hypnogenic properties? Brain Res 637:211-221
34. McGinty DJ, Harper RM (1976) Dorsal raphe neurons: depression of firing during sleep in cats. Brain Res 101:569-575

35. Cespuglio R, Faradji H, Gomez ME, Jouvet M (1981) Single unit recordings in the nuclei raphe dorsalis and magnus during the sleep-waking cycle of semi-chronic prepared cats. Neurosci Lett 24:133-138
36. Cespuglio R, Sarda N, Gharib A, Houdouin F, Jouvet M (1990) Voltammetric detection of the release of 5-hydroxyindole compounds throughout the sleep-waking cycle of the rat. Exp Brain Res 80:121-128
37. Houdouin F, Cespuglio R, Jouvet M (1991) Effects induced by the electrical stimulation of the nucleus raphe dorsalis upon hypothalamic release of 5-hydroxyindole compounds and sleep parameters in the rat. Brain Res 565:48-56
38. Aghajanian G, Bloom FE, Sheard MH (1969) Electron microscopy of degeneration within the serotonin pathway of rat brain. Brain Res 13:266-273
39. Bosler O, Beaudet A (1985) Relations ultrastructurales entre systèmes monoaminergiques et peptidiques de l'hypothalamus. Ann Endocrinol 46:19-26
40. Okomo-Assoumou MC, Geffard M, Daulouède S, Chaugier C, Lemesre JL, Vincendeau P (1995) Circulating antibodies directed against tryptophan-like epitopes in sera of patients with human African trypanosomiasis. Am J Trop Med Hyg 52:461-467
41. Okomo-Assoumou MC (1995) Mécanismes autoimmuns et perturbations du réseau des cytokines dans la physiopathologie de la trypanosomose humaine africaine. Thesis, Bordeaux II, France, p 120
42. Poltera AA (1980) Immunopathological and chemotherapeutic studies in experimental trypanosomiasis with special reference to the heart and brain. Trans R Soc Trop Med Hyg 74:706-715
43. Pentreath VW (1991) The search for primary events causing the pathology in African sleeping sickness. Trans R Soc Trop Med Hyg 85:145-147
44. Vincendeau P, Daulouède S, Veyret B, Dardé ML, Bouteille B, Lemesre JL (1992) Nitric oxide-mediated cytostatic activity on *Trypanosoma brucei gambiense* and *Trypanosoma brucei brucei*. Exp Parasitol 75:353-360
45. Gallais P, Badier M (1952) Recherches sur l'encéphalite de la trypanosomiase humaine africaine. Méd Trop 6:633-675
46. Poltera AA, Owor R, Cox JN (1977) Pathological aspects of human African trypanosomiasis (HAT) in Uganda. A post-mortem survey of fourteen cases. Virchows Arch Path Histol 373:249-265
47. Globus MYT, Prado R, Busto P (1995) Ischemia-induced changes in extracellular levels of striatal cyclic GMP: role of nitric oxide. Neuro Report 6:1909-1912
48. Shibata M, Araki N, Hamada J, Sasaki T, Shimazu K, Fukuuchi Y (1996) Brain nitrite production during ischemia and reperfusion: an *in vivo* microdialysis study. Brain Res 734:86-90
49. Bentivoglio M, Grassi-Zucconi G, Olsson T, Kristensson K (1994) *Trypanosoma brucei* and the nervous system. Trends Neurosci 17:325-329
50. Montmayeur A, Buguet A (1994) Time-related changes in the sleep-wake cycle of rats infested with *Trypanosoma brucei brucei*. Neurosci Lett 168:172-174
51. Schultzberg M, Ambatsis M, Samuelsson EB, Kristensson K, Van Meirvenne N (1988) Spread of *Trypanosoma brucei* to the nervous system: early attack on circumventricular organs and sensory ganglia. J Neurosci Res 21:56-61
52. Pennisi E (1997) Tracing molecules that make the brain-body connection. Science 275:930-931
53. Olsson T, Bakhiet M, Edlund C, Höjeberg B, Van der Meide P, Kristensson K (1991) Bidirectional activating signals between *Trypanosoma brucei* and CD8+ T cells: a trypanosome-released factor triggers interferon-γ production that stimulates parasite growth. Eur J Immunol 21:2447-2454
54. Buguet A, Burlet S, Auzelle F, Montmayeur A, Jouvet M, Cespuglio R (1996) Dualité d'action du monoxyde d'azote (NO) dans la trypanosomose africaine expérimentale. C R Acad Sci Paris 319:201-207
55. Cespuglio R, Burlet S, Marinesco S, Robert F, Jouvet M (1996) Détection voltamétrique du NO cérébral chez le rat. Variation du signal à travers le cycle veille-sommeil. C R Acad Sci Paris 319:191-200
56. Canini F, Bourdon L, Cespuglio R, Buguet A (1997) Voltammetric assessment of brain nitric oxide during heatstroke in rats. Neurosci Lett 231:67-70
57. Moncada S, Higgs A (1993) The L-arginine-nitric oxide pathway. N Engl J Med 329:2002-2011
58. Wong EHF, Kemp JA, Priestley T, Knight AR, Woodruff GN, Iversen LL (1986) The anticonvulsivant MK 801 is a potent N-methyl-D-aspartate antagonist. Proc Natl Acad Sci USA 83:7104-7108
59. Alagarsamy S, Johnson KM (1995) Voltage-dependent calcium channel involvement in NMDA-induced activation of NOS. Neuro Report 6:2250-2254
60. Manzoni O, Prezeau L, Marin P, Deshager S, Bockaert J, Fagni L (1992) Nitric oxide-induced blockade of NMDA receptors. Neuron 8:653-662
61. Chao CC, Hu S, Ehrlich L, Peterson PK (1995) Interleukin-1 and tumor necrosis factor-a synergistically mediated neurotoxicity: involvement of nitric oxide and of N-methyl-D-aspartate receptors. Brain Behav Immun 9:355-365

CHAPTER 12

Electroencephalographic features and evoked potentials in human African trypanosomiasis

F Tabaraud, P Tapie

Introduction

Human African trypanosomiasis (HAT) is an endemic disease which plagues the area between the latitudes 14° North and 29° South. This geographic zone is poorly accessible, undermedicalized, politically unstable, and weak economically; this may explain the endemic persistence and the scarcity of recent electrophysiological studies on such a serious disease.

The illness consists of a meningoencephalitic phase occurring after a phase of trypanosome inoculation and a hemolymphatic phase. The neurological phase is characterized by the combination of daytime somnolence together with poor nocturnal sleep [1], abnormal tone, movement abnormalities, sensory disturbance, neuroendocrine problems (for involvement of the diencephalon), and psychiatric symptoms [2]. These multiple and polymorphic neurological manifestations are the consequence of a diffuse meningoencephalitis predominating in the mesodiencephalic structures and periventricular region; from this, there results cerebral cortical, subcortical, and brainstem dysfunction which can be documented by neurophysiological testing.

The history of this disease is punctuated by three periods of development of electrophysiological techniques.

History

From 1900 to 1950: birth of clinical electrophysiology

In 1903, Annibal Bettencourt wrote in a report presented in Lisbon to the Portuguese Minister of the Seas and Colonies: "In some patients, we have examined the muscles from the perspective of their electrical reactions with the aid of faradic and galvanic excitation. Of the first application… faradic excitation of the nerve trunk of the extremities was shown to be within normal limits" [3].

Martin et al. [3], in 1909, confirmed that the electrical reactions of muscles are little altered in HAT. These little known works are clear antecedents of the next more intensive period of study.

It is during this period that Manson Bahr, in 1942 [1], gave a disease description which remains classic, "somnolent by day and restless by night".

From 1950 to 1988: blossoming of clinical electroencephalography

Doctors, despite difficult field conditions, were able to conduct electroencephalographic exams on patients with trypanosomiasis to obtain diagnostic information, estimate the cerebral repercussions of the illness and their seriousness, find complications and monitor the course or the resistance to treatment, and finally, predict relapses.

Waking and daytime napping electroencephalograms (EEG) have been used to learn about disturbances of vigilance. Even some nocturnal EEGs have been performed to study sleep architecture [4-11]. Waking EEGs were found to be non specific but there is a parallel between the gravity of the clinical signs of the patient and the electroencephalographic alterations; clinical amelioration, with treatment, is associated with an amelioration of the EEG; conversely, relapse produces a renewed deterioration in the tracing. Alterations of the waking EEG consist of (in order of increasing gravity):

– slowing of baseline rhythm reactive alpha activity from the onset of the meningoencephalitic phase which only disappears late in the course of this phase;

– non lateralizing and intermittent theta bursts (4 to 8 Hz): later in disease course, intermittent monomorphic delta runs (1 to 3 Hz) with bifrontal projection that, in the absence of effective treatment, will invade the entire tracing;

– then, disappearance of physiological rhythms which are replaced by diffuse very slow activity marking a major impairment. This slow delta activity can sometimes take on a pseudo-periodic aspect considered by Giordano et al. [12] to be very suggestive of the disease in the endemic zone. This "pseudo-periodic" activity in HAT diminishes during wakefulness and during paradoxical sleep in contrast to the periodic activity of the other leukoencephalitic diseases which remain constant.

Sleep architecture, studied by sleep EEG [11,13], reveals that sleep spindles and "K" complexes diminish with disease severity. Paradoxical sleep tracings are normal, but respiratory irregularities in this stage are more significant.

Dynamic sleep organization assessed according to Rechtschaffen and Kales criteria [14] is made difficult by the presence of slow wave runs with bifrontal projection which increase with depth of slow wave sleep.

During nocturnal and nap tracings, occurrence of slow wave sleep has been described with rapid ocular movements without cataplexy or hypotonia but accompanied by hallucinations.

From 1988 to the present

This study period is still underway; it began in 1989 with the work of Buguet et al. [15] "24 hour polysomnographic evaluation in a patient with sleeping sickness", which will be described in detail in the following section. Increased sophistication is now possible with use of computers; both the EEG and the evoked potentials studies which follow were conducted using recent computerized electrophysiological materials [15-19].

Two studies are detailed here to illustrate the extent and limits of our current understanding of HAT from an electrophysiological perspective. The first consists of electroencephalographic study of patients with HAT in the early stages of the meningoencephalitic phase. These patients and controls underwent extended recordings correlated with recording of other physiological parameters. The second study will then describe evoked potentials done at this same early stage of meningoencephalitic involvement.

Most recent EEG study in HAT

Patients

All patients were examined at Daola, on the Ivory Coast, where the Clinical Research Center on Trypanosomiasis (Dr. Doua) and regional referral hospital of Brazzaville in the Congo (Dr. Mouanga) are situated. All enrolled patients presented clinical signs of trypanosomiasis in the meningoencephalitic phase. The diagnosis was confirmed by specific immunofluorescent tests and the finding of trypanosomes in the blood and in the CSF; other acute or chronic encephalitic conditions and acquired immunodeficiency disease (AIDS) were excluded.

Methods

Paper electroencephalographic recordings were made prior to any specific treatment for a 24 hour period during which patients were kept supine in a temperature-controlled room (24-25° C). The polysomnogram consisted of 8 or 10 channels: one for electrooculogram (EOG), one for electromyogram (EMG) of the chin muscles, and another for registering both anterior tibial muscles in order to track the movements of the lower extremities. Chest movements were collected by a strain gauge and nasal and buccal airflow with Cu-Ct thermocouples. All electrodes were cupular and collodium-fixed. Paper speed was 15 mm/s and Epochs were 20 s.

After disease treatment, outpatients recordings were performed on an Oxford Medilog 9000-II 8-channel recorder reproducing the conditions for the paper tracing except the capture of respiratory and lower extremity movements. These ambulatory recordings were then transferred in entirety to paper and scored by 20 s Epochs.

Polysomnographic recordings of volunteer African subjects in good health (control subjects) were performed in the same manner.

Interpretation criteria

Tracings for all patients and controls were interpreted according to the classical criteria of vigilance states [14]: NREM or sleep stages 1-4 and REM, by consensus of three specialists in sleep tracing reading. Reading technique was adapted according to the criteria of Schwartz and Escande [13], when scoring of patient tracings became impossible because of slow wave invasion of the tracing.

Various parameters were studied:
- during wakefulness:
  - alpha rhythm and its reactivity to opening and closing of the eyes.
- during sleep:
  - spindles or sleep spikes, defined as "a rhythmic group of waves characterized by a progressively increasing and then diminishing amplitude" having a frequency of between 12 and 14 Hz and a duration from 0.5 to 1s. [20];
  - K complexes [21], which correspond to a "slow biphasic wave of large amplitude often accompanied by a sleep spike; they occur during sleep spontaneously or in response to sensorial stimulation; they are not specific to any individual sensory modality" [20];
  - "transient activation phase" (TAP), defined by Schieber et al. [22] as an acceleration of the cardiac rhythm accompanied by a peripheral vasoconstriction and appearance of bursts of muscle action potentials and segmental or complex body movements.
  - "cyclic alternating pattern" (CAP) [23] occurring in slow wave sleep: sequence of two recurrent stereotypic patterns; each of these two sequences having a duration of between 2 and 40 s.

"Slow wakings" are long runs generally of hypersynchronous slow waves of very high voltage preceding a rousing episode, and are also known as "paroxysmal hypnopompic hypersynchrony" [24].

Results

The disorganization of the circadian rhythm of wakefulness and sleep represents the characteristic element of these studies (reported Chapt 11).

Morphological study of EEGs in 18 patients with trypanosomiasis in the early stage of meningoencephalitis revealed that, during wakefulness, the results of prior studies were confirmed. Alpha rhythm is present and reactive in all the patients at onset of meningoencephalitic phase, its frequency vanging from 9 to 11 Hz. Patients display slow wave anomalies with diffuse theta activity and delta bursts, bilateral bursts of slow waves whose duration is inferior to 3 s, predominating in the frontal regions (Fig. 1). These features, occurring on tracings where the baseline rhythms are conserved, should lead to consideration of HAT in an endemic zone. These alterations increase with increasing severity of the clinical signs.

During sleep, bifrontal monomorphic slow waves are found in patients during sleep stages 1-4 and REM; they are most frequent during stages 3 and 4. This frontal intermittent rhythmic delta activity (FIRDA) is not specific to HAT as it has also been reported in organic lesions involving the diencephalon: the thalamus and the superior portion of the brainstem [25]. This tracing feature has been described in diseases as varied as viral encephalitis, metabolic encephalopathy, and deep tumors of the cerebral hemispheres. In the context of established trypanosomiasis, their presence is a sign of brain involvement and of disease severity.

K complexes are abundant and can occur in isolated fashion, but most often occur in salvos of 3 to 6 perhaps reflecting the abundant sensorial stimuli caused by the sensory disturbance which is a constant feature of this disease. Sleep spindles are easily identified in all patients except those with severe involvement; this reduction in sleep spindles in severely-affected patients has previously been described by Bert et al. [11] in cases of severe trypanosomiasis. Besides this sign is not specific and has been described in many chronic neurological conditions such as Parkinson's disease, fatal familial insomnia, neurological manifestations of AIDS... to name just a few.

Patterns of change in arousal are seen, sometimes associated with behavioral or physiological awakening: attenuation phases, beta runs, transient activation phases (TAP). These patterns occur spontaneously or after one or several K complexes; they either lighten sleep stage or

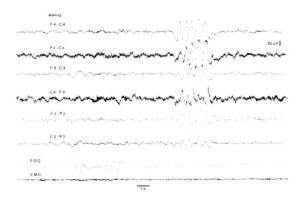

Fig. 1. Delta bursts (2-3 Hz) with frontal predominance

produce a complete wakening (Fig. 2). They do not have a periodicity over a 24 hour period.

TAPs and awakenings have been enumerated during the 24 hours of recording in both patients and control subjects. The number of TAPs is significantly elevated in patients over controls in sleep stages 1 and 2; during REM and stages 3 and 4, there is no significant difference. The importance of sensorial symptomatology in the patient with trypanosomiasis may explain here as well, the increase in TAP during phases 1 and 2 of sleep.

"Slow wakings" (Fig. 3) were noted in some patients, uniquely in sleep stages 3 and 4; they were not seen in the controls.

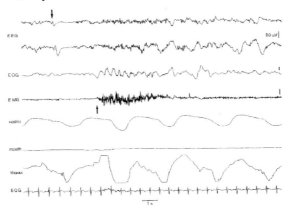

Fig. 2. Paroxysmal hypnopompic hynagogic events with sinusoidal delta activity (1-3 Hz) with a frontal predominance, superimposed with alpha or beta rhythms

Identical anomalies have also been reported in a patient with trypanosomiasis due to *Trypanosoma brucei rhodesiense* [18], in sedated subjects [24] and in subjects with a parasomnic episode (somnambulism, night terrors, and confusional states) [26].

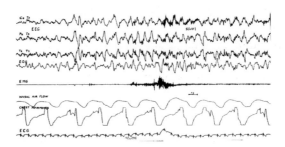

Fig. 3. Transient activation phases (TAP), beginning with a K complex-like pattern, followed successively by an attenuation phase, and slow wave activity superimposed with fast activity, accompanied by a transitory arousal

There is no recent work conducted on later stages of the meningoencephalitic phase; earlier studies have shown that waking and sleep tracings appear to be invaded by slow waves, with progressive disappearance of physiological patterns and of normal reactivity.

Ambulatory polysomnographic follow-up of patients under treatment has shown regression of morphological alterations of the EEG; this finding may be used as a marker of treatment effectiveness.

In conclusion

Sleep and waking EEG results confirm earlier studies but reveal the need to validate study of sleep microstructure with larger numbers of recordings.

The electrophysiological course of this disease and morphological aspects of the EEG at late disease stage or with the therapeutic complication of arsenical encephalopathy still need to be studied.

Evoked potentials

The technique of evoked potentials has been used for many years in neurology to test function of sensorial pathways, and to detect or confirm involvement in the course of many neurological disorders such as multiple sclerosis [27, 28]. Somatosensory evoked potentials (SEP) enable probing of the lemniscal tract; initial auditory evoked potentials (AEP) reflect the auditory pathways in the brainstem, and visual evoked potentials (VEP) the visual tracts from the retina to the occipital cortex. More recently, the technique of motor evoked potential (MEP) has been developed for exploring the corticospinal motor tracts [29].

These techniques currently in use in industrialized countries are, in contrast, rarely used on the African continent, because of the equipment costs necessary for their realization, the scarcity of neurologists and even greater scarcity of neurophysiologists, and finally, because of local conditions. They have, nevertheless, been employed in patients with HAT to study anomalies at the level of the brainstem, the veritable crossroads of studied tracts (auditory, lemniscal, and corticospinal). The technique of VEP may, moreover, permit detection of visual disturbances which have been frequently reported in early observations.

Methods

The study was conducted in 16 hospitalized patients at the PRCT of Daloa on the Ivory Coast [30]. All patients presented an early meningoencephalitic stage of the disease with troubles with sleep and wakefulness in all cases. None of the patients presented with symptomatology of sensory or motor deficiencies. Their ages ranged from 12 to 66 years (mean

32.4 years). None had an associated pathology, except the oldest patient who had diabetes; all had biological stigmata of protein nutritional deficiency.

The evoked potentials were conducted according to the classic methodology for stimulation and data collection [OR logging] described by Mauguière and Fischer [28].

Visual evoked potentials were obtained using a pattern reversal stimulus; the studied parameter was the latency of P100. Auditory evoked potentials were collected after mono aural stimulation by an alternating click; latencies of initial responses at the level of the brainstem were measured (waves I to V).

Somatosensory evoked potentials were studied by stimulation of the median nerve at the wrist and the posterior tibial nerve at the ankle. The latency of the parietal response N20 (upper extremitiy) and P39 (lower extremity) were measured.

Stimulation of the corticospinal (motor) tract was performed with the aid of an electric Digitimer stimulator. Electrical stimulation applied at the level of the scalp and neck enabled a peripheral muscular response (motor evoked potential) to be obtained. The potential was collected at the skin surface at the level of the abducteur brevis muscle of the thumb. The latency of the response obtained after cortical stimulation defined to total conduction time (TCT), that obtained after cervical spinal stimulation the peripheral conduction time (PCT), the difference (TCT-PCT) representing the central conduction time (CCT) (rapid conduction motor fibers of the corticospinal tract).

Data from these different parameters were collected and compared to reference values, and considered abnormal if greater than 2.5 standard deviations from the mean value.

Results

AEP: in all studied patients, bilateral auditory evoked potentials were obtained and latencies and interlatencies I-III, III-V, and I-V were normal in all cases.

VEP: P100 latency was normal bilaterally in all patients studied.

SEP: 2 out of 16 patients (12.5%) had a delayed N20 latency bilaterally, and 5 out of 16 patients (31%) had a significantly delayed P39 latency with bilateral involvement.

MEP: 3 out of 16 patients (18.7%) had a delayed CCT at the level of the upper extremities.

Discussion

There do not appear to be any other published study on the evoked potentials in the disease course of HAT. This study was conducted on patients presenting with early stage neurological disease, which likely explains the results showing relative integrity of the studied tracts explored at this stage.

The lack of abnormality in VEP expresses the integrity of visual pathways. This result is in apparent contradiction with early observations which reported frequent visual disturbance correlated with glial lesions and demyelination along the optic tracts [31]. However, these abnormalities were reported in patients presenting late disease stage with diffuse leucoencephalitic lesions.

The normality of the auditory responses in the brainstem supports, at this stage of the disease, the absence of abnormality in the auditory pathways from the auditory nerve (wave I) up to the level of the inferior colliculus (wave V) which is situated in the cerebral peduncular region. AEP normality in this study corroborates the previously obtained experimental results in a sheep with trypanosomiasis [32].

The N20 response in SEP of the upper extremities is delayed in 2 cases: for one of the patients, this prolongation appears to be connected with arm edema related to IV infiltration; the other patient had diabetes and may have had a diabetic neuropathy (which was not sought out) as the source of the delayed response. The P39 response was delayed in 5 patients who had no clinical particularity compared to the other patients, and in particular did not show clinical signs of lemniscal tract involvement (e.g. had normal vibratory sensation). However, the average age of these 5 patients was 44.2 years compared with only 27 years for the 11 others. One of the 5 had diabetes which, as for the N20 response, was at the source of the delayed somatosensory response from the lower extremities; for the other 4 patients, the delayed somatosensory response could be related to the trypanosomiasis by means of spinal cord or root involvement described in certain observations [31], but at a later disease stage. It should be noted however, that this result is entirely nonspecific and one could not eliminate other potential etiologies such as a folate or B12 nutritional deficiency affecting the posterior spinal columns.

With regard to MEP, only 3 patients had a delayed central conduction time. None of these patients showed signs of pyramidal irritation. As for the somatosensory evoked potentials, the interpretation of this result is difficult. It may be directly related to the trypanosomiasis by means of demyelinating lesions of the corticospinal tract, but may also be related to an associated vascular or lacunar pathology, particularly in the case of the most aged subject who was diabetic. For the 2 other patients, aged 23 and 35 years, there did not appear to be a known associated pathology, and the abnormality of MEP may be related to trypanosomiasis without knowing the precise nature of the lesion nor the site of involvement, cerebrum, brainstem or spinal cord. Pyramidal tract lesions, in fact, have been described in late-stage trypanosomiasis at the level of the pons and anterior horn cells of the spinal cord [31].

In total, the majority of studied patients had normal evoked potentials; these results support a relative sparing of the examined pathways, particularly at the level of the brainstem, at early disease stage. The abnormalities of the sleep-wake cycle caused by the trypanosomiasis at this stage may thus be more related to a nondestructive process involving the sleep centers than from destructive lesions which would be expected to also affect the adjacent sleep centers.

In contrast, it is probable that at a later disease stage, diffuse immune-mediated demyelinating lesions account for significant clinical abnormalities. There are not yet any published studies of evoked potentials at this disease stage, but it is likely that in view of the pathological findings significant abnormality of evoked potentials might be expected; however, it would be particularly difficult to conduct research in view of the abnormal movements, behavioral disturbances, and clinical deterioration at this stage of diffuse leucoencephalitis.

References

1. Manson-Bahr P (1942) Manson's tropical diseases. A manual of diseases of warm climates. Cassel, London, p 1033
2. Dumas M, Boa FY (1988) Human African trypanosomiasis. In: Vinken PJ, Bruyin GW, Klawans HL (eds) Handbook of Clinical Neurology, revised Series. Elsevier Science Publishers, Amsterdam, pp 339-344
3. Martin L, Guillain G, Darré H (1909) Formes nerveuses de la maladie. In: Martin G, Leboeuf A, Roubaud E (eds) Rapport de la mission d'études de la maladie du sommeil au Congo français, 1906-1908. Masson & Cie, Paris, pp 317-339
4. Gallais P, Gastaut H, Gardaire G, Planques L, Pruvost A, Miletto G (1951) Étude électroencéphalographique de la trypanosomiase humaine africaine. Rev Neurol 85: 95-104
5. Gallais P, Badier M (1952) Recherche sur l'encéphalite de la trypanosomiase humaine africaine. Corrélations cliniques, anatomiques, électroencéphalographiques et biolo-giques. Méd Trop 6: 633-673
6. Gallais P, Cros R, Pruvost A, Planques L, Gardaire G, Miletto G, Levy-Cavalleri G, Bert J, Fons R (1953) Étude clinique, biologique, électroencéphalographique, parasitologique de la trypanosomiase d'inoculation. Méd Trop 13: 807-843
7. Radermecker J (1955) Corrélations électrocliniques dans la trypanosomiase humaine africaine et dans la trypanosomiase employée comme thérapeutique des affections psychiatriques graves. Acta Neurol Belg 55: 179-218
8. Radermecker J (1956) Leucoencéphalite à parasites connus: la trypanosomiase systématique et électroencéphalographie des encéphalites et encéphalopathies. Electroencephalogr Clin Neurophysiol 8: 117-124
9. Collomb H, Demarchi J, Miletto G (1958) Trypanosomiase humaine africaine. In: Traité de médecine. Masson, Paris, pp 268-350
10. Collomb H, Bert J, Zwingelstein J (1963) Interêt de l'examen électroencéphalographique dans le diagnostic de la trypanosomiase humaine africaine. In: Proceedings of the First International Symposium on Tropical Neurology. Buenos Aires, pp 216-226
11. Bert J, Collomb H, Fressy J, Gastaud H (1965) Étude électroencéphalographique du sommeil nocturne au cours de trypanosomiase humaine africaine. In: Fischgold H (ed) Le sommeil de nuit normal et pathologique. Études électroencéphalographiques. Masson, Paris, pp 334-352
12. Giordano C, Clerc M, Doutriaux C, Doucet J, Nozais JP, Bureau JP, Piquemal N (1977) Le diagnostic neurologique au cours de différentes phases de la trypanosomiase humaine africaine. Ann Soc Belge Méd Trop 57: 213-225
13. Schwartz BA, Escande C (1970) Sleeping sickness: sleep study of a case. Electroencephalogr Clin Neurophysiol 29: 83-87

14. Rechtschaffen A, Kales A (1968) A manual of standardized terminology, techniques and scanning system for sleep stages of human subjects. Public Health Service, US Government Printing Office, Washington DC, USA
15. Buguet A, Gati R, Sevre JP, Develoux M, Bogui P, Lonsdorfer J (1989) 24 hour polysomnographic evaluation in a patient with sleeping sickness. Electroencephalogr Clin Neurophysiol 72:471-478
16. Buguet A, Bert J, Tapie P, Tabaraud F, Lonsdorfer J, Bogui P, Dumas M (1993) Sleep-wake cycle in human African trypanosomiasis. J Clin Neurophysiol 10:190-196
17. Gati R, Tabaraud F, Buguet A, Bert J, Tapie P, Bittel J, Sparkes B, Breton JC, Doua F, Bogui P, Lonsdorfer A, Lonsdorfer J, Moulin J, Chameaud J, Dumas M (1990) Analyse circadienne du sommeil, de la température rectale et de variables des immunologiques et endocrinologiques dans la maladie du sommeil. Bull Soc Path Ex 83:275-282
18. Montmayeur A, Brosset C, Imbert P, Buguet A (1994) Cycle veille sommeil au décours d'une trypanosomose humaine africaine à *Trypanosoma brucei rhodesiense* chez deux parachutistes français. Deuxième congrès de neurologie tropicale, Limoges, France, 21-23 septembre 1994
19. Tapie P, Buguet A, Tabaraud F, Bogui P, Doua F, Bert J (1996) Electroencephalographic and polygraphic features of 24 hour recordings in sleeping sickness and healthy African subjects. J Clin Neurophysiol 13:339-344
20. Chatrian JE, Bergamini L, Doudey M, Klass D, Lennox-Buchtal M, Peterson M (1974) A glossing of terms most commonly used by clinical electroencephalographers. Electroencephalogr Clin Neurophysiol 37:538-548
21. Roth M, Shaw J, Green J (1956) The form, voltage, distribution and physiological significance of the K complex. Electroencéphalogr Clin Neurophysiol 8:365-385
22. Schieber JP, Muzet A, Ferrière PJR (1971) Les phases d'activation transitoires spontanées au cours du sommeil normal chez l'homme. Arch Sci Physiol 25:443-465
23. Terzano MC, Mancia D, Saleti MR (1985) The cycling alternating pattern as a normal physiologic component of normal MREM. Sleep 8:137-145
24. Erwin CW, Somerville ER, Radtke RA (1984) A review of electroencephalographic features of normal sleep. J Clin Neurophysiol 1:253-274
25. Gastaut H (1980) L'électroencéphalographie clinique en neurologie. Encycl Méd Chir, Paris, Neurologie, 17031:A10-A25
26. Broughton RJ (1968) Sleep disorders: disorders of arousal. Science 159:1070-1078
27. Stockard JJ (1984) Clinically useful applications of evoked potentials in adults. J Clin Neurophysiol 1:159-202
28. Mauguière F, Fischer C (1997) Potentiels évoqués en Neurologie. Encycl Med Chir Neurologie 17031-B-10, Elsevier, Paris, p 40
29. Hugon J, Lubeau M, Tabaraud F, Chazot F, Vallat JM, Dumas M (1988) Potentiels évoqués moteurs. Technique, résultats chez le sujet normal. Rev Neurol 144:91-95
30. Tabaraud F, Hugon J, Tapie P, Buguet A, Lonsdorfer A, Gati R, Doua F, Dumas M (1992) Study of evoked potentials in human African trypanosomiasis. Am J Trop Med Hyg 95:246-252
31. Van Bogaert L, Janssen P (1957) Contribution à l'étude de la neurologie et neuropathologie de la trypanosomiase humaine. Ann Soc Belge Méd Trop 37:379-427
32. Tapie P, Sonan T, Dumas M, Nicolas JA, Tuillas M, Denichoux L, Dubost G, Bouteille B, Ragot J (1988) Potentiels évoqués auditifs précoces chez le mouton sain et le mouton trypanosomé. Bull Soc Path Ex 81:484-489

CHAPTER 13

Clinical aspects of human African trypanosomiasis

M Dumas, S Bisser

Introduction

The host's invasion by a trypanosome produces first a local lesion and then a systemic illness. At the site of inoculation of the trypanosome by the tsetse fly an entry chancre develops, a lesion which constitutes a first cutaneous line of defense that is quickly overcome; rapidly, the multiplication of trypanosomes in this lesion provides an entry point for host invasion via a lymphatic line of defense which in turn is quickly surmounted. The invasion continues its propagation with spread into the bloodstream and invasion of different organ systems, some of which are particularly susceptible such as the heart and the central nervous system (CNS). The precise timing of earliest penetration of the blood-brain barrier (BBB) is still not known but is likely quite early in the course of infection, as use of treatments hematogenously active but which are known not to penetrate the BBB do not cure the patient. This results from lymphatic and hematogenous re-infection by trypanosomes which had been shielded in the CNS; thus, the discovery of clinical and biological clues which herald this early involvement of the CNS assumes major importance in determining the choice of therapy.

Clinical and pathogenic features

The course of illness progresses more or less rapidly depending on the infecting trypanosome; neurological involvement is very rapid in the trypanosomiasis due to *Trypanosoma brucei rhodesiense*, and of slower onset with that due to *T. b. gambiense*. Progressively, and more or less rapidly depending on host defenses which vary from one individual to another and according to the pathogenicity of the parasite, three groups of signs will occur, marking the three major clinical manifestations of the disease: general, cardiac, and most notably neuropsychiatric.

Initial lesion

The initial cutaneous lesion may be practically indistinguishable, particularly in dark-skinned individuals; contrastingly, it can provoke a warm, tense, painful, and pruritic nodule which may be confused with a furuncle, but unlike a furuncle does not progress to suppuration and typically disappears in 1-2 weeks. Identified early on, this lesion permits very early confirmation of the diagnosis by simple cutaneous scarification and examination of this serous fluid by wet mount and/or after fixation and staining; it teems with easily identifiable trypanosomes.

In man, it appears that only *T. b. gambiense* and *T. b rhodesiense* are capable of growing and multiplying. The other trypanosomes responsible for animal trypanosomiasis do not appear to have this capacity (however this fact should not be assumed as it may not be generalizeable to all trypanosome species and still requires further study). Notably, in an endemic zone it is classic to discover, among subjects in apparent good health, specific positive serologies (CATT and IFAT); these may persist for a long time or progressively disappear without development of any clinical disease manifestation. After ruling out false positive reactions, the meaning of positive serologies in this setting is not clear. They may signal host tolerance of a trypanosome strain infectious to man, but having attenuated pathogenicity, or of a trypanosome species uniquely infectious to animals (livestock, domestic or wild animals) which also belong to the group *brucei*. These animal trypanosomes are normally lysed in human serum by lytic factors termed "trypanosome lytic factors" by Mary Rifkin who described them in 1978 [1]. These trypanalytic factors correspond to a haptogloblin [2] which, after linking with a hemoglobin molecule, is internalized by the parasite. There, it first produces a liberation of H_2O_2 which cannot be inactivated by the trypanosome that lacks the necessary catalase and leads to a peroxidation of the lipid membranes of the parasite; then it activates macrophages to release proteolytic enzymes which together with peroxidation destroy the trypanosome. It is conceivable that repeated inoculation of animal trypanosomes, which are inoffensive to man via these host defenses, may be at the origin of a biological response with seropositivity in the absence of clinical disease. Recognition of these facts deserves great caution in the interpretation of positive serologies which may lead to the false assumption of a latent trypanosomiasis and to an unnecessary administration of melarsoprol, a poorly tolerated and at times lethal therapy.

Hemolymphatic phase

During the course of development of the local cutaneous lesion, a regional adenopathy develops . These lymph nodes are generally rubbery and mobile, painful with onset and thereafter nontender, and typically persist over several weeks to months. They can constitute a key sign of the disease provided that they are recognized, but as they are rarely bulky they may easily be overlooked. Thus, it is necessary to search for them attentively, particularly in the lateral cervical region (just posterior to the sternocleidomastoid muscle) where their presence constitutes the Winterbottom sign. Gentle kneading followed by slow aspiration of one of these lymph nodes, with subsequent microscope examination of this material may be sufficient to make the diagnosis. With increased disease duration, these lymph nodes become more firm, making the aspiration and detection of the trypanosome more difficult. These simple procedures well described and codified by authors in the first half of this century [3] are presently too often neglected in favor of serological tests which unfortunately do not make the diagnosis with certainty; isolation of the trypanosome remains necessary for definitive diagnosis.

The lymph tissues appear to be the principal site of the trypanosomes during this stage of infection such that it is notably easier to isolate the trypanosome from the lymph extract of these hypertrophied nodes than from peripheral blood. Pathological examination of these lymph nodes reveals early central hypertrophy that later undergoes sclerotic change which favors the development of lymphoedema.

As this first lymphatic barrier is soon overcome, there is progressive invasion of the bloodstream which is denoted as the hemolymphatic phase or stage 1 of the infection. For some authors, the phase of hematogenous spread is already stage 2; they consider the initial cutaneous lesion to be the actual first stage.

The trypanosome takes advantage of these phases for rapidly multiplying, a strategy which enables it to quickly spread through all systems of the host, including the central nervous system. It achieves its propagation through a double strategy, first of theft, extracting from the host the nutrients it needs for multiplication and movement, then of defense, eluding the human system of counterattack. These two strategies are highly developed. Accompanying this progressive invasion of the host, generalized symptoms are developed– an immunological testament of the host's fight against this shrewd invader – and also more specific cardiac and neurological symptoms which signal attack on these susceptible systems.

Through a strategy of theft of energy supplies, the trypanosome assures its multiplication by longitudinal fission and its movement which are both necessary for its spread throughout the host. To reach these objectives, each trypanosome needs to metabolize in one hour a quantity of glucose equivalent to its own dry weight; this reaches enormous proportions if one considers that a trypanosome can double by fission in around six hours [4]. This sizeable energy consumption is all the more harmful for patients who are often in a precarious state of malnutrition and wasting and have a heavy parasite load. This clinical aspect must not be overlooked at the time of instituting a treatment which is characteristically poorly tolerated. It is preferable to reestablish a nutritional homeostasis before proceeding; in most cases, the initiation of treatment can safely wait several days.

The trypanosome's essential requirement for glucose has led to therapeutic research efforts to inhibit trypanosomal glycolysis [5-7]. This research path appears as the more interesting since glycolysis represents the only route of energy production for the trypanosome which, lacking a mitochondrial cytochrome, does not have a functional Krebs cycle. The antitubercular agent, salicylhydroxamic acid (SHAM), has been shown to be active in arresting the aerobic route of glucose degradation by the trypanosome; however, the arrest of the anaerobic route by addition of glycerol which impedes the trypanosome's glycerolkinase is unfortunately of limited clinical use because glycerol cannot cross the BBB. In addition, the quantities of glycerol required are enormous, beyond what is feasible in clinical practice.

The strategy of defense developed by the trypanosome is itself two-fold; it operates both through a developed strategy of immune escape which enables it to escape the immune recognition by the host, and the neuromediators such recognition would unleash, as well as through a strategy of evasion in finding refuge in protected sites of the host (in particular, the CNS).

The strategy of immune escape [8, 9] resides in the ability of the trypanosome to cloak itself in a thick coat of glycoproteins; these glycoproteins have the highly distinctive feature of being continually replaced as they are recognized by the host's antibodies. The trypanosome is thus always a step ahead and able to escape the host's immune response. The replacement of one variant surface glycoprotein (VSG) by another which is antigenically different is achieved through continual shedding of existing VSGs and recycling their components into antigenically new forms. This property, designated as "antigenic variation", represents a key feature which prevents most hosts coming in contact with a trypanosome from acquiring any truly protective immunity, this problem limiting the creation of a vaccine [10]. For this trait, the parasite dedicates 5-10% of its genetic patrimony for enough coding genes to assure antigenic variation with a large enough repertoire of antigens to permit its proliferation with successive waves of surface changes. The genetic code for VSG, synthesis of its different constituents and their assembly have not yet been fully elucidated. The detailed knowledge of these mechanisms may possibly provide sites for therapeutic targeting. VSGs are anchored in the trypanosome membrane via myristic acids which, if replaced by oxygenated analogs (serving as decoy molecules), might prevent the assembly of these glycoproteins.

Because of such an antigenic turnover, the host remains in a constant state of immune stimulation; there is polyclonal activation of B lymphocytes which become antibody-producing plasmocytes. As a result, a hypergammaglobulinemia occurs which may reach very high levels and be suggestive for trypanosome infection in endemic zones. It is associated with an elevation of circulating immune complexes, immune deposition, hypocomplementemia [11], and effective immunosuppression which predisposes to development of other bacterial and parasitic diseases that in turn aggravate the host response; they cause T lymphocyte activation and cytokine release that synergistically fan immune derangement.

This widespread immune derangement, which marks the host's fight to eliminate the parasite, is the source of the nonspecific clinical signs representing the disease's general manifestations and may itself be suggestive for the disease in endemic zones.

The temperature curve is difficult to schematize but characteristically remains abnormal throughout the months to years of the disease course. The first febrile spikes are typically elevated; they mark the host's invasion by the parasite and the onset of a defense reaction. At this early stage of disease, fever may be one of the most evident disease signs; it is remittent, very irregular, varying over a 24 hour period or in cycles of several days. It is accompanied by systemic signs which are characteristically severe at disease onset with tachypnea, a tachycardia which may exceed 120 bpm (beats per minute) even during afebrile periods, as well as headaches, exhaustion, anorexia, and extreme thirst.

Adenopathy, accompanying immune stimulation as previously highlighted, is of considerable interest for suspecting the diagnosis in a patient from an endemic region when accompanied by fever. Although most commonly found in the subclavicular region, adenopathy may also be found in sub-maxillary, axillary and inguinal region; their consistency is first soft and then becomes firm; their size characteristically is that of a bean or large pea, but may occasionally reach that of a hazelnut. Over time, they tend to shrink and sclerose, rendering their diagnostic puncture more difficult.

Splenomegaly occurs in approximately one quarter of cases. As for the lymph nodes, splenomegaly is typically of fibrous-firm consistency and persists throughout the disease course, becoming even more pronounced at the last stages of disease.

Cutaneous changes are frequent but sometimes hard to distinguish on dark-skinned individuals; they consist of circinate papuloerythematous eruptions visible on the trunk or proximal aspect of the extremities. While often separated by intervals of normal skin, they can at times coalesce into large plaques. The eruptions usually last 10-15 days before spontaneously resolving.

Pruritus, a characteristic disease symptom, very rarely exists at this stage and then tends to be limited in extent; these features help to distinguish the erythematous lesions described above from urticaria with which they can be confused. Later, the pruritus generalizes, and will occur beyond sites of cutaneous erythema.

Edema of the face and extremities are quite common early in the disease course. Sub-orbital edema, for authors at the beginning of this century, was considered to be a key diagnostic feature, present in 75% of patients. This diagnostically suggestive type of edema may still be present after diagnosis but lymph node puncture is no longer possible.

In parallel with these general symptoms which mark the progressive and diffuse invasion of the host by the trypanosome, two target organs are particularly affected by the disease process: the central nervous system and the heart.

The cardiovascular manifestations represent one of the major signs of the American trypanosomiasis due to *T. cruzi*; cardiac involvement is equally present but less severe in human African trypanosomiasis. Beyond the evident tachycardia, clinical manifestations need to be carefully searched for; they may appear as a soft cardiac murmur, a reduced pulse pressure, or at times by more serious signs of myocarditis or pericarditis. Some instances of the edema previously described may also be of cardiac origin. Current methods of cardiovascular investigation (electrocardiogram, Holter monitoring, echography) enable better detection of the cardiac failure which marks the mononuclear infiltration of cardiac muscle. These infiltrates, first perivascular then becoming interstititial, are likely the result of pathological immune response rather than of direct parasitic invasion; in fact, the presence of parasite has never been found in anatomical specimens [12, 13]. Myocarditis of variable severity is the most frequent cardiac lesion to develop with advancing disease. The endocardium is usually little affected by the disease process, but in severe forms, hence advanced stages of the disease, there is a true pancarditis from abundant perivascular and endovascular infiltrates causing focal myocardial necrosis.

The trypanosome's strategy of evasion is achieved by its penetration into the CNS where the trypanosome can take refuge in an inaccessible site, at least for a certain period of time, from the usual defense reactions of the host. This site constitutes a reservoir of parasites sheltered from host defenses, and from suramin and pentamidine, therapies used in the initial stages of infection which unfortunately do not adequately penetrate the BBB.

The trypanosome's means of broaching this barrier is detailed in another chapters 6, 7 and 8 of this manual and hence will only be quickly summarized here. In brief, the infection triggers host release of opposing cytokines (TNF-α, IFN-γ, interleukins) and nitric oxide (NO). These compounds modify the BBB by acting on capillary endothelial cells; they cause pericyte alteration, permitting the penetration of CD8 T lymphocytes and of trypanosomes, and so producing a perivascular proliferation of microglia, astrocytes, lymphocytes and plasmocytes. Among the different cytokines, TNF-α plays a particularly important role, its elevation rising further with disease progression [14]. Its trypanostatic effect is both a direct effect on the trypanosome membrane, and indirect by favoring the production of NO. The role of NO in human African trypanosomiasis is still poorly understood but appears particularly important as an inhibitory mechanism of the trypanosome growth [15]. An ubiquitous mediator NO is normally secreted by endothelial cells and neurons via NO synthases (NOS) which are upregulated in response to many pathological conditions, evidently including trypanosomiasis. Once induced, the activity of the NOS may be further modified by various cytokines [16].

Meningoencephalitic phase

Progressively, the disease evolves toward a stage of neurological attack, with CNS involvement constituting a principal disease sign [52, 53]. This phase occurs in a highly variable time frame depending on the trypanosome type and its pathogenic capacity.

The modification of the BBB, at the source of CNS penetration by *T. b. gambiense* or *T. b. rhodesiense*, occurs first in the most fragile and least protected zones which consist first of the choroid plexus, and then thalamus, postrema, median eminence, pineal and hyphophyseal regions [17]. This selective CNS localization explains in part the principal clinical neurological signs encountered in this disease: difficulty with sleep from involvement of the supraoptic nuclei (which serve to support the mechanisms controlling sleep onset); extrapyramidal signs from involvement of the striatum; deep sensory disturbance with hyperpathia from involvement of the thalamus and related structures; endocrine dysfunction because of hypophyseal-subthalamic involvement.

In the choroid plexus, the exact localization of the trypanosome has been the object of some controversy [18]; it is agreed that it can be found extracellularly in the stroma, but some authors have also described intracellular amastigote forms as in the American trypanosomiasis [19-21]. This intracellular location, if confirmed, may be an important CNS cache leading to relapse and host re-infection.

With disease progression, the trypanosome may be found in the locations previously described, but also in the Virchow-Robin spaces as noted with all experimental models as well as in man. Lesions may occur in the meninges and the cerebral cortex, but also in the subcortical white matter (principally in the regions around the third ventricle), the cerebellum and even the medulla. There is significant perivascular infiltration with involvement of the endothelial cells of the leptomeninges and proliferation of ependymal cells. These perivascular infiltrates are mainly comprised of lymphocytes and plasmocytes, which swell with immunoglobulins taking on the appearance of Mott cells, with rare polynuclear cells. The infiltrates occur in the form of thick mantles particularly around the fine and medium-caliber vessels of the venous system, infiltrating the entire Virchow-Robin space, but also infiltrating open nervous parenchyma. The cortical meningeal lesions are diffuse, extending along the leptomeninges of the entire cortex.

Progressively with disease course, there is significant destruction of myelin in the white matter, particularly in the centrum ovale and the gyri. These demyelinating lesions can appear as severe as the perivascular and inflammatory reactions, which are themselves quite exuberant; however, demyelination is never widespread, massive or confluent. The lesions have a nonspecific appearance, appear to spare the inter-hemispheric association fibers (corpus callosum and commissures), predominating instead in the striatothalamic, putaminocaudal, and pallidonigral white fibers. This myelin destruction appears to always be the result of an inflammatory process; it is associated with a very pronounced glial reaction with microglial proliferation. Additionally, there is cortical cellular thinning which is generally most pronounced in the 3rd and 6th cortical layers in conjunction with changes in the white matter and in the meninges. Identical lesions can be found in the central grey nuclei in the form of lymph-plasmocytic mantles which disrupt the architecture of the nuclei, and especially their connecting fibers [22, 23].

The localization of these different lesions accounts for the different clinical signs whose intensity worsens with disease duration, paralleling the worsening of the neuropathological lesions [24]. These signs remain reversible, however, for a long time with treatment attesting to the predominance of potentially reversible inflammatory lesions over irreversible destructive lesions.

Thalamic involvement accounts for the sensory disturbances which are one of the disease's notable characteristics. These difficulties are always mentioned by patients: the least shock, a simple bump is enough to provoke an episode of memorable pain. Objective sensory defects do not seem to exist, but it is difficult to prove their absence in patients who may be excessively somnolent. Sensory changes may be a very early symptom, appearing within two weeks of trypanosome inoculation, sometimes within only a few days. Useful diagnostically, it has been termed Kerandel's sign after the physician's name who observed and described the disease's effect on himself with great precision in 1910: "I came to take unusual precaution to not touch the edges of doors or furniture, the sides of my bed while sleeping, and the legs of tables near where I was seated" [25, 26].

In addition to thalamic involvement, there occur lesions in the various regions involved with sleep regulation (the postrema, median eminence, and pineal region). The conjunction of all these lesions rapidly cause derangement of the regions involved with sleep regulation; they are described in greater detail in chapter 11 of this manual. These sleep derangements are such a characteristic and constant feature in the neurological phase that they have given name to this (sleeping sickness) disease. Recent study of 24 hour sleep tracings confirms high concordance with the findings of careful clinical observation. Affected individuals sleep frequently, leading to the mistaken attribution of hypersomnia, when in fact dozing episodes are short and, if combined, are practically identical to the total sleep duration of a normal subject. Sleep cycles become more spread out over a 24 hour period the more advanced the disease. The initial sleep episode is most often paradoxical followed by slow wave sleep, but all combinations may be seen. It is interesting to note that these changes, like the other neurological signs discussed, remain reversible for a long time with treatment, attesting the predominance of (potentially reversible) inflammatory lesions over (irreversible) destructive lesions. Secretion of cortisol, renin and prolactin are altered with the disturbance of circadian rhythm, reflecting altered sleep cycling, but their pulsatility remains tightly linked to specific sleep phases [27-29]. Melatonin secretion remains similar to normal subjects, with nocturnal secretion and daytime inhibition [30].

Involvement of the entire tubero-infundibular axis and subthalamic and hypophysial regions account in turn for the additional endocrine derangements which occur throughout the disease course [31]. Impotence, amenorrhea and infertility are frequent patient complaints, which in women may be accompanied by an atrophy of external genitalia and diminishment of uterine size, and in men by prostatic and testicular atrophy with a fall in testosterone levels. In 50% of men and women, there is a decrease in testosterone and estradiol levels, respectively, though pituitary gonadotropin levels (FSH and LH) remain normal; the results of gonad adenohypophyseal tests suggest that the dysfunction may be of supra- or extra-hypophyseal origin [32]. Excessive chilliness, which oblige patients to seek out the warmth of sunlight, and facial myxedema mark thyroid insufficiency with reduced free T3 and total T3 levels. Sensations of hunger and thirst are disturbed – often excessive – in contrast to the degree of malnutrition of the patients which otherwise tends to blunt these sensations.

Disorders of tone and mobility, and abnormal movements are consistent features of the disease. They vary in their distribution, intensity, and onset. They also reflect lesion localization in the diencephalon and superior mesencephalon. However, lesions may occur anywhere throughout these regions, reflecting symptom variability, and hence the difficulty of general characterization. Parkinsonian-like rigidity with paratonia can occur. At times, it may give way to episodes of flaccidity, or be obscured by abnormal movements. In association with disturbed wakefulness, these signs may lead to the characteristic frozen and apathetic appearance of some patients.

The abnormal movements are characterized by their great variability. They do not fit any classical description, but assume different characteristics from patient to patient and within a single patient with respect to location and timing. Amidst slow writhing athetoic movements of the hands, sudden brief and rapid – even explosive – movements supervene, reminiscent of chorea. Involvement is more common in the distal rather than proximal portions of the upper extremity. The mixture of these different components may produce a picture of incessant gesticulation marked by motor instability, grimaces, and clumsiness. For some patients, these exist in stereotypic patterns. Amidst these signs or independently, a diffuse fine tremor or resting myoclonus may be observed.

The neuropathological lesions in man and animal hosts have been described by different authors as perivascular infiltrates with occasional necrosis occurring in multiple deep cerebral sites [22, 33]. These lesions are probably connected with significant presence of circulating antibodies crossreacting with the tryptophan epitope found by Okomo-Assoumou et al. [34] in sera of patients with sleeping sickness.

Sometimes wide-based ataxia may be the dominant symptom, reflecting inflammatory cerebellar lesions, although the ataxia is not generally a pure cerebellar syndrome. Determining the full clinical extent of cerebellar contribution to problems of coordinated body movements can be difficult, but the cerebellar lesions themselves are typical of the neuropathology seen at other affected sites. They consist, as for cerebral cortical lesions, of a meningeal infiltrate comprised of plasmocytes, lymphocytes and macrophages. All cerebellar regions may be affected, the vermis as well as the hemispheres. While the granular layer is typically preserved, the Purkinje cells are quite altered and reduced in number with barely visible dendrites.

Despite all these movement derangements, it is rare for true paralysis to occur. When cases do occur, they may be indirect, the result of a neuropathy due to a vitamin deficiency associated with trypanosomiasis. This, however, is highly unusual, despite the often advanced state of malnutrition found in these patients. Radiculoneuritis in association with the meningoencephalitis has been described by Van Bogaert and Janssen [33]; it results from a panvasculitis with preferential involvement of dorsal roots near the spinal ganglion, and may result in unusual patterns of mononeuritis multiplex, ascending polyneuritis or isolated nerve paralysis.

Very rarely at this disease stage, hemiplegia may occur. Hemiparesis and hemiplegia may occur with intracranial hypertension. Their pathological basis is inflammatory infiltration or demyelination of the white matter of the centrum ovale; they may also show reversibility with adequate treatment.

Mental difficulties are a key disease feature. Of very variable appearance, they can manifest in the foreground or develop in the shadow of more serious neurological and general signs. Appearing sometimes at the onset of the malady, sometimes at a more advanced stage, but always before the terminal phase, they can be the source of major diagnostic errors and lead to the mistaken belief of a primary psychiatric illness or criminally prosecutable misconduct. The interpretation of these mental difficulties is not always clear-cut: some may be the direct result of the (sleeping sickness) disease process; others may be more influenced by the patient's personality and sociocultural milieu. One can distinguish:

– purely psychiatric problems: mood disorders with lability and indifference, character problems of aggression, antisocial behaviour, and alteration in activity level with apathy and nonchalance or, contrastingly, with impulsive actions;

– forms with mental confusion and/or dementia with stereotypic behavior, impulsive actions, and fugue states. On the background course of progressive dementia, there may be added: manic episodes, melancholia, and states of delirium with themes of grandeur or of persecution, accompanied by hallucinations.

A given severity of mental disturbance may be appreciated at differing degrees depending on the observer and the tolerance of the family milieu. This frequency and variability should lead one to note that in an endemic zone, any mental disorder needs to lead to consideration of and search for a trypanosome. Such mental disturbances may be the first manifestation of a long-standing trypanosomiasis or, contrastingly, may appear in the course of a disease which has already been detected, diagnosed, and treated.

Beginning with the hemolymphatic phase, there can be subtle or not so subtle mood and behavioral problems with anxiety, mood instability, manic and depressive episodes, apathy, indifference, dishevelment, insouciance, reduced activity, and negligence with social responsibilities: in effect, a complete rupture with the person's habitual way of living. The importance of these signs needs to be emphasized; because of their early onset, they may sometimes be the only signs to attract attention during this disease phase. In the course of the meningoencephalitic phase, psychiatric disturbance may dominate. The classic picture is of disinterest, indifference to external circumstances and apathy. Somnolence, sluggishness, moral apathy, psychological fatigue, stupor: these all seem linked to the problem of altered arousal. However, on this calm base, as a general rule there are violent unfounded episodes of irritability and explosions of unexpected rage which signal a global and profound alteration of the affective sphere. Despite these episodes of delirium and agitation, the intellectual faculties of these patients remain nonetheless intact. Dementia does not develop until the terminal end-stage of the disease, where it is quite evident.

Phase of demyelinating encephalitis

The last stage corresponds to the terminal phase of the disease. It is marked in particular by the destruction of myelin. Lesions do not cause large areas of demyelination, but are instead diffuse, particularly affecting the white matter which links the central grey nuclei; they generally appear at sites of earlier infiltrative lesions. The discovery of anti-galactocerebroside auto-antibodies in the serum and cerebrospinal fluid (CSF) of patients may be one of the explanations of this demyelinating encephalitis as galactocerebrosides are one of myelin's major constituents [35, 36]. These antibodies may constitute a good biological marker of CNS involvement. The mechanisms leading to their formation have not yet been elucidated; they may be the result of cross-reactivity between epitopes common to the trypanosome and the nervous system, as it has been observed in *T. cruzi*. Alternatively, they may be the consequence of a secondary immune reaction to an original myelin lesion caused by trypanosome antigen fixation to glial cells, or more simply be the consequence of cytotoxic events secondary to the direct action of the parasite, causing the formation of auto–antibodies. In addition to the auto–antibodies directed against lipid constituents, there has also been evidence of auto–antibodies directed against protein constituents of the nervous system; these are anti-neurofilament antibodies of essentially IgM type, detectable in the serum and CSF of patients [37]. Their production may result either from a nonspecific stimulation of B lymphocytes, or from a cross-reaction to common epitopes between the trypanosome and neuronal proteins. This also holds for the anti-tryptophan-like antibodies which may also represent good markers of neurological involvement [34].

Without appropriate treatment, this terminal phase of the disease progressively develops. Clinically, this stage is characterized by an increase in the disturbances of consciousness and the development of dementia marked by incoherence, incontinence, and epileptic crises. Finally, there is deterioration in the general state of the patient who then succumbs to progressive cachexia ending in death. Pathologically, cardinal features of this phase are the irreversibility of lesions as a result of major demyelination and atrophy, observed in the past with encephalography [51], and more recently by CT imaging.

Clinical forms

The clinical picture just described is the most common and represents the most classical one. The development as subdivided into four stages, corresponds to the chronological events. These may be assessed in particular by the way in which immunological changes, electroencephalographic anomalies and anatomical lesions become manifest, but the true chronological duration of the stages can never be established precisely. Certain forms follow a subacute or acute course, developing over some weeks or months; others, however, develop more slowly with latent periods of several years. This may lead to incorrectly believe that there is recovery. These episodes of clinical silence must be appreciated so as not to fail to recognize the disease before the patient deteriorates abruptly or decompensates due to intercurrent disease. Sometimes, latent periods may be induced when drugs are used which are sufficient to initially improve the patient's general condition but insufficient for cure because they do not cross the BBB, and thus do not destroy the trypanosomes in the brain. Some rare hemiplegic and pseudotumor forms associated with intracranial hypertension and papilloedema may also confuse the diagnosis [38].

In children, the clinical picture of the disease may be just as variable. Essentially, it presents as a neurological condition in which disturbances of tone, involuntary movements, and disturbances of sleep and consciousness (often coma) are prominent. Minor behavioral and character disturbances are constant features, as well as fever which is frequent but rarely very pronounced. Infestation occurs either due to the bite of the tsetse fly or via the placenta [39-42].

Diagnosis

Keeping in mind the polymorphism of clinical signs, sleeping sickness may simulate a great number of other diseases. Therefore, a discussion of diagnosis needs to be very extensive, covering the whole range of its symptomatology.

A key danger exists of missing the diagnosis at the beginning of the disease during the hemolymphatic stage when the neurological signs are still too atypical to necessitate lumbar puncture. Here, the history of a stay in tropical Africa with test results revealing changes in serum globulins, positive flocculation tests, or increase in IgM and IgG, should suggest the diagnosis. The clinical association of pruritus, adenopathy, and fever may at first suggest malignant myeloma or Waldenstrom's macroglobulinemia, but one also has to think of sleeping sickness.

During the neurological phase of the disease, in endemic areas, the diagnosis is readily established from the clinical signs. However, whether in endemic regions or elsewhere, the diagnosis is less easy to establish in the early phases. The signs which mark cerebral involvement may be rarely apparent in the early stage of neurological attack. It is necessary to attentively seek out the existence of primitive brainstem reflexes which mark loss of superimposed higher cortical function (frontal release signs):

reappearance of palmo-mental, grip, rooting, and sucking reflexes which normally disappear several months after birth. These carry greater significance than a simple exaggeration of the osteotendinous reflexes or the very inconstant Babinski sign, whose interpretation is made all the more difficult because many patients walk without shoes and thus have a thick and superficially insensitive plantar surface. Of greater value is the appearance of sensory difficulty characterized by either spontaneous diffuse pain or a pain provoked by an apparently trivial contact. As with all disorders of vigilance, patients are able to rouse their attention but may be only after some delay; this begins with small attentional difficulties with some aspects of indifference, then a tendency toward some periodic daytime somnolence develops which contrasts with the tendency for nocturnal insomnia. These problems develop gradually and tend to attract little attention from the patient's entourage who do not attach importance to these small perturbations of attention or displays of the patient's indifference.

The problem of diagnostic suggestion is compounded if one considers the associated constellation of possible symptoms: impotence or amenorrhea, alterations in eating habits (often of bulimic type), and pruritus are always difficult to interpret; excess chilliness is likely more characteristic than the existence of choreoathetoic movements and extrapyramidal signs whose appearance is much later in the disease course. This is similarly the case for behavioral problems with dreamlike delirium, aggressivity or, constrastingly, total indifference which all may evoke thought of a primary psychiatric disorder.

Thus, any of this grouping of clinical neurological signs described above should guide the decision for prompt lumbar puncture; the CSF will generally be consistent with a subacute or chronic meningitis characterized by a moderate lymphocytic-predominant pleocytosis; the presence of plasma or Mott cells is sufficient for a presumptive diagnosis. However, isolation of the trypanosome, necessary for certain diagnosis, even in the presence of CSF changes is not straightforward.

The extraction of CSF must not be parsimonious; it is better to remove the quantity needed to conduct all analyses rather than having to perform a second lumbar puncture; the 150 mls of CSF contained in the CNS are renewed 3 times per day; the removal of 15 mls and even 20 mls needed to perform the necessary exams will not increase the severity of post-lumbar puncture headaches if they have to occur. These headaches are the consequence of a CSF leak secondary to perforation of the dura mater and/or of the psychological experience of the puncture more than a consequence of the quantity of CSF removed. The advantage of a sufficient sample is the ability to improve the detection rate of the trypanosome, but the procedure is involved: a differential centrifugation is conducted first at 2000 rpms for at least 5 min, then the resulting pellet is drawn into 2 microhematocrit tubes, where it is respun at 12000 rpms for 2 min, and this second pellet is then directly examined under microscope. These steps require an adequate time allocation: 15 or even 20 min may be necessary, without certitude of finding the trypanosome. This point was already underscored in 1909 by Martin et al. [3] who wrote: "the trypanosomes are more abundant in the cerebrospinal fluid of patients with advanced disease, but this is subject to numerous exceptions. One could be perfectly likely to only find very occasional parasites in patients with advanced disease or even in those in the final stages of illness, while one can sometimes observe numerous [trypanosomes] in some men who were merely suspected [of having the disease] in whom it is impossible clinically to make certain diagnosis".

Given the difficulty of this parasite search, even if no trypanosome is isolated, the presence of suggestive signs or symptoms in a patient who has stayed even only a few hours in tropical Africa should lead one to consider other systematic diagnostic testing. Diagnosis of trypanosomiasis is established by finding a four-fold increase in serum IgM or increase in CSF IgM to 10% of CSF total protein. The finding of an increase greater than 30% in CSF total gammaglobulins is also good presumptive evidence; increases of this magnitude are only found in few other diseases: tuberculosis, syphilis, or subacute sclerosing panencephalitis. Cultures and/or serological tests can discriminate between these possibilities.

When the trypanosome is found in a blood or lymph node sample, diagnosis of trypanosomiasis is established with certainty; the only remaining problem is then to determine disease stage. The choice of treatment regimen (pentamidine or melarsoprol) depends on whether the patient's disease is in a hematological or neurological phase. In the absence of the trypanosome in the CSF, the precise differentiation between these two stages can be very problematic, and often the therapeutic choice reflects what is best adapted to the treatment setting (e.g. in a hospital setting or in the bush). Classically, a pleiocytosis (greater than 5 nucleated cells) in the CSF of a patient with trypanosomiasis suggests the neurological phase has been reached, hence requiring the use of melarsoprol which is able to cross the BBB, but quite toxic, proving lethal in around 5% of treated subjects. One can easily grasp the seriousness of making such a treatment decision, compounded by the fact that for certain authors the diagnosis of trypanosomiasis does not absolutely require the discovery of the trypanosome in the blood and/or lymph tissue but can be made based solely on positive serologies (CATT, IFAT). These serologies, although credited with a certain specificity, have the possibility of both false positivity and false negativity in field usage. If serology alone is used to make the diagnosis, then the finding of more than 5 nucleated cells in the CSF could lead to the usage of the quite toxic melarsoprol. This is all the more problematic because trypanosomiasis exists in regions plagued with many other diseases which can also produce meningoencephalitis: tuberculosis, AIDS, fungal infections, filariasis, schistosomiasis, and other parasite infestations which can coexist with trypanosomiasis.

These concerns have led certain authors to raise the limit of cellular elements from five to twenty, which may be an improved but not fully successful approach. It is for this reason that immunological tests to differentiate the two phases and randomized therapeutic trials are currently being pursued. The discovery of anti-galactocerebroside and anti-neurofilament antibodies in the serum and CSF, antibodies directed against the lipid and protein constituents of nervous tissue, may turn out to be a good marker of nervous system involvement [35-37].

Conclusion

Although human African trypanosomiasis has undergone considerable resurgence in recent years, it is therefore troubling to report that this dilemma of therapeutic choice still exists; from a therapeutic perspective, this disease remains one of the neglected step-children of infectious pathologies. HAT has the mistake, because of the biology of its conducting vector, to only occur in tropical black Africa (thus in underdeveloped regions) and to not have a sexual transmission which would grant its special consideration. And so, although many of the nitro-imidazole derivatives are endowed with some efficacy in killing trypanosomes, there is a certain disinterest, when not from the pharmaceutical industry itself, due to a theoretical concern of mutagenicity and carcinogenicity (which has never been demonstrated in man), and their development has not been pursued. It is then surprising to note that metronidazole (a 5-nitroimidazole derivative) is used daily, with success, as treatment of numerous infections, while multiple studies have confirmed the mutagenic risk of this molecule in the laboratory animal; this risk, however, has never been demonstrated by its widespread usage in man. Used alone, this medication is, unfortunately, ineffective for trypanosomiasis; used in conjunction with suramine, it has some effect, temporary and debatable.

Other 5-nitroimidazole derivatives are associated with some activity, alone or in combination with suramin, on chronic *T. b. brucei* infections in the mouse and monkey [43-45]. The combination of these derivatives with melarsoprol permits significant reduction in the dose of this arsenical compound, which is not without considerable interest. Ro 15-0216, similar to benznidazole (Radanil®), used alone has produced definitive cure in mice and sheep with nervous phase infection by *T. b. brucei* [46], and the efficacy of this molecule has even been reported in multiresistant strains of *T. b. brucei* [47] and in mice infected by *T. b. gambiense* [48]. Megazol (CL 64 855), active against *T. cruzi*, has been shown to be effective in conjunction with suramin in chronic *T. b. brucei* infections in mice [49, 50].

It may appear revolting or at least surprising that, possessing molecules which have shown their efficacy against the nervous phase of the disease due to *T. b. brucei* in animal models, there has been refusal to pursue their research and development because of a possible mutagenicity or carcinogenicity which has not been proven. At the same time, one continues to accept killing, with full impunity, about 5% of patients by use of melarsoprol, the sole molecule which is still available, and perhaps not for long as its production is of prohibitive expense, considering the weak market supporting this cost.

References

1. Rifkin MR (1978) Identification of the trypanocidal factor in normal human serum: high density lipoprotein. Proc Natl Acad Sci (USA) 75:3450-3454
2. Smith AB, Esko JD, Hadjuk SL (1995) Killing of trypanosomes by the human haptoglobin related protein. Science 268:284-286
3. Martin G, Leboeuf A, Roubaud E (1909) Rapport de la Mission d'Études de la Maladie du Sommeil au Congo Français, 1906-1908. Masson & Cie, Paris, p 721
4. Fairlamb A (1982) Biochemistry of trypanosomiasis and rational approaches to chemotherapy. Trends Biochem Sci 7:249-253
5. Opperdoes FR, Aarsen PN, Van Der Meer C, Borst P (1976) *Trypanosoma brucei*: an evaluation of salicylhydroxamic acid as a trypanocidal drug. Exp Parasitol 40:198-205
6. Evans DA, Brightman CAJ (1980) Pleomorphism and the problem of recrudescent parasitaemia following treatment with salicylhydroxamic acid (SHAM) in african trypanosomiasis. Trans R Soc Trop Med Hyg 74:601-604
7. Van Der Meer C, Verluis-Broers JAM, Opperdoes FR (1979) *Trypanosoma brucei*: trypanocidal effect of salicylhydroxamic acid plus glycerol in infected rats. Exp Parasitol 48:126-134
8. Borst P, Gommers-Ampt JH, Ligtenberg MJL, Rudenko G, Kieft R, Taylor MC, Blundell PA, Van Leuwen F (1993) Control of antigenic variation in African trypanosomes. Cold Spring Harbor Symposia on Quantitative Biology 18:105-114
9. Doering TL, Raper J, Buxbaum LU, Adams SP, Gordon JI, Hart GW, Englund PT (1991) An analog of myristic acid with selective toxicity for African trypanosomes. Science 252:1851-1854
10. Pays E (1990) Variation antigénique des trypanosomes africains. Ann Parasitol Hum Comp 65 (Suppl 1):5-10
11. Greenwood BM, Whittle HC (1980) The pathogenesis of sleeping sickness. Trans R Soc Trop Med Hyg 74:716-723
12. Poltera AA, Cox JN, Owor R (1976) Pancarditis affecting the conducting system and valves in human African trypanosomiasis. Brit Heart J 38:827-837
13. Adams JH, Haller L, Boa FY, Doua F, Dago A, Konian K (1986) Human African trypanosomiasis (*T. b. gambiense*): a study of 16 fatal cases of sleeping sickness with some observations on acute reactive arsenical encephalopathy. Neuropathol Appl Neurobiol 12:81-94
14. Okomo-Assoumou MC, Daulouède S, Lemesre JL, N'Zila-Mouanda A, Vincendeau P (1995a) Correlation of high serum levels of tumor necrosis factor-α with disease severity in human African trypanosomiasis. Am J Trop Med Hyg 53:539-543
15. Vincendeau P, Daulouède S, Veyret B, Dardé ML, Bouteille B, Lemesre JL (1992) Nitric oxide-mediated cytostatic activity on *Trypanosoma brucei gambiense* and *Trypanosoma brucei brucei*. Exp Parasitol 75:353-360
16. Buguet A, Burlet S, Auzelle F, Montmayeur A, Jouvet M, Cespuglio R (1996) Dualité d'action du monoxyde d'azote (NO) dans la trypanosomose africaine expérimentale. C R Acad Sci Paris (Sciences de la vie) 319:201-207
17. Schultzberg M, Mabatsis M, Samuelsson EB, Kristensson K, Van Meirvenne N (1988) Spread of *Trypanosoma brucei* to the nervous system: early attack on circumventricular organs and sensory ganglia. J Neurosci Res 21:56-61
18. Van Marck EAE, Gigase PLJ, Beckers A, Wéry M (1981) Experimental infections of laboratory rodents with recently isolated stocks of *Trypanosoma brucei gambiense*. 2. Histopathological investigations. Z Parasitenkd 64:187-193

19. Abolarin MO, Evans DA, Tovey DG, Ormerod WE (1982) Cryptic stage of sleeping-sickness trypanosome developing in choroid plexus epithelial cells. Brit Med J 285:1380-1382
20. Mattern P, Mayer G, Felici M (1972) Existence de formes amastigotes de *Trypanosoma gambiense* dans le tissu plexuel choroïdien de la souris infectée expérimentalement. C R Acad Sci Paris 274:1513-1515
21. Ormerod WE, Venkatesan S (1971) An amastigote phase of the sleeping sickness trypanosome. Trans R Soc Trop Med Hyg 65:736
22. Bertrand I, Bablet J, Sicé J (1935) Lésions histologiques des centres nerveux dans la trypanosomiase humaine africaine (à propos de deux cas mortels non traités). Ann Inst Pasteur 54:91-147
23. Collomb H, Lemercier G, Girard PL, Dumas M (1973) Aspects neuro-pathologiques de la trypanosomiase humaine africaine à *Trypanosoma gambiense*. 10th International Congress of Neurology (8-15 September), Barcelona, Spain (oral communication)
24. Gallais P, Badier M (1952) Recherches sur l'encéphalite de la trypanosomiase humaine africaine: corrélations cliniques, anatomiques, électro-encéphalographiques, biologiques. Med Trop 12:633-673
25. Kerandel P (1910) Un cas de trypanosomiase chez un médecin. Bull Soc Path Ex 3:642-662
26. Nozais JP (1988) Mais où est donc passée la "clé" de Kérandel? Bull Soc Path Ex 81:477-479
27. Brandenberger G, Buguet A, Spiegel K, Stanghellini A, Mouanga G, Bogui P, Montmayeur A, Dumas M (1994) Maintien des relations entre la sécrétion pulsatile des hormones et la structure interne du sommeil dans la trypanosomiase humaine africaine. Bull Soc Path Ex 87:383-389
28. Buguet A, Bert J, Tapie P, Bogui P, Doua F, Mouanga G, Stanghellini A, Sarda J, Tabaraud F, Gati R, Montmayeur A, Chauffard F, Lonsdorfer J, Dumas M (1994) Distribution du sommeil et de la veille dans la trypanosomose humaine africaine. Bull Soc Path Ex 87:362-367
29. Radomski MW, Buguet A, Bogui P, Doua F, Lonsdorfer A, Tapie P, Dumas M (1994) Disruptions in the secretion of cortisol, prolactin, and certain cytokines in human African trypanosomiasis patients. Bull Soc Path Ex 87:376-379
30. Claustrat B, Buguet A, Geoffriau M, Montmayeur A, Bogui P, Muanga G, Stanghellini A, Dumas M (1994) Le rythme nycthéméral de la mélatonine (MLT) est conservé dans la trypanosomose humaine africaine. Bull Soc Path Ex 87:380-382
31. Noireau F, Apembet JD, Frézil JL (1988) Revue clinique des troubles endocriniens observés chez l'adulte trypanosomé. Bull Soc Path Ex 81:464-467
32. Hublart M, Lagouche L, Racadot A, Boersma A, Degand P, Noireau F, Lemesre JL, Toudic A (1988) Fonction endocrine et trypanosomiase africaine. Bilan de 79 cas. Bull Soc Path Ex 81:468-476
33. Van Bogaert L, Janssen P (1957) Contribution à l'étude de la neurologie et neuropathologie de la trypanosomiase humaine. Ann Soc Belge Méd Trop 37:379-427
34. Okomo-Assoumou MC, Geffard M, Daulouède S, Chaugier C, Lemesre JL, Vincendeau P (1995b) Circulating antibodies directed against tryptophan-like epitopes in sera of patients with human African trypanosomiasis. Am J Trop Med Hyg 52:461-467
35. Jauberteau MO, Ben Younes-Chenoufi A, Amevigbe M, Bouteille B, Dumas M, Breton JC, Baumann N (1991) Galactocerebrosides are antigens for immunoglobulins in sera of an experimental model of trypanosomiasis in sheep. J Neurol Sci 101:82-86
36. Jauberteau MO, Bisser S, Ayed Z, Brindel I, Bouteille B, Stanghellini A, Gampo S, Doua F, Breton JC, Dumas M (1994) Détection d'autoanticorps anti-galactocérébrosides au cours de la trypanosomose humaine africaine. Bull Soc Path Ex 87:333-336
37. Ayed Z, Brindel I, Bouteille B, Van Merveinne N, Doua F, Houinato D, Dumas M, Jauberteau MO (1997) Detection and characterization of autoantibodies directed against neurofilament proteins in human African trypanosomiasis. Am J Trop Med Hyg 57:1-6
38. Collomb H, Baylet R, Lacan A (1968) La trypanosomiase humaine africaine. In: Epidemiologie et prophylaxie des endémies dominantes en Afrique Noire. Masson, Paris, pp 27-46
39. Darré H, Mollaret P, Tanguy Y, Mercier P (1937) Hydrocéphalie par trypanosomiase congénitale. Démonstration de la possibilité du passage transplacentaire dans l'espèce humaine. Bull Soc Path Ex 20:159-176
40. Debroise A, Debroise-Ballereau C, Stage P, Rey M (1968) La trypanosomiase africaine du jeune enfant. Arch Fr Pediatr 25:703-720
41. Burke JA, Bengosi ME, Diantet NL (1974) Un cas de trypanosomiase africaine (*T. gambiense*) congénitale. Ann Soc Belge Med Trop 54:1-4
42. Olowe SA (1975) A case of congenital trypanosomiasis in Lagos. Trans R Soc Trop Med Hyg 69:57-59
43. Jennings FW, Urquhart GM, Murray PK, Miller BM (1983) Treatment with suramin and 2-substituted 5-nitroimidazoles of chronic *Trypanosoma brucei* infections with central nervous system involvement. Trans R Soc Trop Med Hyg 77:693-698
44. Sayer PD, Onyango JD, Gould SS, Waitumbi JN, Raseroka BH, Akol GWO, Ndung'u JM, Njogu AR (1987) Treatment of African trypanosomiasis with combinations of drugs with special reference to suramin and nitroimidazoles. International Scientific Council for trypanosomiasis Research and Control. 19th Meeting, Lomé, Togo, pp 205-210
45. Cluckler AC, Malanga CM, Conroy J (1990) Therapeutic efficacy of new nitroimidazoles for experimental trichomoniasis, amebiasis and trypanosomiasis. Am J Trop Med Hyg 19:916-925

46. Bouteille B, Dardé ML, Pestre-Alexandre M, Dumas M, Breton JC, Nicolas JA, Catanzano G, Munoz M (1988) Traitement de la trypanosomiase expérimentale du mouton à *Trypanosoma brucei brucei*: efficacité du Ro 15-0216 (dérivé 2-nitroimidazolé). Bull Soc Path Ex 81:616-622
47. Borowy NK, Nelson RT, Hirumi H, Brun R, Waithaka HK, Schwartz D, Polak A (1988) Ro 15-0216: a nitroimidazole compound active in vitro against human and animal pathogenic African trypanosomes. Ann Trop Med Parasitol 82:13-19
48. Richard-Lenoble D, Kombila MY, Félix H (1988) Efficacité du Ro 15-0216 sur les souches humaines de *Trypanosoma gambiense* entretenues sur les rongeurs. Étude préliminaire. Bull Soc Path Ex 81:609-615
49. Marie-Daragon A, Rouillard MC, Bouteille B, Bisser S, De Albuquerque C, Chauvière G, Périé J, Dumas M (1994) Essais d'efficacité sur *Trypanosoma brucei brucei* de molécules franchissant la barrière hématoméningée et du mégazol. Bull Soc Path Ex 87:347-352
50. Bouteille B, Marie-Daragon A, Chauvière G, De Albuquerque C, Enanga B, Dardé ML, Vallat JM, Périé J, Dumas M (1995) Effect of megazol on *Trypanosoma brucei brucei* acute and subacute infections in Swiss mice. Acta Trop 60:73-80
51. Heches P, Girard PL, Dumas M (1972) Fractional gas encephalography in trypanosomiasis. Bull Soc Med Afr Noire Lang Fr 17:12-15
52. Collomb H (1957) Encéphalite de la trypanosomiase humaine africaine. Gaz Méd Fr 64:1064-1078
53. Borremans P, Van Bogaert L (1933) Les manifestations extra-pyramidales de la trypanosomiase chez l'européen. (Syndrome d'inhibition avec stéréotypies, pigmentations cutanées symétriques et anneau cornéen). J Belge Neurol Psychiatr 33:561-588

CHAPTER 14

Biological diagnosis of human African trypanosomiasis

N Van Meirvenne

Introduction

Human African trypanosomiasis or sleeping sickness is an endemic disease in many sub-Saharan countries. The causative agent is the unicellular haemoflagellate parasite *Trypanosoma brucei*, which is cyclically transmitted through the saliva of blood sucking tsetse flies. The more chronic "Gambian" form of sleeping sickness, which evolves fatally over a period of several months or years, predominates in West and Central Africa. It is caused by trypanosomes of the subspecies *T. b. gambiense*. The more fulminant "Rhodesian" form, due to the morphologically identical *T. b. rhodesiense*, is merely found in East Africa, where a large variety of game and domestic animals act as reservoir hosts. This parasite, which sometimes strikes safari tourists, can kill the patient within weeks or months.

T. brucei is an extracellular parasite dwelling in various body and tissue fluids. The spindle-shaped cell (20-30 by 1.5-3.5 µm) multiplies by longitudinal binary fission. The live trypanosome is easily recognizable through its very active wiggling movements, brought about by undulatory motion of the flagellum, which runs along the cell body and ends in front in a free tip. In stained preparations the locomotory apparatus, a central nucleus and a subterminal DNA containing particle, the kinetoplast, are easily seen. The only possible source of confusion are the South American trypanosomes *T. cruzi* and *T. rangeli*.

At the site of inoculation by an infective tsetse fly bite, a transient skin lesion, the trypanosomal chancre, often develops, especially in case of *T. b. rhodesiense*. During the first or haemato-lymphatic stage of the disease the trypanosomes invade lymph, blood, bone marrow and various other tissues, where they multiply and elicit vigorous immune responses resulting in cyclical partial clearance of the parasite burden but not in cure. Sooner or later, a question of weeks or months, the infection spreads to the central nervous system, thus initiating the second or meningo-encephalitic stage, which requires much more delicate chemotherapeutic intervention.

A variety of unspecific and inconstant clinical symptoms, including lymphadenopathy, skin rash, periodic fever, headache, joint and muscle pains, oedema, cardiac problems and endocrinological disorders, may be present from an early stage on [1]. Onset is commonly most rapid and severe in case of *T. b. rhodesiense* infection. As the disease progresses, emaciation and neurological symptoms become apparent. It should be stressed that patients infected with *T. b. gambiense* or less virulent variants of *T. b. rhodesiense* may have vague complaints only, for some period of time.

The pathogenesis of the disease is largely due to inflammatory and immunopathological reactions caused by continuous activation of cellular and humoral immune responses. The main governing factor is the outstanding capacity of the trypanosomes for antigenic variation, a process regulated by gene switching and recombination [2]. The parasites are coated with a variant surface glycoprotein (VSG): it is strongly immunogenic and elicits trypanocidal antibodies which efficiently eliminate trypanosomes of corresponding variable antigen type (VAT). In the meantime, however, some new VATs have already been generated and this alternation of parasite proliferation and destruction continues throughout infection. The ever repeated contact with variable and invariant antigens induces a number of cytotoxic and physiopathological processes. Blood and serum anomalies commonly present, include: low haematocrit, increased autoagglutinability and sedimentation rate of the red blood cells, decreased albumin, strongly increased IgM, increased levels of plasma kinins, immune complexes and fibrinogen degradation products, hypocomplementaemia, various kinds of autoantibodies (see other chapters in this book).

Demonstration of the parasite, usually in blood or lymph fluid, is imperative for unequivocal diagnosis. This may be quite difficult to achieve, especially in case of *T. b. gambiense* infection. Since the number of trypanosomes circulating in the blood fluctuates considerably, repeated daily examination may be required. Much depends on the sensitivity of the parasitological technique used. Serodiagnostic tests are indispensable complementary tools. Simplified field tests have been developed for screening the population under risk in *T. b. gambiense* endemic areas, but more research is needed for the development of more reliable tests based on antibody, antigen and DNA detection.

Different drugs and schemes are used for treatment of first and second stage patients. Stage determination and post treatment follow-up are based on examination of cerebrospinal fluid.

Parasitological methods

The standard approach for parasitological diagnosis is meticulous microscopic examination of lymph aspirate and blood. Finding of a single trypanosome with a simple technique is good enough. However, since the parasitaemia is occasionally very low, the use of more sensitive concentration or cultivation techniques is indicated. Details about various methods can be found in the literature [1, 3-7]. The microscopist should be aware of the pleomorphism of *T. brucei*, ranging from long slender to short stumpy forms.

Direct methods

Lymph

Lymph aspirate is classically obtained from enlarged cervical lymph nodes. A swollen gland is immobilized between thumb and fingers, punctured with a large bore needle and gently massaged to harvest a little bit of lymph. The needle is withdrawn while closing the free opening with a finger tip. It is then mounted on a small syringe containing some air. The juice is expelled onto a microscope slide, a cover slip is put and the wet preparation immediately examined under the microscope (10 x 40) for the presence of motile trypanosomes. Each microscopic field should be looked at carefully, and total reading time may be 10-15 min. If desired, a permanent lymph smear preparation can be made and further processed as described for thin blood smears.

Gland palpation followed by wet lymph examination is currently used in field surveys for Gambian sleeping sickness. However, not all trypanosome carriers are presenting swollen lymph nodes and the parasite may be missed in the aspirate. Upon single examination the overall sensitivity of the procedure varies between 40 and 80%.

Blood

Most techniques are using freshly collected anticoagulant blood (heparin or EDTA), obtained by finger prick or venepuncture. In a refrigerated blood sample, trypanosomes can survive for several hours. Other motile organisms possibly present in blood are *T. cruzi*, *T. rangeli*, the more thiny male gametes of *Plasmodium* sp, the much larger microfilariae and, not to forget, bacteria.

Wet blood film

On a microscope slide, a small drop (up to 10 µl) of plain or anticoagulated blood is mounted with a relatively large cover slip, so as to obtain a monolayer of red blood cells. The cover slip can be slightly pressed, so that it does not float. This wet preparation is immediately examined under the microscope (10 x 40), carefully screening each field for the presence of a motile trypanosome, which often attracts the attention by agitation of the surrounding red blood cells. Under phase contrast or dark ground illumination the trypanosomes are more easily seen as contrasting refringent vermicules, even at a 10 x 25 magnification. Attentive reading of the entire preparation may take up to 20 min. The sensitivity of the technique is relatively low. A parasitaemia of less than 10,000 trypanosomes per ml of blood, corresponding to about 1 parasite per 200 microscope fields, will often be missed.

A simple trick to improve the visibility of scarce trypanosomes is the addition of sodium dodecyl sulfate (SDS), a haemolytic detergent which does not affect the motility of the parasites for quite some time [8]. The SDS wet blood clarification technique is as follows. Haemolytic solution: dissolve 100 mg of SDS (toxic!) in 100 ml of isotonic TRIS-NaCl-glucose buffer, pH 7.5 (Trizma-Base® 14.0 g, NaCl 3.8 g, glucose 10.8 g, dissolve in 750 ml H_2O; add 90-100 ml 1N HCl and H_2O to 1000 ml); can be stored in small vials at ambient temperature for many months. Blood and SDS solution are mixed in equal volumes. To avoid killing of the trypanosome by the detergent, both reagents should have a temperature of at least 15°C. On a microscope slide, put 5-10 µl of blood. With a calibrated plastic inoculating loop, add 10 µl of haemolytic solution, mix. Put a sufficiently large cover slip (24 x 24 or 24 x 36). Immediately examine the preparation under an ordinary or phase contrast microscope at a 10 x 25 magnification. White blood cells, unlysed and ghost red blood cells serve for focusing. After some experience, passing microscopic fields can be examined at a glance.

Thin blood smear

Its permanent character and excellent quality makes this preparation ideal for morphological examination. However, it often requires a long reading time and is unsuitable for detecting parasitaemias of less than 10,000 trypanosomes per ml. Basically, the technique is the same as that used for white blood cell formula. A minute drop of blood (up to 5 µl) is put at the end of a clean microscope slide and a thin smear made in the usual way. The slide is air-dried and further processed by one of the various possible fixation and staining procedures, including May-Grünwald-Giemsa, Giemsa, Field's stain or Dade® Diff Quicke®. Microscopic examination is done without a cover slip, at 10 x 40, or with a x100 oil immersion objective.

Thick blood smear

This is a general technique for finding microfilariae, malaria parasites and trypanosomes. A thick layer of blood is dried on a slide and stained without fixation, leaving a haemolysed permanent preparation showing white cells and parasites. As compared with thin smears there is a concentration factor of about 10 times and parasitaemias of 5,000 trypanosomes per ml become detectable.

A large drop (5-10 µl) of plain or anticoagulated blood is put in the centre of a clean microscope slide and spread over an area of about 1 cm in diameter, by circular movements with a tooth pick or the corner of a slide. Correct thickness is important. Newsprint should just be readable when the wet film is put on top. Let air-dry in a horizontal position for a few hours, in a dust free place. Avoid heating, even by sunlight. Staining is done with Giemsa or Field's stain as for thin smears but without previous fixation; washing with tap water should be done very gently. The preparation is examined with a x100 oil immersion objective. Checking 200-300 microscopic fields may take 20 min.

Buffy coat/centrifugation techniques

Upon centrifugation of infected anticoagulated blood, most of the trypanosomes get concentrated at the buffy coat-plasma interface. When the centrifugation is performed in capillary tubes, as in the microhaematocrit technique (mHCT), the parasites can be searched under the microscope directly *in situ* or in a preparation made from the buffy coat layer.

Heparinized capillary tubes (75 x 1 mm) are filled 3/4 full with fingerprick or venous blood. The dry ends are firmly sealed by rotating in plasticine and the tubes are spun for 6-8 min in a microhaematocrit centrifuge. Make sure that the temperature of the blood remains below 40°C. The centrifuged tubes are temporarily kept in an upright position and examined as soon as possible.

For direct examination, the capillaries can be secured on a microscope slide with adhesive tape or plasticine. A better microscopic image is obtained by using a specially devised mounting slide with grooves and viewing chambers that can be filled with water and covered with a cover slip. The buffy coat-plasma area is examined at a 10 x 10 or 10 x 20 magnification, moving from the buffy coat edges towards the plasma layer, thus covering 4-5 overlapping microscopic fields. Per horizontal field, 4-5 different vertical levels are brought into focus with the microscrew. A total of only 20-25 microscopic fields have to be examined in this way, which takes less than 2 min. Since the trypanosomes to be detected under these conditions are very small, some experience is required. Careful application of this technique is very rewarding. Using a few capillary tubes per blood sample, a parasitaemia of less than 5,000 trypanosomes per ml is detectable in a relatively short period of time. With a 12 Volt driven centrifuge, the technique becomes perfectly applicable under field conditions.

When direct reading of the capillary tube proves impossible by abundant presence of microfilariae, a wet preparation can be made. With a diamond or a thiny grinding disk as used by dentists, scratch the tube just below the buffy coat layer, carefully break the tube (risky!) and dab the buffy coat onto a slide. Put a cover slip and examine as for wet blood. Otherwise, make a thin smear and process it as for whole blood.

Sensitivity can be increased by previous centrifugation of a larger volume of blood in a narrow tube (700 g, 10 min), followed by a second centrifugation of the buffy coat in capillary tubes. The more laborious triple centrifugation technique [6] has gone out of use.

Recently, an elaborated and most sensitive version of the microhaematocrit technique, designated QBC® (Quantitative Buffy Coat method), has been introduced. This technique is using capillary tubes precoated with EDTA and the fluorescent dye acridine orange, and containing a small floating cylinder that automatically comes to rest on top of the red blood cell sediment. The expanded and fractionated buffy coat layer thus obtained is examined under an epi-illuminated fluorescence microscope. Various blood parasites can be detected in this way. Trypanosomes show a bright yellow-green color and remain motile for about 20 min. They accumulate just above the blood platelets. Very satisfactory results have been recorded by some authors [9-11]. All materials, including centrifugation and microscope equipment for laboratory and field use, are commercially available. Cost price could be a limiting factor for large-scale application.

Mini-Anion-Exchange-Column-Technique (mAECT)

Trypanosomes are less negatively charged than blood cell elements and can selectively be eluted through a DEAE-cellulose anion exchange column. Phosphate-saline-glucose buffer (PSG), pH 8, of well–defined ionic strenght has to be used [12]. For the separation of *T. brucei* from human blood, a suitable PSG composition of ionic strenght 0.181 is as follows: Na_2HPO_4 13.48 g, NaH_2PO_4 $2H_2O$ 0.78 g, NaCl 4.25 g, glucose 20 g, distilled water to 2000 ml.

DEAE-cellulose powder (Whatman DE52, preswollen) is equilibrated with PSG and the slurry stacked to a height of 3-4 cm in a column with porous bottom and adjustable tubing outlet. A syringe barrel plugged at the bottom with a little bit of cotton-wool will also do. The heparinized blood sample is carefully put on top of the slurry and allowed to seep into. The maximal volume of blood that can be processed is about 1/5 of that of the slurry. Elution of the trypanosomes is achieved by progressive addition of PSG, in a total volume of twice that of the packed slurry. The eluate is collected in a conical tube and centrifuged at 700 g for 10 min. The supernatant liquid is removed by aspiration, leaving a few drops of "deposit" to be examined under the microscope as wet preparation. A much more convenient method is to transfer a somewhat larger volume of eluate deposit into a small, finely drawn-out Pasteur pipette. The pipette is sealed, put into a protective holder, consisting of a tube and a conical plastic cap, and submitted to a second centrifugation. Using an appropriate viewing chamber the very tip of the pipette is then examined under the microscope (10 x 10) for the presence of trypanosomes, in a similar manner as for mHCT capillaries.

A miniaturized field and laboratory version, allowing processing of 0.2-0.3 ml of blood, has been introduced [4, 13, 14]. Ready-for-use mAECT kits can be obtained at reasonable cost price. Using mAECT, a parasitaemia of less than 100 trypanosomes per ml becomes detectable.

Cerebrospinal fluid (CSF)

In some patients with long lasting *T. b. gambiense* infection trypanosomes become more easily demonstrable in CSF than in lymph or blood. Clinical suspicion of late stage sleeping sickness, therefore, may urge to do a lumbar puncture. CSF examination is required anyway for stage determination before chemotherapy and for post treatment follow-up.

Put a few ml of freshly collected CSF into a conical tube. Centrifuge at 700 g for 10 min. Separate the supernatant by decantation or aspiration, saving it for other tests. Resuspend the sediment in the few drops left at the bottom of the tube by tapping it with the finger. Make a wet preparation between slide and cover slip and examine this under the microscope (10 x 25 or 10 x 40) for the presence of motile trypanosomes.

In the more sensitive double centrifugation system, the sediment is sucked into one or more plain microhaematocrit capillary tubes, leaving a dry end of about 1 cm [15]. Seal the dry ends in a flame; centrifuge in a microhaematocrit centrifuge for 2 min; put the capillaries on a microscope slide, cover the sealed tips with water and a cover slip. Examine the sediments directly under the microscope (10 x 10 or 10 x 16).

Other biopsy material

During the first few weeks of infection, the trypanosomal chancre is a candidate site for parasitological diagnosis. Juice from a suspected lesion can be obtained by superficial scarification and microscopically examined as a wet, or fixed and Giemsa stained, preparation. In some exceptional cases of persistent diagnostic incertitude, examination of a Giemsa stained preparation of bone marrow may reveal the presence of trypanosomes.

Details of staining procedures

Giemsa stain

a) Stock solution
   In a dark bottle containing some glass beads: put 3.8 g of Giemsa powder and 250 ml of methanol p.a. Close the bottle and shake for 3 min. Add 250 ml of glycerol p.a. Leave the solution for a few days and vigorously shake from time to time. Filter and store in a tightly closed dark bottle. Keep the solution perfectly clean, avoiding any trace of moisture.
b) Staining of smears

1. For single day use: make a 1/30 dilution of the stock solution in phosphate buffered distilled water, pH 7.2 (Na_2HPO_4 3.0 g, KH_2PO_4 0.6 g/l or something similar, possibly making use of buffer tablets; correct pH is important);
2. fix with methanol for 3 min, briefly dry and continue or store for some time (note: thick blood smears should not be fixed!);
3. stain for 25 min;
4. gently wash with tap water. Let dry.

Field's stain

Is a combination of buffered solutions A (methylene blue and azure I dye) and B (yellow eosin). Prepared powders are commercially available and processed as follows. Powder A: 5.9 g, dissolve in 600 ml hot distilled water, filter when cool. Powder B: 4.8 g, same procedure.
Staining of thick blood smear:
1. Dip into solution A for 3 s;
2. gently wash by dipping once into water;
3. dip into solution B for 3 s;
4. gently wash as in step 2. Let air dry.

Culture methods

When inoculated into a suitable medium, viable trypanosomes multiply and become detectable after some time. In principle, any type of freshly collected wet sample can be inoculated. Sensitivity of these methods may be very high but waiting-time is a matter of several days or even weeks. Large-scale field application would be rather expensive and laborious.

Animal inoculation

T. b. rhodesiense easily grows in mice. Up to 0.5 ml of heparin or EDTA anticoagulated blood is i.p. inoculated into each of two mice. From day 2-3 on, tail wet blood films of the animals are examined at 1-2 day intervals, for a period of 10 days. Field samples of *T. b. gambiense* often fail to infect mice or just cause a transient infection with low parasitaemia. Other animals, including nursling rats, have proved more susceptible.

In vitro culture

Recently, a ready-for-use kit for *in vitro* isolation (KIVI) has been developed [16]. The kit contains vials with medium and antibiotics, vials with the anticomplementary anticoagulant polyanethol-sulfonate, syringes and needles, parafilm stretching foils. Reagents can be stored in the refrigerator for many months. In this medium at 25°C bloodstream form trypanosomes first transform into large "procyclic forms", a developmental stage that is also found in the tsetse fly. These forms then proliferate by cell division.

Bleeding and further manipulations should be done aseptically. 9.5 ml of blood is collected on 0.5 ml anticoagulant solution and 2-5 ml of this mixture inoculated through the desinfected rubber stopper into each of two culture vials. Caps are sealed with parafilm and culture vials incubated in the dark in an upright position at a temperature of about 25°C. The bottom layer of the medium is aseptically sampled and microscopically examined once or twice a week for a period of 3-4 weeks.

The recorded sensitivity of KIVI for spotting *T. b. gambiense* infections proves very high [9, 11, 17]. New developments are to be expected, including miniaturization of the technique and possibly application to lymph and CSF [18].

Immunological methods

Trypanosomes have a complex antigenic structure and elicit production of a large spectrum of antibodies. Since clinical symptoms are highly variable and parasitological methods rather laborious and insensitive, diagnosis can be aided by serological tests. Serological screening of the entire population in endemic areas has become a routine strategy in field surveys on Gambian sleeping sickness.

Trypanosomal antigens and antibody response

Two main categories of trypanosomal antigens can be distinguished: variable and invariant antigens [19].

Variable antigens

The trypanosomes evolving in the patient, including the initial metacyclic forms inoculated by a tsetse fly, are coated with a layer of variant surface glycoprotein (VSG). Each trypanosome has the genetic potential to express several hundreds of distinct VSGs, corresponding to a repertoire of as many variable antigen types (VATs) [2]. Serotyping of VATs is currently done by immune lysis or direct agglutination tests using clone populations and specific antisera [20, 21]. In the course of the infection, sets of different VATs succeed each other in a loosely defined hierarchical order, starting with a series of so-called predominant VATs and followed by less and less predictable ones. Trypanosomes of different genotype may express distinct VAT repertoires, with varying degrees of overlap due to serologically similar IsoVATs. In this respect, *T. b. rhodesiense* shows much more diversity than *T. b. gambiense* [20, 21].

The strongly immunogenic VSG represents about 10% of the total protein content of a trypanosome. It has a molecular weight of 60-65 kD and contains about 550 amino acids. A NH_2-terminal domain of about 400 amino acids shows extensive sequence diversity, whereas there is more sequence conservation in the C-terminal domain which is positioned on the cell membrane.

Whereas only a few "external" VSG epitopes are exposed at the outer surface of a live trypanosome, a variety of additional "internal/cryptic" VSG epitopes are exposed by detached molecules and by dead cells.

Invariant antigens

Trypanosomes probably contain hundreds or thousands of invariant antigens. Only few of these have been studied in detail. In buffer extracts a few tens of components can be revealed by immunoelectrophoresis with hyperimmune antisera [19]. In addition, various non hydrosoluble antigens and a lot of minor protein components have been reported, including cytoskeleton elements, invariant surface glycoproteins (ISGs), enzymes and receptor proteins [22-24].

Antibody response

Efficient trypanocidal IgM antibodies are mounted against the external VSG epitopes of each of the VATs successively generated in the course of the infection and these antibodies, possibly replenished with IgG counterparts, accumulate in the serum [20, 21]. As antibody coated VATs become lysed or phagocytosed, the immune response expands to internal/cryptic VSG epitopes and to a large spectrum of invariant antigens. The levels of circulating antibodies directed against common epitopes of VSGs or invariant antigens, may periodically fluctuate as a result of immune complex formation and elimination.

Antibody detection tests

Of the numerous antibody test systems that have been published, only some have been put into practical use for some time and very few are still currently used nowadays. Insiders believe that more accurate tests could be developed through the selection of adequate antigens. In the meantime, rigorous standardization of existing tests is a first requirement. Antigen preparations are routinely derived from trypanosomes growing in the blood of mice or rats. The entire parasite cells or buffer extracts thereof, apart from some host components, represent a cocktail of variable and invariant antigens, including intracellular VSG precursor material. Since both groups of antigens can react with antibodies from infection sera, trypanosome populations of well defined VAT should always be used as starting material. Procyclic culture forms of any *T. brucei* stock, on the other hand, can serve as a source of invariant antigens which, in general, are better suited for *T. b. rhodesiense* than for *T. b. gambiense* serology.

How sensitive are the tests? Theoretically, the earliest antibodies that can be demonstrated are those directed against the external surface epitopes of the metacyclic trypanosomes inoculated by the fly, followed by analogous highly specific antibodies to predominant bloodstream form VATs. Production of antibodies to some invariant antigens and internal VSG epitopes, probably starts during the first few weeks of infection. Presumedly, current antibody tests become positive after 3-4 weeks of infection, which is a relatively long period of time in case of *T. b. rhodesiense*. Another complicating factor, especially with the latter parasite, may be the periodic fluctuations in antibody levels, due to *in vivo* consumption. Finally, false seronegative results ascribed to a state of immunosuppression have sporadically been recorded in late stages of Gambian sleeping sickness. In summary, whenever possible, it may be wise to use a combination of two different test systems and to retest seronegative individuals one or two weeks later.

How specific are the tests? False or undesired seropositive results do occur to some extent. Repeated inoculation by tsetse of non infective animal trypanosomes might play a role. Some cross-reactions with other Trypanosomatidae (visceral leishmaniasis, Chagas' disease) are to be expected. Much more embarassing is the tendency of malaria to give false positive results in some test systems, particularly at low cut-off values. After successful chemotherapy most antibody tests remain positive for long periods of time.

Under laboratory conditions, where combined parasitological and serological tests can be repeatedly applied on individual patients, correct diagnosis will eventually be made. During field surveys, however, where hundreds of people have to be examined daily by rapid screening methods, a certain per cent of infected individuals will unavoidably be missed. Surveyors should be aware of the classical formula to calculate the predictive value of positive and negative serological test results. The higher the prevalence rate, the higher the positive predictive value and the more useful serological screening tests become. The choice of a suitable test system depends on various factors: feasibility, cost price, sensitivity, specificity, prevalence rate, frequency of visits.

Specialized laboratories can be contacted for the supply of reagents and accessory materials. Samples of serum, dried blood or cerebrospinal fluid can also be sent for more detailed analyses.

Tests using live trypanosomes

When live trypanosomes are incubated with infection antiserum, surface epitopes may react with antibodies, resulting in agglutination or lysis.

Working with bloodstream form trypanosomes, grown in mice or rats, the tests are strictly VAT specific, which raises complicated questions about the VATs to be used. Working with procyclic culture form trypanosomes, invariant surface antigens only come into play.

All these test systems remain confined to specialized laboratories.

Immune trypanolysis

This test needs to be done with serotyped clone populations of bloodstream form trypanosomes of well–defined VAT. Briefly, a set of VATs is incubated in serum dilutions in the presence of guinea pig complement. After 1-2 hours the preparations are examined under the phase contrast microscope and the per cent of lysis recorded. False positive results virtually do not occur. Sensitivity, however, depends on the VATs used. VAT recognition patterns are relatively well–known for *T. b. gambiense* infection sera but less predictable for *T. b. rhodesiense* [20, 21, 25].

Direct agglutination

Bloodstream form clone populations of defined VAT can be used for direct agglutination assays. Technical details can be found in the literature, including several publications of A. R. Gray. The immune lysis test is probably easier to standardize.

A delicate procyclic agglutination trypanosomiasis test (PATT), using culture forms of *T. brucei*, has been proposed for serodiagnosis of Gambian and Rhodesian trypanosomiasis [26, 27].

Tests using fixed trypanosomes

These tests, although far from perfect, have the advantage of using stable reagents and requiring few laboratory facilities. Antigen preparations are currently derived from bloodstream form trypanosomes grown in mice or rats. In most test versions both variable and invariant antigens can take part in the reaction, often without knowing their relative contribution to the final result.

Direct agglutination

Large quantities of pure trypanosomes can be separated from infected rat blood through a DEAE-cellulose anion exchange column [12] and further processed to obtain a reagent for a rapid card agglutination test for trypanosomiasis, known as CATT [28]. A field kit of CATT/*T. b. gambiense* is currently available. It contains a freeze-dried antigenic reagent consisting of formaldehyde fixed and Coomassie stained trypanosomes, together with control sera and small accessory materials and, as permanent apparatus, a 12/220 Volt card shaker. In Africa the test is widely used to facilitate field surveys on Gambian sleeping sickness, for which purpose it has been specially devised. The test takes only five minutes and is quite easy to perform, so that it can be combined with simultaneous clinical and parasitological examinations.

The easiest CATT version, allowing screening of several hundreds of people per day, is by testing a drop of heparinized blood. This detects the majority of infected patients, with reported figures on sensitivity ranging from 75 to 100%. However, some false positive results, in the range of 1-3%, are equally obtained. When applied on a 1/4-1/8 dilution of blood, serum or plasma, the test becomes more specific with little loss of sensitivity. In areas with high prevalence rates, say 1% or more (this is quite common), systematic screening of the population with CATT reduces the workload of parasitological examinations to be made and considerably increases the detection rate. Parasitological confirmation remains required anyway. Further developments aiming at improved sensitivity of CATT, are conceivable. The present reagent is prepared from trypanosomes of a predominant though not ubiquitous VAT of *T. b. gambiense*, i.e. type LiTat 1.3. It might be supplemented with a few other predominant VATs. Another possibility to increase the sensitivity of the test is the addition of Na_2-EDTA, which counteracts a prozone phenomenon due to complement. Finally, it would be attractive to have a broad spectrum CATT that also serves for *T. b. rhodesiense*. Trials to develop such a test with fixed procyclic trypanosomes have not been successful. Maybe, bloodstream form trypanosomes can be manipulated for this purpose. There is good evidence that CATT processed trypanosomes expose a much broader spectrum of surface epitopes than the live organisms, presumably internal/cryptic VSG epitopes and maybe some invariant surface antigens as well.

Immunofluorescence

This elegant test system has been widely used in field surveys on Gambian sleeping sickness. It can be applied on buffer eluates of dried blood spots collected on filter paper, serum and cerebrospinal fluid samples.

The test procedure is simple. Briefly, slides are coated with trypanosomes and subdivided into small reaction areas. Isotonic phosphate buffered saline (PBS), pH 7.2-7.4, serves as diluent and washing liquid (0.1% sodium azide can be added as preservative). A drop of suitably diluted test sample is allowed to react for 20-30 min, remnants then washed off with PBS. Following a 20-30 min incubation with FITC conjugated anti-human-IgG antibodies, slides are washed again and mounted with glycerol-PBS (50-50 mixture, can be stored in the refrigerator for 1-2 weeks; pH should be 7.2 at least) and a cover slip. Reading is done under an epifluorescence microscope (10 x 40). Processed slides can be stored in the refrigerator for a few days or in the freezer for several weeks.

Antigen preparations commonly used are thin smears of heparinized blood from mice or rats infected with *T. brucei* since 2-4 days. Smears are air-dried for 1-2 hours and fixed with water-free acetone for 15 min. Smears can be hermetically packed with silica gel and stored at -20 to -70°C for several months. Reaction zones are demarcated with nail varnish. Prior to adding the test samples, slides are dehaemoglobinized and rehydrated by giving them a PBS dipping and a 10-15 min soaking bath. From now on, reaction zones should remain wet throughout. Before adding reagents, excess PBS is shaken off and absorbed with tissue paper. A convenient starting dilution of test serum commonly is 1/100-200. The choice of the fluorescent conjugate is a critical factor. According to our experience an IgG (gamma chain) specific reagent should be used. Its optimal working dilution is 2-4 times less than the end-titre found upon titration with positive control sera. Evans blue, in a final concentration of 0.01%, can be added as a counterstain fluorochrome (a 0.1% stock solution can be kept in the refrigerator for a few weeks).

For some obscure reasons, different stocks and populations of *T. brucei* have been used as antigenic reagent by different laboratories. The least one can do for standardization is always to use the same serotyped VATs. Moreover, as regards *T. b. gambiense*, the use of selected wide-spread predominant VATs is an extra advantage [21, 25]. Freeze-dried trypanosomes of defined VAT can be obtained for this purpose [29]. The freeze-dried reagent is reconstituted with PBS and set out onto multispot slides. A fixation procedure for bloodstream form trypanosomes, using 80% acetone and 0.25% formaline in saline, has also been proposed [30]. The final reagent can be stored at 4°C for at least one year and serves for coating multispot slides. However, it still contains a lot of VSG (personal observations).

To circumvent the problems related to VAT diversity, an immunofluorescence test system based on recognition of invariant antigens and suitable for both *T. b. gambiense* and *T. b. rhodesiense* would be welcome. Unfortunately, preparations of procyclic culture form trypanosomes yield a low test sensitivity [30].

Tests using solubilized antigens

Hydrosoluble extracts of trypanosomes are currently made by ultrasonication or by X-Press treatment. The pellet obtained after centrifugation probably contains interesting serodiagnostic antigens but little research has been devoted to this aspect. Surprisingly, only two major types of soluble antigen preparations have been put into practical use: total buffer extracts and purified VSG fractions.

Immunodiffusion

Precipitation tests in membrane or gel media, allowing visualization and appreciation of individual immune complexes, are potentially useful.

Taylor and Smith [31] have described a countercurrent-immunoelectrophoresis method in agarose gel using extracts of cloned bloodstream forms of *T. brucei*, but only few serum samples were tested. Proper antigen selection is a key problem [32]. In a recent study [33], using Ouchterlony double diffusion tests in agarose gel, serum samples from 50 *T. b. gambiense* infected and 40 non trypanosomiasis patients were combined with buffer extracts of respectively procyclic *T. brucei*, four bloodstream form populations of distinct VAT and purified VSG fractions thereof. Invariant antigens yielded a lot of false positive results. VSGs were only recognized by trypanosomiasis sera and, when used in combination, yielded a test sensitivity of 96%.

Enzyme linked immunosorbent assay (ELISA)

Since 1975 several publications have been dealing with this test system. There are as many test versions as authors, but no wide scale applications so far. Once again, the choice of the antigen preparation is of critical importance. Invariant antigens of procyclic and bloodstream form trypanosomes might be suitable reagents for *T. b. rhodesiense* serodiagnosis but, as regards *T. b. gambiense*, such preparations do not give satisfactory results and crude extracts of bloodstream form trypanosomes may give very different results, depending on the VAT population used [25]. This observation has led to the development of ELISA/ *T. b. gambiense* versions based on the use of selected variable antigens [34, 35].

An innovative experimental test system is the anti-trypanosome-enzyme-specific ELISA [36].

Indirect agglutination

Well–established procedures exist for coating red blood cells or latex particles with trypanosomal antigens, to be used in relatively simple agglutination tests. The freeze dried reagents are remarkably stable under tropical conditions. All these features make indirect agglutination tests most practical tools for laboratory and field use. However, further research is needed to develop standardized and optimized tests using well –defined antigens. It should be stressed that existing test versions have been devised for *T. b. gambiense* serology only.

Boné and Charlier [37] have proposed an elegant capillary haemagglutination test, using erythrocytes coated with a buffer extract of uncloned bloodstream. The test has been produced for some time as Testryp, by Smith Kline-RIT. At present, a Cellognost®, microplate version using a similar reagent, is commercially available from Bœhring. All these test versions have been successfully applied in various African countries. Attempts to develop an analogous test with invariant antigens from procyclic *T. brucei* have failed and there is good experimental evidence that VSG antigens are the decisive reactants [25].

A rapid latex/*T. b. gambiense* card agglutination test, using particles covalently coated with selected VSG, has been presented [38]. More recently a field kit with freeze dried multi-VSG coated reagent has been prepared by the same authors. Preliminary evaluation results are quite encouraging.

Antigen and nucleic acid detection tests

During the last decade efforts have been made to develop antigen detection tests that would make clear-cut distinction between ongoing and cured infections. Apart from ELISA versions, a latex agglutination test has also been presented [26, 39]. The working principle and reliability of these pilot test systems remain controversial.

Diagnostic tests based on trypanosomal DNA or mRNA amplification by PCR are still at the research stage [40, 41; Burke unpublished].

Staging of the disease and post treatment follow-up

Detection of trypanosomes in blood or lymph must be followed by examination of cerebrospinal fluid (CSF) to determine whether the patient is already in the meningoencephalitic stage of the disease and requires special treatment. The conventional criteria for second stage infection are: elevated number of white cells (more than 5 per µl), increased total protein concentration (more than 37 mg per 100 ml) and presence of trypanosomes [1]. This arbitrary and rudimentary system causes a great deal of controversy. Some specialists believe that 20 cells would be a better threshold. Others claim that the presence of a few trypanosomes alone, without other abnormalities, is not a sufficient indication. If facilities permit, supplementary analyses should be made. The presence of IgM in CSF is a significant parameter. Moreover, classical indications of impairment of the blood-brain barrier are highly suggestive of advanced infection. Comparative antibody tests on serum and CSF and various other markers are now being explored in some research laboratories. Successful treatment should rapidly be followed by complete negativation of the most sensitive parasitological tests. CSF should remain or progressively become normal. Complete follow-up after second stage treatment normally takes 1-2 years, with regular check-up at 3-6 months intervals. Antibody tests remain positive for very long periods after cure.

References

 1. Anonymous (1997) Epidemiology and control of African trypanosomiasis. World Health Organization, Technical Report Series, Geneva
 2. Barry JD, Turner CMR (1991) The dynamics of antigenic variation and growth of African trypanosomes. Parasitol Today 7:207-211
 3. Anonymous (1980) Manual of basic techniques for a health laboratory. World Health Organization, Geneva
 4. Anonymous (1987) Sleeping Sickness Control. African Medical and Research Foundation, Nairobi
 5. Anonymous (1991) Basic laboratory methods in medical parasitology. World Health Organization, Geneva
 6. Baker JR (1970) Techniques for the detection of trypanosome infections. In: Mulligan HW (ed) The African Trypanosomiases. George Allen and Unwin Ltd, London, p 67
 7. Fleck SL, Moody AH (1988) Diagnostic techniques in medical parasitology. Butterworth & Co Ltd, London
 8. Van Meirvenne N, Büscher P, Aerts D (1989) Use of SDS for detection of trypanosomes in haemolysed blood samples. Poster presentation, 25th Trypanosomiasis Seminar, British Society for Parasitology, Glasgow
 9. Truc P, Bailey JW, Doua F, Laveissière C, Godfrey DG (1994) A comparison of parasitological methods for the diagnosis of gambian trypanosomiasis in an area of low endemicity in Côte d'Ivoire. Trans R Soc Trop Med Hyg 88:419-421
10. Bailey JW, Smith DH (1994) The quantitative buffy coat for the diagnosis of trypanosomes. Tropical Doctor 24:54-56
11. McNamara JJ, Bailey JW, Smith DH, Wakhooli S, Godfrey DG (1995) Isolation of *Trypanosoma brucei gambiense* from northern Uganda: evaluation of the kit for *in vitro* isolation (KIVI) in an epidemic focus. Trans R Soc Trop Med Hyg 89:388-389
12. Lanham SM, Godfrey DG (1970) Isolation of salivarian trypanosomes from man and other mammals using DEAE-cellulose. Exp Parasitol 28:521-534
13. Lumsden WHR, Kimber CD, Evans DA, Doig SJ (1979) *Trypanosoma brucei*: miniature anion-exchange centrifugation technique for detection of low parasitaemias: adaptation for field use. Trans R Soc Trop Med Hyg 73:312-317
14. Lumsden WHR, Kimber CD, Dukes P, Haller L, Stanghellini A, Duvallet G (1981) Field diagnosis of sleeping sickness in the Ivory Coast. I. Comparison of the miniature anion-exchange/centrifugation technique with other protozoological methods. Trans R Soc Trop Med Hyg 75:242-250
15. Cattand P, Miezan BT, De Raadt P (1988) Human African trypanosomiasis: use of double centrifugation of cerebrospinal fluid to detect trypanosomes. Bull WHO 66:83-86
16. Aerts D, Truc P, Penchenier L, Claes Y, Le Ray D (1992) A kit for in vitro isolation of trypanosomes in the field: first trial with sleeping sickness patients in the Congo Republic. Trans R Soc Trop Med Hyg 8:394-395
17. Truc P, Aerts D, McNamara JJ, Claes Y, Allingham R, Le Ray D, Godfrey DG (1992) Direct isolation *in vitro* of *Trypanosoma brucei* from man and other animals, and its potential value for the diagnosis of gambian trypanosomiasis. Trans R SocTrop Med Hyg 86:627-629
18. Truc P (1996) A miniature kit for the *in vitro* isolation of *Trypanosoma brucei gambiense*: a preliminary field assessment on sleeping sickness patients in Côte d'Ivoire. Trans R Soc Trop Med Hyg 90:246-247
19. Le Ray D (1975) Structures antigéniques de *Trypanosoma brucei* (Protozoa, Kinetoplastida). Analyse immuno-électrophorétique et étude comparative. Ann Soc Belge Méd Trop 55:129-311
20. Isharaza WK (1990) Variant-specific trypanolytic antibodies in sera from patients infected with *Trypanosoma brucei rhodesiense*. Bull WHO 68:33-37
21. Van Meirvenne N, Magnus E, Büscher P (1995) Evaluation of variant specific trypanolysis tests for serodiagnosis of human infections with *Trypanosoma brucei gambiense*. Acta Trop 60:189-199
22. Overath P, Chaudi M, Steverding D, Ziegelbauer K (1994) Invariant surface proteins in bloodstream forms of *Trypanosoma brucei*. Parasitol Today 10:53-58
23. Imboden M, Müller N, Hemphill A, Mattioli R, Seebeck T (1995) Repetitive proteins from the flagellar cytoskeleton of African trypanosomes are diagnostically useful antigens. Parasitology 110:249-258
24. Pays E, Berberof M (1995) Antigènes variables et non variables des trypanosomes africains. Médecine Science 11:261-267
25. Vervoort T, Magnus E, Van Meirvenne N (1983) Serological tests for sleeping sickness: importance of antigen selection. In: Crooy PJ (ed) Symposium on the diagnosis of African sleeping sickness due to *T. gambiense*. Smith Kline-RIT, Rixensart, p 47
26. Liu MK, Cattand P, Gardiner IC, Pearson TW (1989) Immunodiagnosis of sleeping sickness due to *Trypanosoma brucei gambiense* by detection of anti-procyclic antibodies and trypanosome antigens in patients' sera. Acta Trop 46:257-266

27. Ngaira JM, Olaho-Mukani W, Omuse JK, Tengekyon KM, Mbwabi D, Olado D, Njenga JN (1992) Evaluation of procyclic agglutination trypanosomiasis test (PATT) for the immunodiagnosis of *Trypanosoma brucei rhodesiense* sleeping sickness in Kenya. Trop Med Parasitol 43:29-32
28. Magnus E, Vervoort T, Van Meirvenne N (1978a) A card-agglutination test with stained trypanosomes (C.A.T.T.) for the serological diagnosis of *T. b. gambiense* trypanosomiasis. Ann Soc Belge Méd Trop 58:169-176
29. Magnus E, Van Meirvenne N, Vervoort T, Le Ray D, Wéry M (1978b) Use of freeze-dried trypanosomes in the indirect fluorescent antibody test for the serodiagnosis of sleeping sickness. Ann Soc Belge Méd Trop 58:103-109
30. Katende JM, Nantulya VM, Musoke AJ (1987) Comparison between bloodstream and procyclic form trypanosomes for serological diagnosis of African human trypanosomiasis. Trans R Soc Trop Med Hyg 81:607-608
31. Taylor AER, Smith V (1983) Microcounter immuno electrophoresis: a rapid screening technique for trypanosomiasis. Trans R Soc Trop Med Hyg 77:481-486
32. Van Meirvenne N, Le Ray D, Wéry M, Afchain D (1975) Serological and parasitological diagnosis of sleeping sickness. Ann Soc Belge Méd Trop 55:545-549
33. Poortmans A (1995) Serodiagnostic evaluation of variable and invariant antigens of *Trypanosoma brucei gambiense* in gel precipitation tests (in dutch). Thesis, Institute of Tropical Medicine, Antwerp
34. Vervoort T, Magnus E, Van Meirvenne N (1978) Enzyme-linked immunosorbent assay (ELISA) with variable antigen for serodiagnosis of *T. b. gambiense* trypanosomiasis. Ann Soc Belge Méd Trop 58:177-183
35. Büscher P, Depla E, Van Meirvenne N, Magnus E (1995) A serodiagnostic ELISA using variable antigens of *Trypanosoma brucei gambiense*. In: Sones KR (ed) Proceedings 22nd meeting ISCTRC, Kampala 1993. OAU/STRC, Nairobi, p 46
36. Borowy NK, Schell D, Schefer C (1991) Diagnosis of human African trypanosomiasis and visceral leishmaniasis based on the detection of anti-parasite-enzyme antibodies. J Inf Dis 164:422-425
37. Boné G, Charlier JL (1975) Hémagglutination en tubes capillaires: une méthode de diagnostic de la trypanosomiase applicable sur le terrain. Ann Soc Belge Méd Trop 55:559-569
38. Büscher P, Draelants E, Magnus E, Vervoort T, Van Meirvenne N (1991) An experimental latex agglutination test for antibody detection in human African trypanosomiasis. Ann Soc Belge Méd Trop 71:267-273
39. Nantulya VM, Doua F, Molisho D (1992) Diagnosis of *Trypanosoma brucei gambiense* sleeping sickness using an antigen detection enzyme-linked immunosorbent assay. Trans R Soc Trop Med Hyg 86:42-45
40. Shares G, Mehlitz D (1996) Sleeping sickness in Zaire: a nested polymerase chain reaction improves the identification of *Trypanosoma brucei gambiense* by specific kinetoplast DNA probes. Trop Med Int Health 1:59-70
41. Penchenier L, Dumas V, Grebaut P, Reifenberg JM, Cuny G (1996) Improvement of blood and fly gut processing for PCR diagnosis of trypanosomosis. Parasite 3:387-389

CHAPTER 15

Present strategies in the treatment of human African trypanosomiasis

S Van Nieuwenhove

Introduction

Almost a century after the introduction of Atoxyl® in 1905, the ideal antitrypanosomal drug, that should be cheap, effective in early and advanced sleeping sickness, of low toxicity, with a long duration of action after oral absorption and not rapidly inducing resistance, has yet to be discovered. Although trypanosomes, because of their intriguing biological characteristics, have become favourite study objects for scientists and several of their features are sufficiently different from those of humans to form attractive targets for antitrypanosomal therapy, the current drugs are few, toxic, generally expensive and often difficult to obtain.

The situation in the first half of this century was even worse. Atoxyl® cured only early *Trypanosoma brucei gambiense* infections, thus advanced disease was lethal until the discovery of tryparsamide in 1919. Patients with advanced *T. b. rhodesiense* infections remained incurable until melarsoprol became available in 1949 and by that time *T. b. gambiense* had already developed widespread resistance against tryparsamide. Patients with tryparsamide-resistant sleeping sickness were depicted by Friedheim [1] as the "doomed left overs of mass treatment, dragging on for a number of years through the repeated sequence of treatment and relapse to the unavoidable fatal end". Atoxyl®, tryparsamide, melarsonyl potassium (Mel W) and several other drugs are no longer produced [2-5].

Sleeping sickness therapy is complicated by differences in sensitivity to antitrypanosomal drugs between *T. b. gambiense* and *T. b. rhodesiense* (which is generally more refractory) and by the existence of two disease stages. During the first or haemolymphatic stage (stage I), trypanosomes are present in lymph, blood, bone marrow and extravascular spaces of various organs, but not in the central nervous system (CNS). In the second or meningoencephalitic stage (stage II), they have also penetrated in the CNS, where they are (at least partially) protected by the selective permeability of the blood-brain barrier against several drugs that can cure stage I disease. Hence, the need to distinguish between stages in order to prescribe the right type of treatment.

Stage definition is based on cerebrospinal fluid (CSF) examination. A patient is considered to be in stage I when trypanosomes cannot be detected in the CSF, the CSF-leukocyte count does not exceed 5/µl and the CSF-protein content is not higher than 37 mg/100ml (dye binding method) or 25 mg/100ml (trichloracetic acid turbidometry method). However, these empirically defined criteria are quite unsatisfactory: stage I drugs are able to cure a proportion of patients with stage II sleeping sickness and show synergism with other antitrypanosomal compounds against CNS disease; trypanosomes detected after double centrifugation [6] in otherwise normal CSF can generally be eliminated with stage I drugs; CSF-leukocytes and protein are also raised in the course of several other diseases, which may complicate stage definition and interpretation of post-treatment follow-up results. Research to identify more satisfactory CSF parameters is continuing [7], but it is likely that the CSF itself is only an imperfect yardstick for trypanosome penetration in the cerebral parenchyma.

Post treatment follow-up is tedious: a patient is considered cured if he is clinically well 2 years after therapy, no trypanosomes can be detected in CSF, lymph and blood and CSF-leukocyte count and protein content are normal. Normalization of CSF parameters may be slow, thus as long as they progress towards normal there is no reason to suspect a relapse.

Therapy of stage I sleeping sickness has not changed in decades: suramin was introduced in 1922, pentamidine in 1941 and diminazene in the mid-1950s. Treatment of stage II disease still relies almost exclusively on melarsoprol. None of these drugs would easily pass current registration regulations, but, fortunately for hundreds of thousands who have been cured by them, they got on the market in the first half of the century.

About two decades ago, it was shown that nifurtimox and eflornithine are able to cure stage II *gambiense* sleeping sickness, including melarsoprol-refractory infections, but (because of its high cost, complicated IV-regimen and limited availability) hardly 2000 patients have so far received eflornithine (mostly during clinical trials) and nifurtimox has been used on an even more limited scale, mainly as compassionate treatment for melarsoprol-refractory *gambiense* sleeping sickness.

In this chapter, suramin, pentamidine, diminazene, melarsoprol, nifurtimox and eflornithine will be reviewed, and ways to optimise therapy suggested.

Suramin (Bayer 205, Germanin®)

Chemistry and mode of action

Suramin is a sulphonated naphtylamine with a high molecular weight that is strongly negatively charged at physiological pH and is taken up by trypanosomes by pinocytosis, as a plasma-protein-bound complex [8, 9]. It is an inhibitor of many dehydrogenases and kinases [10], of thymidine kinase

[11] and dihydrofolate reductase [12]. It is also a potent inhibitor of glycolytic enzymes [13]. Nine glycolytic enzymes are in trypanosomes confined within glycosomes, organelles enclosed by a phospholipid bilayer membrane [14, 15] that most likely prevents glycosomal uptake of suramin. As glycolysis provides the only source of energy for bloodstream trypanosomes [16], a quick lysis would be expected if any of these enzymes became inhibited, but suramin has a slow antitrypanosomal effect [9]. This pleads against direct inhibition of any of these enzymes inside the glycosomes. Wang [17] has postulated that suramin may bind to the glycolytic enzymes during the few minutes they are present in the cytoplasm of trypanosomes, between their synthesis and their import into the glycosomes, but this hypothetical mechanism of action has yet to be proven experimentally.

Pharmacokinetics

The pharmacokinetic profiles vary from one individual to another [18]. Serum levels are usually high immediately after intravenous injection, but fall rapidly within the first couple of hours and subsequently more slowly over the next few days. CSF penetration is negligible. The elimination half-life after a course of 6.2 g suramin was very long (44 to 54 days), 99.7% of the drug was bound to plasma proteins, total plasma levels remained higher than 100 µg/ml for several weeks and urinary excretion accounted for most of the elimination [19].

Effectiveness and indications

Because of its poor CSF penetration, suramin monotherapy is restricted to stage I infections. Trypanosomes disappear within 12 to 36 h from blood and lymph [10]. It is the drug of choice for *T. b. rhodesiense*, but is nowadays rarely used as monotherapy in *T. b. gambiense* sleeping sickness. There is great variation in treatment regimens. Cure rates in stage I disease of close to 100% have been reported [3, 4, 20], but failure rates of 25-35% have also been observed [21, 22]. More recently, relapse rates of 6.9% [23] and 31% [24] were reported in stage I *T. b. rhodesiense* infections.

A combination with pentamidine has quite extensively been used for treatment of stage I *gambiense* sleeping sickness and was considered more effective and less toxic than monotherapy with any drug [2]. A relapse rate of 3% was reported by Neujean [22], as compared to 16% with pentamidine monotherapy. This combination is still administered in some areas and Pépin and Milord [25] recently noticed 9.6% relapses.

Suramin (a single or several injections) is often given prior to melarsoprol, mainly for stage II *T. b. rhodesiense* infections and has also been used for chemoprophylaxis [26]. During field surveys, a single injection is sometimes administered to newly detected patients to clear their blood of trypanosomes between the time of diagnosis and the moment they arrive at a health service for further treatment.

Notwithstanding its low CSF penetration, cures have been reported after suramin monotherapy of patients with stage II disease [27] and there is experimental evidence [28-34] clinical indications [35, 36] that suramin shows synergism with other drugs (including nifurtimox, nitroimidazoles and eflornithine) against stage II infections. Thus, it may be assumed that it reaches some parts of the CNS, as was suggested by animal experiments [29].

Adverse effects

As sometimes lethal idiosyncratic reactions (anaphylactic shock) may occur, most frequently when there is concomitant onchocerciasis, it is recommended to start with a test dose of 100-200 mg slowly IV. Other adverse effects are: fever (very common after the first injection), nausea, vomiting, conjunctivitis, photophobia, palpebral oedema, pruritus, urticaria, cutaneous eruptions (including exfoliative dermatitis), stomatitis, diarrhoea, arthralgia, neurotoxic reactions (paresthesia, hyperesthesia, peripheral polyneuropathy), changes in the blood picture (leukocytopenia, thrombocytopenia, haemolytic anaemia and agranulocytosis). Most commonly, some degree of renal damage occurs [3], but treatment should only be discontinued when there is high albuminuria or when casts or red blood cells appear in the urine sediment. Cheson et al. [37] have noticed liver dysfunction, adrenal insufficiency, thrombocytopenia and neutropenia during administration to AIDS patients. Because of its long elimination half-life, suramin toxicity is cumulative [10]. Adverse effects during sleeping sickness therapy are most likely underreported. Veeken et al. [24] registered a mortality rate of 1.3% (2 patients) during treatment of stage I *T. b. rhodesiense* infections and Pépin et al. [38] observed 4 deaths (0.64%) among patients with stage II *gambiense* sleeping sickness within days after a single injection (before melarsoprol could be started). Burri and Blum [39] reported a lethal encephalitic reaction (5 days after the first suramin) in a patient with stage I *T. b. gambiense infection*. Prospective studies to update knowledge of suramin toxicity during sleeping sickness therapy seem warranted.

Recommended regimens, availability and cost

Suramin is produced by Bayer under the trade name Germanin® and is available in vials containing 1g of whitish powder. A 10% solution in distilled water should be prepared for immediate IV injection (IM administration is too painful). Usually, 5 injections are administered at intervals of 5-7 days (20 mg/kg with a maximum of 1 g per injection). One vial costs about 5.4 US$, thus a 5-injection course amounts to 27 US$ (added expenses for injection materials, and hospitalisation not included).

Pentamidine (pentamidine isethionate BP, Pentacarinat®)

Chemistry and mode of action

Pentamidine is an aromatic diamidine that is positively charged at physiological pH. Until recently, two closely related molecules were available: pentamidine isethionate (Pentacarinat®) and methanesulfonate (Lomidine®), but the latter is no longer produced. The mode of action is unknown. Mechanisms that may play a role have been reviewed by Wang [17]: various bindings to nucleic acids, disruption of kinetoplast DNA, inhibition of RNA-editing in trypanosomes and inhibition of mRNA trans-splicing. Pentamidine has been shown to reversibly inhibit trypanosomal S-adenosylmethionine decarboxylase, which plays a role in the biosynthesis of polyamines [40], but recent findings suggest that inhibition of this enzyme is not involved in the antitrypanosomal action [41]. The drug is transported into bloodstream forms of *T. b. brucei* via a carrier-mediated process, which raises the concentration within trypanosomes to many times the plasma level [42]. Resistance can easily be induced in the laboratory [41, 43] and is primarily due to a diminished ability of resistant parasites to import pentamidine.

Pharmacokinetics

Several specific high performance liquid chromatographic methods (HPLC) have been developed during the last 10 years [44-46]. Conte et al. [47] measured a 3 times higher peak plasma concentration after IV than after IM administration and found a renal clearance of 2.5% and 5%, respectively. Accumulation occurred when multiple doses were given, without achieving steady state levels after 13 injections [48]. The estimated elimination half-life after the last injection was 12 days and after 6 weeks pentamidine could still be detected in almost all patients. Bronner [49] studied the pharmacokinetics in three groups of patients with sleeping sickness and one with AIDS, who received 1-10 doses of 1.7-4.5 mg pentamidine base/kg as IV infusions or IM injections every day or on alternate days. There was accumulation during repeated dosing, and metabolism was found to be the major route of drug elimination. The elimination half-life after the last dose was in the order of one to several weeks. There was a 3-fold variation in plasma clearances, which may reflect individual differences in metabolic capacity. Tissue binding is high and plasma protein binding is estimated at 70%. In sleeping sickness patients treated with 10 IM injections of methanesulfonate (3.5-4.5 mg/kg on alternate days), maximum plasma levels were generally reached within 1h after injection and varied extensively (420-13420 nmol/l), presumably due to varying absorption from the injection site [50]. The median plasma concentration after the last dose was about 5 times higher than after the first. CSF levels after the

10th injection were very low (0.5-0.8% of plasma concentrations), but pentamidine generally persisted in CSF for at least 30 days. *In vitro*, quite low concentrations have a trypanocidal effect during prolonged exposure [51]. The role played by metabolites needs further elucidation.

There is pharmacokinetic evidence that 3 injections may be equally effective and less toxic than 7-10 injections [49]. TDR (UNDP/WB/WHO Special Programme for Research and Training in Tropical Diseases) is planning a randomised trial to compare effectiveness and toxicity of a 3-day with a 7-day regimen (4 mg isethionate/kg/day IM).

Effectiveness and indications

Pentamidine is the drug of choice for stage I *gambiense* sleeping sickness. Van Hoof et al. [52] detected trypanosomes in blood and lymph up to 48 h after the first injection. It gives less reliable results in *T. b. rhodesiense* [4, 53, 54]. Relapse rates of about 7% have been reported [55-57], but higher rates (16%) were also observed [22]. Patients who were not cured usually responded to melarsoprol [22-56], but little is known about the present sensitivity of pentamidine refractory trypanosomes. Recent rumours about considerable relapse rates in a few areas need confirmation.

A combination with suramin is sometimes used against stage I *T. b. gambiense* infections (see under suramin) and 1-2 pentamidine injections are often administered as pre-treatment prior to melarsoprol in stage II *gambiense* sleeping sickness.

Cures with pentamidine monotherapy of stage II infections have been reported [20, 58, 59], but the proportion of relapses increased in parallel with the CSF-leukocyte count. Doua et al. [60] found recently only 6% relapses in 52 patients with stage II infections (CSF-leukocytes up to 20/µl), but preliminary results of a similar trial in Uganda suggest a higher relapse rate (Legros, unpublished). Prospective randomised studies on a larger scale are needed before any modification of routine treatment strategies could be recommended.

Six-monthly injections have been used for "chemoprophylaxis" campaigns [26, 52]. In view of the limited sensitivity of microscopic techniques, it must be assumed that quite a number of already infected individuals were given intermittent subcurative treatment instead of prevention and progressed undiagnosed to advanced stage II disease. It has been suggested that mass chemoprophylaxis may have induced increased resistance, but this is not well substantiated.

Adverse effects

Adverse effects have been more extensively studied in AIDS patients than in individuals with sleeping sickness, where there is most likely underreporting. Nephrotoxicity is the most common complication, but is generally mild to moderate and reversible [61]. Other adverse effects include: syncope (adrenaline should be at hand), hypotension, breathlessness, pallor, fever, hypersalivation, headache, nausea, vomiting, abdominal pain, diarrhoea, dizziness, pruritus, cutaneous eruptions, peripheral polyneuropathy, cardiotoxicity (tachycardia, ventricular arrhythmias), pancreas toxicity (hypoglycaemia, hyperglycaemia and diabetes), hepatic dysfunction; changes in the blood picture (leukocytopenia, anaemia, thrombocytopenia) and hypocalcaemia. IM injections are quite painful, may cause severe local inflammation ("sterile abscess"), sciatic nerve damage and occasionally gangrene. Doua and Boa Yapo [62] noticed 1 case of convulsions, 1 sudden death and 1 diabetes mellitus among 150 patients with stage I *T. b. gambiense* infections. The occurrence in 1993 of 4 deaths after a single injection has prompted cessation of pentamidine pre-treatment (prior to melarsoprol) in Uganda (John, unpublished), where also hepatic dysfunction, leukocytopenia and thrombo-cytopenia have been reported. There is no laboratory evidence of mutagenicity [63] or of adverse effects on pregnancy [54, 64]. Prospective studies to update knowledge of pentamidine toxicity during sleeping sickness therapy seem warranted.

Recommended regimens, availability and cost

Courses of 7-10 IM injections of 4 mg pentamidine isethionate/kg (daily or on alternate days) are usually given. Patients should lie down for some time after each injection. Although slow IV infusion (over 1-2 h) is better tolerated, IM administration is, for practical and budgetary reasons, likely to remain the major route in areas with high sleeping sickness prevalence.

Pentamidine isethionate is produced under the trade name Pentacarinat® by May and Baker (a subsidiary of Rhône Poulenc Rorer) for treatment of *Pneumocystis carinii* pneumonia in AIDS patients. One 300 mg vial costs about 30 US$ (although the drug was very cheap in the pre-AIDS era). Fortunately, the manufacturer agreed to provide pentamidine free of charge to WHO for sleeping sickness treatment, but shortages have been a recurrent problem. WHO charges 1.5 US$ per 200 mg vial for handling and transport, thus a 10-day IM regimen amounts to 15 US$ (added expenses for injection materials, and hospitalisation not included). The same course with commercial Pentacarinat® costs about 200 US$. Depending on the results of forthcoming clinical trials, a short 3-day regimen may be recommended in the future.

Diminazene aceturate (Berenyl®)

Diminazene aceturate is a pentamidine analogue, developed as a trypanocide for veterinary use. Although it is not registered for sleeping sickness therapy, several thousands of patients with stage I *T. b. gambiense* and *T.*

b. rhodesiense infections have been treated with IM diminazene [65-70]. It has occasionally been given orally [71] and has also been used as pretreatment prior to melarsoprol. Relapse rates were low (3-15%). Adverse effects are quite similar to those of pentamidine: fever, nausea, vomiting, cutaneous eruptions (including exfoliative dermatitis) and albuminuria. Abaru et al. [72] reported 1% coma and 6% peripheral polyneuropathy during treatment of 99 patients with stage I *T. b. rhodesiense* infections, but all recovered. De Raadt [73] observed "reactive encephalopathy" after administration in stage II *rhodesiense* sleeping sickness (prior to melarsoprol). Diminazene has several advantages over pentamidine: injections are less painful, treatment is shorter (better suited for ambulatory therapy), it is very cheap, regularly produced and easily obtainable. However, because of poor profit prospects, the manufacturer is not pursuing the registration procedure for human therapy. As long as it is not registered, diminazene can not be recommended for routine use, but further toxicological studies and randomised trials to compare its effectiveness and adverse effects with those of pentamidine seem long overdue.

Diminazene is marketed by Hoechst under the trade name Berenyl® as sealed sachets containing 1.05 g diminazene aceturate to be dissolved in distilled water. Treatment regimens for human trypanosomiasis have varied widely: 1-3 IM injections of 5-7 mg/kg, usually on alternate days. A 3-injection course costs about 2 US$ (added expenses for injection materials not included).

Melarsoprol (Mel B, Arsobal®)

Chemistry and mode of action

Melarsoprol is a combination of the trivalent organic arsenical melarsen oxide with the heavy metal chelator BAL (British Anti-Lewisite, dimercaprol) [74]. The adduct is about 100 times less toxic than melarsen oxide, but only 2.5 times less trypanocidal. The mechanisms of action are not well understood. Trypanothione, that is believed to be a major cofactor in the dithiol-disulfide redox balance in trypanosomes and that forms the adduct MelT with melarsen oxide, has been proposed as the primary target [75]. However, as melarsen oxide-treated trypanosomes become already swollen and nonmobile when the bulk of intracellular reduced trypanothione is still intact, lysis of trypanosomes is more likely caused by blocked glycolysis [17]. Resistance is probably due to reduced drug uptake into the parasite [43, 76].

Pharmacokinetics

Melarsoprol is mainly eliminated in the faeces [77]. Current treatment regimens were generated by trial and error: 2-4 series of 3-4 daily injections, separated by at least 7 days. With an unspecific bioassay (determining antitrypanosomal activity), Burri and Brun [78], Burri et al. [79] and Burri [80] calculated an elimination half-life of about 35 h and found no evidence of drug accumulation. Serum levels were high enough to eliminate blood trypanosomes in a short time, but dropped to almost zero in between series of injections. CSF concentrations were 50-100 times lower than serum levels, increased very slowly and varied considerably among patients. Variation could be due to differences in drug metabolism, altered protein binding, different degrees of impairment of the blood-brain barrier and variation in active drug transport. CSF levels after the first series of injections may be too low in some cases to clear the CNS of all trypanosomes. Low melarsoprol levels are likely to be trypanocidal if they persist for a sufficiently long time, which may not always be the case when series of injections are separated by 7-day intervals.

With a specific HPLC-method much lower plasma levels were measured (except immediately after injection), less than 1% was found in the urine and melarsoprol could not be detected in plasma or CSF 24h after the 4th injection [81]. The discrepancy between results obtained with HPLC and bioassay suggests that the drug is transformed into active metabolites. More research is needed to fully elucidate its kinetics.

Towards the end of his life, Friedheim and Distefano [82], who had in the past successfully treated a number of patients with 1.8 mg/kg for 10 days, strongly recommended consecutive treatment with lower melarsoprol doses. Good results were reported in monkeys with 1.8 mg/kg/day [31] and during preliminary field trials, promising results were obtained with 1.8 mg/kg/day (Van Nieuwenhove, unpublished) and with 2.2 mg/kg/day [83] during 10 consecutive days. Shorter and cheaper regimens would have invaluable operational and financial advantages during control campaigns in highly infected areas and are expected to be equally effective and possibly less toxic. Randomised trials with lower dose consecutive melarsoprol have started in Angola (Tropical Institute Basel) and in the DRC (Institute of Tropical Medicine Antwerp).

Effectiveness and indications

Melarsoprol is still the most powerful trypanocide and is able to cure both stages of *gambiense* and *rhodesiense* sleeping sickness, but is nowadays rarely used for stage I infections. Treatment regimens vary enormously: total doses of 1.26-2.16 g (35-60 ml), intervals between series of 7-14 days, gradually increasing doses or maximum dose from the onset. The number of series usually increases in parallel with the CSF-leukocyte count [21], but some experts recommend maximum treatment as soon as the CNS is involved. Melarsoprol is generally preceded by a single or several injections of stage I drugs (which may complicate assessment of effectiveness because of possible synergism) and is often associated with corticosteroids. Examples of various treatment regimens can be found elsewhere [84].

Melarsoprol remains remarkably effective, notwithstanding the wide range of treatment regimens. It would be interesting to figure out if pre-treatment with stage I drugs may have slowed down the development of resistance. Primary refractoriness (trypanosomes persisting in the CSF throughout treatment) is rare and relapses are estimated at less than 10% [23, 25]. However, even a low relapse rate can represent an important number of patients, e. g. of the individuals treated with melarsoprol in the Democratic Republic of Congo (DRC) in 1997, about 1500 are expected to relapse. A second melarsoprol course may cure some [22, 23, 25, 85-87], but the majority will need alternative treatment to get cured. Recent rumours about considerable relapse rates in a few areas need confirmation.

Adverse effects

Injections of stage I drugs prior to melarsoprol may complicate assessment of toxicity, as they have a long elimination half-life and are quite toxic themselves. Adverse effects may be severe [3, 4, 25, 62]: CNS complications ("reactive arsenical encephalopathy"), peripheral polyneuropathy (possibly due to direct drug toxicity), subjective sensations of heat, disturbance of smell, headache, fever, nausea, tremors, chest pains, abdominal pain, vomiting, diarrhoea, myalgia, cutaneous eruptions (including exfoliative dermatitis), Lyell syndrome (sometimes lethal), thrombophlebitis (due to propylene glycol), cardiotoxicity (likely responsible for some of the mortality), renal dysfunction (albuminuria), hepatic dysfunction and agranulocytosis. The teratogenicity is unknown, but the risk of untreated trypanosomiasis for mother and foetus (congenital trypanosomiasis) may outweigh the potential danger of melarsoprol treatment. Overall mortality rates during therapy vary from 2% [57] to 9.8% [88] in *gambiense,* and from 3.4% [89] to 12% [86] in *rhodesiense* sleeping sickness.

Little attention has been paid to the solvent propylene glycol, that was found to considerably increase melarsoprol toxicity in mice [90], nor to the potential adverse effects of dimercaprol, the other part of the melarsoprol adduct. Dimercaprol during CNS complications has also been given as "antidote" against arsenic toxicity, but is no longer recommended for that purpose.

"Reactive arsenical encephalopathy" is an ill-defined set of CNS complications. The severe form usually starts with a rapid deterioration of the patient's level of consciousness, often accompanied by fever, convulsions or status epilepticus and frequently leading to pulmonary oedema, coma and death (often within 48 h). Abnormal behaviour and psychotic reactions are sometimes considered as a less severe form, with a more favourable prognosis. The distinction between "reactive" and "haemorrhagic" encephalopathy [91, 92] is of non operational use, as it is based on post-mortem findings. Most CNS reactions occur towards the end of the first series of injections, during the first interval or in the course of the second series. Mortality rates attributed to "reactive arsenical encephalopathy" vary from 0.95% [93] to 9.4% [24], but the determination of the exact cause of death is often the result of educated guess work, particularly in retrospective studies.

There is no consensus about the best way to reduce frequency and severity of CNS complications. Interruption of melarsoprol and high doses of parenteral corticosteroids and anticonvulsive drugs are probably the most useful measures that should be taken as soon as CNS complications are suspected. Mannitol and subcutaneous adrenaline [94] have been tried, but their potential benefit has never been evaluated in controlled trials and their administration in rural health services may be dangerous. Resumption of therapy, after patients have sufficiently recovered, generally does not trigger off further CNS complications [3, 24, 38, 91, 93], but recurrence of encephalitic reactions has occasionally been observed (John, unpublished).

It is generally assumed that elimination of trypanosomes from blood and lymph by slower acting stage I drugs [86, 89, 91] or by starting with a low dose of melarsoprol [86] may reduce CNS complications, presumably as a result of a more gradual release of trypanosomal antigens. Schmidt and Bafort [95] and Jennings et al. [96] postulate, on the contrary, that "reactive arsenical encephalopathy" may be induced by initial subcurative treatment and recommend more aggressive antitrypanosomal therapy from the start, associated with immunosuppressive drugs [96-98]. However, it needs to be proven that their observations in animal experiments can be extrapolated to humans before routine treatment regimens ought to be altered.

Concomitant administration of corticosteroids remains controversial. Pépin et al. [38] observed a non-significantly lower proportion of encephalopathy-related deaths (2.8%) in patients with stage II *gambiense* sleeping sickness who received concomitant corticosteroids, as compared to 6.2% when only melarsoprol was given. However, the overall mortality showed an even less significant difference between both groups: 5.2% vs. 7.5% and as the trial was done in an area with low HIV-seroprevalence, it would appear judicious to study the impact of concomitant corticosteroids on relapse rate and frequency of intercurrent infections in immunocompromised HIV-seropositive patients before routine use could be recommended. It has been shown that drug-induced immunosuppression in mice causes more frequent relapses and a faster selection of arsenical-resistant trypanosomes [99]. Moreover, tuberculosis, more frequent in areas with high HIV-seroprevalence, is likely to be exacerbated by corticosteroids, as well as strongyloidiasis and amoebiasis.

Treatment-induced encephalopathy

The cause of "reactive arsenical encephalopathy" is still a matter of speculation. If CNS complications were due to direct drug toxicity, one would expect them to increase towards the end, but they are rare after the second series of injections. Their frequent occurrence during the first interval and the second series could, in view of the very slow CSF penetration of melarsoprol [79, 80] theoretically be explained in several ways. Melarsoprol might only reach sufficiently high levels towards the end of the first or during the second series to eliminate a large number of parasites in the CNS, causing an important release of trypanosome products that could trigger off Jarish-Herxheimer type reactions or immune-mediated CNS complications [100, 101]. On the other hand, concentrations at the end of the first series may be too low in some patients to clear the brain of all trypanosomes and the remaining parasites could trigger off an immune-mediated exacerbation of the pre-existing meningoencephalitis [96-98]. Schmidt and Bafort [95] postulated that subcurative treatment triggers off an invasion of surviving trypanosomes into the CNS, and Haller et al. [102] suggested that melarsoprol may bind as a hapten to parasite proteins or immune complexes causing a type of delayed hypersensitivity reaction. It is most likely that "reactive arsenical encephalopathy" is an immune-mediated phenomenon, but further research is obviously needed to elucidate the exact underlying mechanisms. The role of increased cytokine production is currently being investigated [103]. Golden [104] postulated that low levels of selenium, that has a protective effect *in vitro* against arsenic-induced chromosomal damage in mammalian cells [105], may contribute to the occurrence of encephalopathy, but this needs *in vivo* validation. Alcohol intake, heavy physical efforts and concomitant administration of thiabendazole [106] may exacerbate CNS complications. The role of concomitant metronidazole has apparently never been investigated, although this drug has inherent CNS toxicity and some antitrypanosomal activity.

Van Nieuwenhove [107] has drawn attention to the occurrence during nifurtimox and eflornithine therapy of sometimes lethal CNS complications (convulsions, status epilepticus, coma and severe psychotic reactions), clinically indistinguishable from "reactive arsenical encephalopathy". Pépin and Milord [100] reported 2 lethal encephalitic reactions (7.7%) during combined IV/PO eflornithine, and convulsions and/or status epilepticus during eflornithine have been reported by many others [62, 108-111]. Encephalitic reactions have also been observed during treatment of stage I infections with melarsoprol [93, 102, 112] and after administration of diminazene [72] or suramin [39] to patients with stage I disease and suramin or diminazene [73] to patients with stage II infections.

Lethal CNS complications will most likely continue to occur during treatment of sleeping sickness with any effective drug, particularly in individuals with advanced stage II infections, who have often severe meningoencephalitis. Their frequency was found to increase in parallel with the CSF-leukocyte count [22, 38, 59], and is higher in patients with trypanosomes in the CSF [100] and in those admitted in bad conditions [3, 91, 106]. Thus, "treatment-induced encephalopathy" seems a much more appropriate label than "reactive arsenical encephalopathy". Differences in frequency and severity of CNS complications triggered off by melarsoprol, nifurtimox or eflornithine may be related to the stronger and faster antitrypanosomal action of melarsoprol. The potential benefit of concomitant corticosteroids, immunosuppressive and anti-inflammatory drugs, a. o. chloroquine [89], deserves further investigation.

Recommended regimens, availability and cost

The drug is produced under the trade name Arsobal® by Rhône Poulenc Rorer as a 3.6% solution in propylene glycol, delivered in 5 ml ampoules containing 180 mg melarsoprol. Because of the highly irritating solvent, it has to be injected strictly intravenously, usually at a dose of 3.6 mg/kg/day in 2-4 series of 3-4 injections, depending on the degree of alteration of the CSF. Each series is separated by at least one week and the maximum dose per injection is 180 mg (5 ml). Stage I sleeping sickness is sometimes treated with 1 series of 3-4 injections. The price is about 5.3 US$ per ampoule, thus a 9-injection course amounts to 48 US$ (added expenses for injection materials, and hospitalisation not included).

Depending on the results of ongoing trials, consecutive administration during 10 days of lower doses may be recommended in the future. Topical application as a gel gave promising results in mice [113], but more research is needed before this interesting route could be recommended for human treatment.

Nifurtimox (Bayer 2505, Lampit®)

Chemistry and mode of action

Nifurtimox is a 5-nitrofuran introduced for treatment of American trypanosomiasis (Chagas' disease) in the late 1960's. Its mode of action needs further elucidation. The activity of 5-nitrofurans relies on the toxicity of free radicals generated by nitrofuran reduction and their binding to proteins and DNA [114]. Radical scavenging by trypanothione (a conjugate of spermidine and glutathione) probably plays a role in the trypanosome's defence against nifurtimox-induced radical damage, but trypanothione activity may be reduced by nifurtimox, that serves as a substrate for trypanothione reductase, thus leading to futile redox cycling [115-118].

Pharmacokinetics

Initial pharmacokinetic studies [119] showed that serum levels of unchanged nifurtimox after oral administration were very low and peaked after 1-3 h, followed by a rapid decline, so that no unchanged drug was detected after 24 h. Only 0.5% of unchanged nifurtimox is excreted in the urine. In healthy volunteers, Paulos et al. [120] found peak levels on average 3.5 h after a single dose of 15 mg/kg and the mean elimination half-life was 2.95 h. Low serum concentrations of unchanged nifurtimox are presumably due to an almost complete metabolisation by the liver, which raises the question of the therapeutic activity of metabolites. Measurements of radioactivity after administration of nifurtimox labelled with radioactive sulphur-^{35}S to rats showed relatively high concentrations (representing unchanged substance and its metabolites) throughout the animal, including brain and spinal cord [121].

Effectiveness and indications

Nifurtimox is able to cure both stages of *gambiense* sleeping sickness, including melarsoprol-refractory disease. Trypanosomes have been detected in blood and lymph up to 3 days after the onset of treatment (Van Nieuwenhove, unpublished). Virtually nothing is known about its effectiveness in *T. b. rhodesiense* infections.

Janssens and De Muynck [122] obtained encouraging results with oral nifurtimox in 3 not previously treated patients with stage II *gambiense* sleeping sickness (120 mg 3 times/day during 40-120 days, preceded by 1.5 g suramin IV).

In an open trial, oral nifurtimox was given during 14-45 days to 20 patients with stage II *gambiense* sleeping sickness (15 mg/kg/day for adults and 20 mg/kg/day for children, divided in 3 intakes, approximately every 8 h) [107, 123, 124]. This series included 75% arsenical-refractory patients (relapses after melarsoprol or melarsonyl potassium), mostly in very poor condition, who would inevitably have died of terminal sleeping sickness. Nifurtimox was preceded by a single injection of IV suramin (1.5 g in adults and 20 mg/kg in children) or 7 mg/kg IM diminazene (when veins had been obliterated by melarsoprol). The results were excellent: 75% cures (follow-up at least 24 months), 10% probable cures (follow-up between 12 and 24 months), 1 favourable evolution (5%) (follow-up 4 months), 1 relapse (5%) and 5% mortality (1 patient died with cachexia 10 days after discontinuation of treatment).

In another open trial, the same dosage of oral nifurtimox monotherapy (without prior administration of stage I drugs) was given during 14-45 days to 75 patients with stage II, mostly arsenical-refractory *gambiense* sleeping sickness [107]. The results were very good: 60% cures, 12% probable cures, 5.3% possible cures (follow-up between 6 and 12 months), 1 favourable evolution (1.3%) (follow-up 4 months), 8% no follow-up beyond the post-treatment CSF examination (when no trypanosomes were found and CSF parameters had considerably improved), 6.7% relapses and 6.7% mortality. The role of nifurtimox in the mortality remains conjectural.

Moens et al. [125] obtained excellent results with oral nifurtimox monotherapy (4-5 mg/kg 3 times/day during 60 days) in 15 patients with stage II *gambiense* sleeping sickness (follow-up 30 months): 87% cures, 1 relapse (13%) and no mortality in 8 patients with arsenical-refractory infections and 72% cures, 1 relapse (14%), 1 possible relapse (14%) (only slightly increased CSF-leukocyte count) and no mortality in 7 not previously treated individuals.

Pépin et al. [126] reported deceiving results with oral nifurtimox (5 mg/kg 3 times/day during 60 days) in 25 patients with stage II arsenical-refractory *gambiense* sleeping sickness. All initial responses were favourable, but 11 relapses (44%) were diagnosed 1-9 months post-treatment. Trypanosomes were detected in only 4 CSFs, while all other relapses only had increased CSF-leukocyte count and/or elevated CSF-protein. Follow-up was too short to allow definitive conclusions in 36% of patients and 20% had no follow-up at all. Nifurtimox was well tolerated and there was no mortality. Regional differences in sensitivity of *T. b. gambiense* were suggested to explain the high relapse rate, but an equally plausible explanation might be a lack of drug compliance, as the evening dose was left with patients suffering of advanced melarsoprol-refractory disease (Milord F, personal communication) who are often extremely confused or frankly psychotic, and of whom compliance can hardly be expected. The adverse effects here were mildest and the relapse rate highest of all trials, which casts further doubt on the regularity of drug intake.

In a study with high-dose oral nifurtimox (30 mg/kg/day during 30 days) in 30 patients with stage II *gambiense* sleeping sickness, Pépin et al. [127] observed 30% relapses, 53% were considered cured (follow-up 1-38 months), 13.3% had no follow-up and 1 patient (3.3%) died. High-dose nifurtimox was considered more effective, but significantly more toxic. Although complications were more pronounced than during the 15 mg/kg/day trial, one would expect them to be even more impressive after regular ingestion of 3-4 times the recommended dose for Chagas' disease. A cure rate of over 50% in previously incurable patients should be considered a breakthrough, but is in this publication presented as a failure.

Nowadays, nifurtimox is used with good results for compassionate treatment of melarsoprol-refractory *T. b. gambiense* infections in the DRC (Sema Hurugo N, unpublished), and a randomised trial to compare its effectiveness and toxicity with melarsoprol is going on (Institute of Tropical Medicine, Antwerp).

The effectiveness of nifurtimox against *rhodesiense* sleeping sickness has never adequately been investigated. In view of the proven activity of another nitrofuran, nitrofurazone [3, 4, 128], and the lack of satisfactory alternative treatment for patients with melarsoprol-refractory *T. b. rhodesiense* infections, controlled trials seem long overdue.

Adverse effects

Many patients so far treated with nifurtimox were in very bad condition at the onset of treatment, which complicates assessment of drug toxicity. Frequency and severity of complications increased with dose and duration of treatment, but generally reversed rapidly after discontinuation: CNS toxicity (vertigo, convulsions, status epilepticus, psychotic reactions), peripheral polyneuropathy, anorexia, weight loss, gastrointestinal discomfort (nausea, vomiting, epigastric pain), skin rash and exacerbation of pre-existing symptoms (headache, arthralgia). As the tablets have an intense yellow colour, the sclera may turn yellow, which should not be confounded with jaundice. Convulsions generally did not recur after resumption of treatment. Cerebellar syndrome (ataxia, nystagmus, tremors and vertigo) was seen during high dose nifurtimox (30 mg/kg/day for 30 days), but reversible upon cessation of therapy [127]. Acute haemolytic anaemia, reported in G6PD-deficient patients during nitrofurazone treatment [129], has not been observed. Moens et al. [125] and Pépin et al. [126] had no lethal complications with 15 mg/kg/day. The 6.7% mortality during monotherapy of mostly advanced arsenical-refractory patients [107] was similar to the 5.9% reported with 14-day IV eflornithine [108], and lower than the 11.3% and 19.2% respectively seen during oral [107] and combined IV/oral eflornithine [109]. The increase of severe complications and mortality with duration of treatment, prompted Van Nieuwenhove [107] to reduce nifurtimox administration to 14-21 days. While effectiveness remained high, toxicity and mortality decreased considerably.

Recommended regimens, availability and cost

Nifurtimox is supplied as yellow 120 mg tablets under the trade name Lampit®. The active substance is produced by Bayer Germany and the final product by Bayer Argentina. Lampit® can currently be ordered from Bayer Belgium, where a stock is available. One bottle with 100 tablets costs about 12 US$.

For stage II *T. b. gambiense* infections, including melarsoprol-refractory disease, we recommend 5 mg/kg PO 3 times/day for adults and 7 mg/kg 3 times/day for children during 14-21 days. The potential therapeutic benefit of higher doses or longer administration does not outweigh the increased risk for severe complications. A 21-day nifurtimox regimen amounts to about 20 US$ (plus expenses for hospitalisation), which is only 2% of the cost for drug and intravenous fluids for a 14-day IV eflornithine course (approximately 1000 US$, expenses for hospitalisation not included).

Eflornithine (DFMO, Ornidyl®)

In the early 1980s, eflornithine was a rationally designed cytostatic drug in search of a disease, as its expected activity in tumour chemotherapy had turned out to be unsatisfactory. Its pharmacokinetics and toxicity were known and clinical trials in cancer patients had been done. At that time, there was no satisfactory treatment for melarsoprol-refractory sleeping sickness and oral eflornithine was first tried in the field in 1981 [130], after Bacchi et al. [131] had shown that it could eliminate trypanosomes *in vivo*.

Chemistry and mode of action

Eflornithine (DL-α-difluoromethylornithine) is a specific, enzyme-activated, irreversible inhibitor of ornithine decarboxylase (ODC). Decarboxylation of ODC is an obligatory step in the biosynthesis of polyamines, which are thought to play important roles in cell division and differentiation. Eflornithine can effectively inhibit ODC activity and deplete polyamines in trypanosomes, which brings them into a dormant state that is vulnerable to the host's immune attack, thus a sufficiently active immune system is required to achieve cure [132, 133]. A decrease in trypanothione content (a conjugate of glutathione and spermidine), presumably due to a depletion of spermidine [134], has also been postulated as mechanism of action. However, the achieved reduction in trypanothione levels does not lead to enhanced sensitivity to oxidative stress and no subsequent lysis of trypanosomes occurs, as they only turn dormant [135]. Moreover, eflornithine-treated bloodstream trypanosomes are still able to differentiate into procyclic forms *in vitro* [136]. These findings suggest that, except for the inability to proliferate, few other abnormalities are associated with eflornithine-treated parasites [17].

Pharmacokinetics

After oral intake, peak plasma-levels are reached about 4 h after ingestion, the mean plasma elimination half-life is 3.3 h and the bioavailability is 54% [137]. Following IV administration, the mean elimination half-life is approximately 3 h and 80% is excreted unchanged in the urine within 24 h. Eflornithine does not bind significantly to plasma proteins. It produces CSF/plasma ratios between 0.13 and 0.51 (Marion Merrell Dow Inc, Product Information, Sept. 1991). Milord et al. [111] found higher ratios at the end of a 14-day IV regimen: 0.91 in adults and 0.58 in children less than

12 years of age. The mean steady state serum concentration in children was only half as high as in adults, while their mean CSF concentration was only one third, presumably due to a higher renal drug clearance. CSF/serum ratios were higher in patients with melarsoprol-refractory infections, possibly because severe impairment of the blood-brain barrier due to chronic meningoencephalitis increases its permeability.

Effectiveness and indications

Eflornithine is able to cure both stages of *gambiense* sleeping sickness, including melarsoprol-refractory disease. Trypanosomes have been detected in blood and lymph up to 4 days after the start of oral eflornithine (Van Nieuwenhove, unpublished). The results in *T. b. rhodesiense* infections are deceiving. In view of its high cost, complicated administration and limited availability, eflornithine should be reserved for melarsoprol-refractory infections.

During the first field trial, 53 patients with arsenical-refractory stage II *T. b. gambiense* infections (who would inevitably have died of terminal sleeping sickness) were given oral eflornithine monotherapy, 100 mg/kg every 6 h, during 21-49 days [107, 138]. The results were very satisfactory: 49.1% cures (follow-up at least 24 months), 9.4% probable cures (follow-up between 12 and 24 months), 1 possible cure (1.9%) (follow-up 7 months), 1 favourable evolution (1.9%) (follow-up 3 months), 9.4% no follow-up beyond post-treatment CSF examination (when no trypanosomes were found and CSF parameters had considerably improved), 17% relapses (mostly in children) and 11.3% mortality. The high relapse rate in children was presumably due to a higher renal drug clearance and might be reduced by administration of proportionately higher doses.

Up to March 1991, 711 individuals with *gambiense* sleeping sickness (94.9% stage II), including the 53 previously described patients, had world-wide been treated with eflornithine. An update was presented by Hardenberg et al. [108], based on data provided by various investigators [109, 138-142; and others]. Only an abstract was published, but the below reported data come straight from a copy of Hardenberg's manuscript. Various regimens had been applied: 135 patients (81.5% arsenical-refractory) received oral eflornithine (100 mg/kg every 6 h during 21-45 days), 252 (70% arsenical-refractory) received 100 mg/kg IV every 6 h during 14 days followed by 75 mg/kg PO every 6 h during 21-30 days, and 324 (43.5% arsenical-refractory) were given 100 mg/kg IV every 6 h or 200 mg/kg IV every 12 h during 14 days. The results in this heterogeneous group of 711 patients were: 26.9% cures (follow-up at least 24 months), 22.1% probable cures (follow-up between 12 and 24 months); 38% follow-up less than 12 months; 5.3% relapses and 6.9% mortality. As follow-up was inferior to 24 months in almost 70% of enrolled patients, no definitive conclusions on effectiveness could be drawn. However, the number of reported relapses was higher in patients who received the oral regimen alone (14.8% vs. 6.9%) and in children up to 12 years of age (35.7% vs. 6.7%).

Milord et al. [110] administered various eflornithine regimens to 207 patients with stage II *T. b. gambiense* infections (100 arsenical-refractory, 107 not previously treated). There were 9% relapses, but the follow-up was too short to allow definitive conclusions. Relapses were more common among not previously treated patients, after oral eflornithine alone and in children below 12 years of age. The 200 mg/kg 12-hourly IV regimen was less effective than 100 mg/kg 6-hourly.

Eflornithine gave deceiving results, presumably due to deficient immune responses, in 4 HIV-seropositive patients with *gambiense* sleeping sickness, of whom 2 relapsed and 1 died shortly after treatment [143].

Results of a randomised TDR trial comparing 7-day and 14-day IV regimens should be available shortly. Preliminary data suggested that relapses were more frequent after the 7-day course.

Eflornithine (200 mg/kg every 6 h, even in combination with suramin) gave unreliable results in *rhodesiense* sleeping sickness [144; Kuzoe, unpublished], but does sometimes achieve cure [36].

Adverse effects

Eflornithine is a cytostatic drug, thus considerably toxic. The following adverse effects were reported with 14-day 6-hourly and 12-hourly IV regimens in 324 individuals [108]: 44.4% anaemia, 26.5% leukocytopenia, 11.7% diarrhoea, 7.4% convulsions, 7.1% vomiting, 6.8% thrombocytopenia, 5.9% fever, 2.5% abdominal pain, 2.2% headache, 1.9% alopecia, 1.2% anorexia, 0.3% dizziness and 5.9% mortality. Loss of hearing acuity was also observed. Diarrhoea was more frequent during oral administration, but not necessitating cessation of treatment. Most episodes of convulsions (80%) occurred during the first week and generally did not recur after resumption of treatment. The contribution of eflornithine to the mortality was judged conjectural.

Milord et al. [110] observed 2% mortality and 4% convulsions during 14-day IV eflornithine given to 207 patients with stage II *gambiense* sleeping sickness (about 50% arsenical-refractory), while Doua and Boa Yapo [62] reported 3.2% mortality and 6.7% convulsions during IV or combined PO/IV eflornithine administration to 126 stage II patients.

The drug has contragestational effects in rodents [145], but little is known about its effect on human pregnancy.

Recommended regimens, availability and cost

Eflornithine was originally supplied as a whitish powder (in sealed sachets containing 1 g each) that was dissolved in water and drunk. It was until recently produced by Marion Merell Dow under the trade name Ornidyl® and may nowadays be obtained from ILEX (via WHO), at about 60 US$ per 100 ml vial containing 200 mg eflornithine/ml. The content is hypertonic and must be diluted (at least 4 parts of water per part of eflornithine concentrate) and injected as a slow IV infusion. Opened vials should be used the same day.

The recommended regimen for adults is 100 mg/kg IV every 6 h during 14 days, while children should receive 150 mg/kg IV. Similar doses may be given during 14-21 days for oral treatment. The potential therapeutic benefit of longer regimens does not outweigh the increased risk for severe complications. A treatment course for adults costs approximately 1070 US$ (800 US$ for eflornithine, 200 US$ for intravenous fluids and 70 US$ for hospitalisation).

As large scale IV administration is practically unfeasible in rural areas, TDR is considering a randomised trial to compare effectiveness and adverse effects of the oral and IV routes and it is possible that oral eflornithine therapy will be reintroduced in the future (provided the necessary funds are found).

Combination therapy with current drugs

Combinations containing stage I drugs

The combination of suramin and pentamidine for treatment of stage I *T. b. gambiense* infections has been mentioned under suramin. Controlled trials to evaluate if there is synergism and if this regimen has advantages over monotherapy are long overdue.

Monotherapy with suramin or pentamidine is able to cure some patients with stage II sleeping sickness and there is experimental evidence for synergism between suramin and several other compounds (including nifurtimox, eflornithine, nitroimidazoles and anticancer drugs) [28-34].

Pre-treatment with suramin, pentamidine or diminazene prior to melarsoprol is, in view of their long elimination half-life, in fact combination therapy, but synergism with melarsoprol has never adequately been investigated, as it was assumed that stage I drugs do not cross the blood-brain barrier.

An association of suramin and metronidazole was able to temporarily clear the CSF of trypanosomes and to induce clinical improvement in a patient with multirefractory *T. b. rhodesiense* infection [35] and suramin together with high dose eflornithine can cure advanced *rhodesiense* sleeping sickness [36]. Although some associations containing suramin (and possibly pentamidine or diminazene) may cure stage II disease, combination therapy with stage II drugs is likely to be more effective.

Combination therapy with several stage II drugs

Compassionate treatment with a combination of full dose nifurtimox and oral eflornithine of patients with multirefractory *gambiense* sleeping sickness (melarsoprol-refractory cases who had also relapsed after monotherapy with nifurtimox and/or eflornithine) gave good results in the Sudan (Van Nieuwenhove, unpublished). An association of full dose melarsoprol and nifurtimox gave good results in patients with melarsoprol-refractory *T. b. gambiense* infections (Vandenweghe, unpublished), and Simarro and Ndong Asumo [146] cured a patient with melarsoprol-refractory *gambiense* sleeping sickness, who had also relapsed after one oral eflornithine course, with a combination of melarsoprol and oral eflornithine.

These clinical observations are consistent with experimental evidence of synergism between concomitantly administered stage II drugs [34]. Melarsoprol, eflornithine and nifurtimox all interfere, at different levels, with trypanothione synthesis and activity in trypanosomes [115-118], but it has yet to be proven that this is the cause of synergism.

Associations of lower doses of melarsoprol and nifurtimox, melarsoprol and eflornithine or nifurtimox and eflornithine are likely to be more effective, possibly less toxic and shorter than monotherapy. Controlled trials should allow to establish optimum treatment regimens. However, eflornithine is expensive, difficult to obtain and its IV-regimen is not suitable for large scale field use. Lower dose oral eflornithine would be perfect for combination therapy but still expensive, and availability may remain problematic. Thus, an association of melarsoprol and nifurtimox is likely to be the only realistic option in the near future.

More research seems warranted on the use that could be made of benznidazole and metronidazole in combination therapy. There is experimental evidence for synergism between eflornithine and benznidazole (Radanil®), a nitroimidazole registered for treatment of Chagas' disease [147] and metronidazole has shown synergism with suramin [29, 35]. Both nitroimidazoles are commercially available and, although they are probably less active against African trypanosomiasis than nifurtimox, they deserve attention, as the mechanism of action of nitroimidazoles and nitrofurans depends on different pathways [114].

Conclusions

Currently, suramin and pentamidine are used to treat stage I *gambiense* sleeping sickness (diminazene is not registered), while melarsoprol is generally restricted to stage II infections and nifurtimox and eflornithine to melarsoprol-refractory disease (provided they are available). The choice for *T. b. rhodesiense* sleeping sickness is practically limited to suramin for stage I and melarsoprol for stage II disease, as pentamidine and eflornithine are unreliable and virtually nothing is known about the effectiveness of nifurtimox. Melarsoprol-refractory *T. b. rhodesiense* infections may sometimes get cured with a combination of high dose eflornithine and suramin, but associations of several stage II drugs are likely to be more effective.

Suramin and pentamidine are often given in ambulatory during field surveys, while it is recommended to hospitalise patients for melarsoprol treatment, although hospital facilities and nursing care are frequently substandard where sleeping sickness is most endemic. When hospitalisation is not possible, patients should be treated as soon as possible, in the best possible way (ambulatory treatment if necessary), in their own interest and in order to rapidly reduce the trypanosome reservoir and prevent further spread.

Apart from the quest for new and better antitrypanosomal drugs, there definitely remains a huge immediate need for more research and well designed trials to clarify many ill understood and often controversial issues: a. o. CSF criteria to distinguish between disease stages; mechanisms of action and pharmacokinetics of current drugs and their metabolites; drug effectiveness and toxicity (updates through prospective studies are badly needed); gradual or aggressive treatment; concomitant use of corticosteroids, immunosuppressive and anti-inflammatory drugs; treatment of CNS complications and other adverse effects; drug synergism; optimum combination therapy regimens; reduction of treatment duration and alternative treatment for melarsoprol-refractory infections.

The probability that new drugs will be registered for sleeping sickness in the near future is small, thus optimisation of therapy will depend on the development of more rational treatment regimens for both monotherapy and combination therapy with current drugs. Associations of lower doses of several compounds are likely to be more effective and possibly less toxic than monotherapy and may allow shorter treatment courses. Particular attention should be paid to alternative treatment for patients with melarsoprol-refractory disease, many of whom are at present left to die, as eflornithine is often unavailable and nifurtimox is, for obscure reasons, only used on a limited scale.

In view of the present dramatic resurgence of sleeping sickness in central Africa, regular production and permanent availability of current drugs need to be guaranteed by all means. As the high cost of antitrypanosomal therapy can not be borne by poor local communities in highly endemic rural areas, governments and the international community need to assume their responsibility.

References

1. Friedheim EAH (1951) Mel B in the treatment of tryparsamide resistant *T. gambiense* sleeping sickness: observations on drug resistance in the trypanosomes of the French Cameroun. Am J Trop Med Hyg 31:218-227

2. Williamson J (1970) Review of chemotherapeutic and chemoprophylactic agents. In: Mulligan HW (ed) The African Trypanosomiases. George Allen and Unwin, London, p 126
3. Apted FIC (1970) Treatment of human trypanosomiasis. In: Mulligan HW (ed) The African Trypanosomiases. George Allen and Unwin, London, p 684
4. Apted FIC (1980) Present status of chemotherapy and chemoprophylaxis in the eastern hemisphere. Pharmac Ther 11:391-413
5. Wéry M (1994) Drugs used in the treatment of sleeping sickness (human African trypanosomiasis: HAT). Int J Antimicrob Agents 4:227-238
6. Cattand P, Miézan BT, De Raadt P (1988) Human African Trypanosomiasis: use of double centrifugation of cerebrospinal fluid to detect trypanosomes. Bull World Health Org 66:83-86
7. Lejon V, Moons A, Büscher P, Magnus E, Van Meirvenne N (1995) Trypanosome specific antibody profile in serum and cerebrospinal fluid of *T. b. gambiense* patients. In: Proceedings 23rd meeting ISCTRC, Banjul. OAU/STRC, Nairobi, p 78
8. Fairlamb AH, Bowman IBR (1980a) *Trypanosoma brucei* maintenance of concentrated suspensions of bloodstream trypomastigotes in vitro using continuous dialysis for measurement of endocytosis. Exp Parasitol 49:366-380
9. Fairlamb AH, Bowman IBR (1980b) Uptake of the trypanocidal drug suramin by bloodstream forms of *Trypanosoma brucei* and its effect on respiration and growth rate *in vivo*. Mol Biochem Parasitol 1:315-333
10. Hawking F (1978) Suramin with special reference to onchocerciasis. Adv Pharmacol Chemother 15:289-322
11. Chello PL, Jaffe JJ (1972) Comparative properties of trypanosomal and mammalian thymidine kinases. Comp Biochem Physiol B 43:543-562
12. Jaffe JJ, McCormack JJ, Meymariam E (1972) Comparative properties of schistosomal and filarial dihydrofolate reductases. Biochem Pharmacol 21:719-731
13. Willson M, Callens M, Kuntz DA, Perié J, Opperdoes FR (1993) Synthesis and activity of inhibitors highly specific for the glycolytic enzymes from *Trypanosoma brucei*. Mol Biochem Parasitol 59:201-210
14. Opperdoes FR, Borst P (1977) Localization of nine glycolytic enzymes in a microbody-like organelle in *Trypanosoma brucei*: the glycosome. FEBS Lett 80:360-364
15. Opperdoes FR, Baudhuin P, Coppens I, De Roe C, Edwards SW (1984) Purification, morphometric analysis and characterization of the glycosomes (microbodies) of the protozoan haemoflagellate. J Cell Biol 98:1178-1184
16. Clarkson AB, Brohn FH (1976) Trypanosomiasis: an approach to chemotherapy by the inhibition of carbohydrate catabolism. Science 194:204-206
17. Wang CC (1995) Molecular mechanisms and therapeutic approaches to the treatment of African trypanosomiasis. Annu Rev Pharmacol Toxicol 35:93-127
18. Hawking F (1940) Concentration of Bayer 205 (Germanin) in human blood and cerebrospinal fluid after treatment. Trans R Soc Trop Med Hyg 34:37-52
19. Collins JM, Klecker RW, Yarchoan R, Lane HC, Fauci AS, Redfield RR, Boder S, Myers CE (1986) Clinical pharmacokinetics of suramin in patients with HTLV-III/LAV infection. J Clin Pharmacol 26:22-26
20. Harding RD (1945) Late results of treatment of sleeping sickness in Sierra Leone by Antrypol, tryparsamide, pentamidine and propamidine singly and in various combinations. Trans R Soc Trop Med Hyg 39:99-124
21. Neujean G (1950) Contribution à l'étude des liquides rachidiens et céphaliques dans la maladie du sommeil à *Trypanosoma gambiense*. Ann Soc Belge Méd Trop 30:1225-1236
22. Neujean G, Evens F (1958) Diagnostic et traitement de la maladie du sommeil à *T. gambiense*. Bilan de dix ans d'activité du centre de traitement de Léopoldville. Mém Acad R Sci Colon Cl Sci Nat Méd Coll in-8° Tome VII, fasc 2
23. Wellde BT, Chumo DA, Reardon MJ, Abinya A, Wanyama L, Dola S, Mbwabi D, Smith DH, Siongok TA (1989) Treatment of Rhodesian sleeping sickness in Kenya. Ann Trop Med Parasitol 66:7-14
24. Veeken HJGM, Ebeling MCH, Dolmans WMV (1988) Trypanosomiasis in a rural hospital in Tanzania. A retrospective study of its management and the result of treatment. Trop Geogr Med 41:113-117
25. Pépin J, Milord F (1994) The treatment of human African trypanosomiasis. Adv Parasitol 33:2-49
26. Waddy BB (1970) Chemoprophylaxis of human trypanosomiasis. In: Mulligan HW (ed), The African Trypanosomiases. George Allen and Unwin, London, p 711
27. Keevill AJ (1934) Subsequent histories of six cases of *Trypanosoma rhodesiense* infection treated with "Bayer 205" or "Fourneau 309". Trans R Soc Trop Med Hyg 28:101-102
28. Clarkson AB, Bienen EJ, Bacchi CJ, Mc Cann PP, Nathan HC, Hutner SH, Sjoerdsma A (1984) New drug combination for experimental late-stage African trypanosomiasis: DL-alpha-difluoromethylornithine (DFMO) with suramin. Am J Trop Med Hyg 33:1073-1077
29. Raseroka BH, Ormerod WE (1985) Suramin/metronidazole combination for African sleeping sickness. Lancet ii:784-785
30. Raseroka BH, Ormerod WE (1986) The trypanocidal effect of drugs in different parts of the brain. Trans R Soc Trop Med Hyg 80:634-641

31. Sayer PD, Gould SS, Waitumbi JN, Murray PK, Njogu AR (1985) The successful treatment of *Trypanosoma rhodesiense* infected monkeys with low doses of melarsoprol or with suramin and nitroimidazole. In: Proceedings 18th meeting ISCTRC, Harare. OAU/STRC, Nairobi, p 131
32. Sayer PD, Gould SS, Waitumbi JN, Raseroka BH, Akol GWO, Ndungu JM, Njogu AR (1987) Treatment of African trypanosomias with combinations of drugs with special reference to suramin and nitroimidazoles. In: Proceedings 19th meeting ISCTRC, Lomé. OAU/STRC, Nairobi, p 205
33. Bacchi CJ, Nathan HC, Clarkson AB, Bienen EJ, Bitonti AJ, Mc Cann PP, Sjoerdsma A (1987) Effect of ornithine decarboxylase inhibitors DL-alpha-difluoromethylornithine and alpha-monofluoromethyldehydroornithine methyl ester alone and in combination with suramin against *Trypanosoma brucei brucei* central nervous system models. Am J Trop Med Hyg 36:46-52
34. Jennings FW (1993) Combination chemotherapy of CNS trypanosomiasis. Acta Trop 54:205-213
35. Arroz J, Djedje M (1988) Suramin and metronidazole in the treatment of *Trypanosoma brucei rhodesiense*. Trans R Soc Trop Med Hyg 82:421
36. Taelman H, Clerinx J, Bogaerts J, Vervoort T (1996) Combination treatment with suramin and eflornithine in late stage rhodesian trypanosomiasis: case report. Trans R Soc Trop Med Hyg 90:572-573
37. Cheson BD, Levine AL, Mildvan D, Kaplan LD, Wolfe P, Rios A, Groopman J, Gill P, Volberding PA, Poiesz BJ, Gottlieb M, Holden H, Volsky DJ, Silver SS, Hawkins MJ (1987) Suramin therapy in AIDS and related disorders. Report of the US suramin working group. JAMA 258:1347-1351
38. Pépin J, Milord F, Guern C, Mpia B, Ethier L, Mansinsa D (1989a) Trial of prednisolone for the prevention of melarsoprol-induced encephalopathy in gambiense sleeping sickness. Lancet 333:1246-1250
39. Burri C, Blum J (1996) A case of reactive encephalopathy after treatment with suramin of stage I sleeping sickness. Trop Med Int Health 1:A36
40. Bitonti AJ, Dumont JA, Mc Cann PP (1986a) Characterization of *Trypanosoma brucei brucei* S-adenosyl-L-methionine decarboxylase and its inhibition by Berenil, pentamidine and methylglyoxal bis(guanylhydrazone). Biochem J 237:518-521
41. Berger BJ, Carter NS, Fairlamb AH (1993) Polyamine and pentamidine metabolism in African trypanosomes. Acta Trop 54:215-224
42. Damper D, Patton CL (1976) Pentamidine transport and sensitivity in *brucei*-group trypanosomes. J Protozool 23:349-356
43. Frommel TO, Balber AE (1987) Flow cytofluorimetric analysis of drug accumulation by multidrug-resistant *Trypanosoma brucei brucei* and *T. b. rhodesiense*. Mol Biochem Parasitol 26:183-192
44. Dickinson CM, Navin TR, Churchill FC (1985) High-performance liquid chromatographic method for quantification of pentamidine in blood and serum. J Chromatogr 345:91-97
45. Berger BJ, Hall JE (1989) High-performance liquid chromatographic method for the quantification of several diamidine compounds with potential chemotherapeutic value. J Chromatogr 494:191-200
46. Ericsson O, Rais M (1990) Determination of pentamidine in whole blood, plasma and urine by high-performance liquid chromatography. Ther Drug Monit 12:362-365
47. Conte JE, Upton RA, Phelps RT, Wofsy CB, Zurlinden E, Lin ET (1986) Use of a specific and sensitive assay to determine pentamidine pharmacokinetics in patients with AIDS. J Infect Dis 154:923-929
48. Conte JE (1991) Pharmacokinetics of intravenous pentamidine in patients with normal renal function or receiving haemodyalisis. J Infect Dis 163:169-175
49. Bronner U (1994) Pharmacokinetics of pentamidine. Focus on treatment of *Trypanosoma gambiense* sleeping sickness. PhD Thesis. Karolinska Institute, Stockholm
50. Bronner U, Doua F, Ericsson O, Gustafsson LL, Miézan TW, Rais M, Rombo L (1991) Pentamidine concentrations in plasma, whole blood and cerebrospinal fluid during treatment of *Trypanosoma gambiense* infection in Côte d'Ivoire. Trans R Soc Trop Med Hyg 85:608-611
51. Miézan TW, Bronner U, Doua F, Cattand P, Rombo L (1994) Long term exposure of *Trypanosoma brucei gambiense* to pentamidine *in vitro*. Trans R Soc Trop Med Hyg 88:332-333
52. Van Hoof L, Henrard C, Peel E (1944) Pentamidine in the prevention and treatment of trypanosomiasis. Trans R Soc Trop Med Hyg 37:271-280
53. Nash TAM (1960) A review of the African trypanosomiasis problem. Trop Dis Bull 57:973-1003
54. Schneider J (1963) Traitement de la trypanosomiase humaine africaine. Bull World Health Org 28:763-786
55. Le Rouzic (1949) Nouveaux trypanocides en expérimentation. Bull Méd AOF 47:63-72
56. Jonchère H, Gomer J, Reynaud R (1951) Traitement par les diamidines de la phase lymphatico-sanguine de la trypanosomiase humaine en Afrique Occidentale Française. Bull Soc Path Ex 44:603-625
57. Dutertre J, Labusquière R (1966) La thérapeutique de la trypanosomiase. Méd Trop 26:342-356
58. Lourie EM (1942) The treatment of sleeping sickness in Sierra Leone. Ann Trop Med Parasitol 36:113-131

59. Duggan AJ, Hutchinson MP (1951) The efficacy of certain trypanocidal compounds against *Trypanosoma gambiense* infection in man. Trans R Soc Trop Med Hyg 44:535-544
60. Doua F, Miézan TW, Sanon Singaro JR, Boa Yapo F, Baltz T (1996) The efficacy of pentamidine in the treatment of early-late stage *Trypanosoma brucei gambiense* trypanosomiasis. Am J Trop Med Hyg 55:586-588
61. Sands M, Kron MA, Brown RB (1985) Pentamidine: a review. Rev Infect Dis 7:625-634
62. Doua F, Boa Yapo F (1993) Human trypanosomiasis in the Ivory coast: therapy and problems. Acta Trop 54:163-168
63. Stauffert I, Paulini H, Steinman U, Sippel H, Estler CJ (1990) Investigations on mutagenicity and genotoxicity of pentamidine and some related trypanocidal diamidines. Mutation Res 245:93-98
64. Gall D (1954) The chemoprophylaxis of sleeping sickness with the diamidines. Ann Trop Med Parasitol 48:242-358
65. Hutchinson MP, Watson HJC (1962) Berenil in the treatment of *Trypanosoma gambiense* infection in man. Trans R Soc Trop Med Hyg 56:227-230
66. De Raadt P, Van Hoeve K, Bailey NM, Kenyanjui EN (1966) Observations on the use of Berenil in the treatment of human trypanosomiasis. In: Annual Report 1965. East African Trypanosomiasis Research Organisation, p 60
67. Onyango RJ, Bailey NM, Okach RW, Mwangi EK, Ogada T (1970) The use of Berenil for the treatment of early cases of human trypanosomiasis. In: Annual Report 1969. East African Trypanosomiasis Research Organization, p 120
68. Temu SE (1975) Summary of cases of human early trypanosomiasis treated with Berenil at EATRO. Trans R Soc Trop Med Hyg 69:277
69. Ruppol JF, Burke J (1977) Follow-up des traitements contre la trypanosomiase expérimentés à Kimpangu (République du Zaïre). Ann Soc Belge Méd Trop 57:481-494
70. Abaru DE, Matovu FS (1984) Berenil in the treatment of early stage human trypanosomiasis cases. Bull Trim Inf Gloss Trypanosom 7:150-151
71. Bailey NM (1968) Oral Berenil in the treatment and prophylaxis of human trypanosomiasis. Trans R Soc Trop Med Hyg 62:122
72. Abaru DE, Liwo DA, Isakina D, Okori EE (1984) Retrospective long-term study of effects of Berenil by follow-up of patients treated since 1965. Trop Med Parasitol 35:148-150
73. De Raadt P (1967) Reactive encephalopathy occurring as a complication during treatment of *T. rhodesiense* with non-arsenical drugs. In: Annual Report 1966. East African Trypanosomiasis Research Organisation, p 85
74. Friedheim EAH (1949) Mel B in the treatment of human trypanosomiasis. Am J Trop Med Hyg 29:173-180
75. Fairlamb AH, Henderson GB, Cerami A (1989) Trypanothione is the primary target for arsenical drugs against African trypanosomes. Proc Natl Acad Sci USA 86:2607-2611
76. Yarlett N, Goldberg B, Nathan HC, Garofalo J, Bacchi CJ (1991) Differential sensitivity of *Trypanosoma brucei rhodesiense* isolates to *in vitro* lysis by arsenicals. Exp Parasitol 72:205-215
77. Cristeau B, Placidi M, Legait JP (1975) Étude de l'excrétion de l'arsenic chez le trypanosomé traité au melarsoprol. Méd Trop 35:389-401
78. Burri C, Brun R (1992) An in vitro bioassay for quantification of melarsoprol in serum and cerebrospinal fluid. Trop Med Parasitol 43:223-225
79. Burri C, Baltz T, Giroud C, Doua F, Welker HA, Brun R (1993) Pharmacokinetic properties of the trypanocidal drug melarsoprol. Chemotherapy 39:225-234
80. Burri C (1994) Pharmacological aspects of the trypanocidal drug melarsoprol. Ph D Thesis. Philosophisch-Naturwissenschaftlichen Fakultät der Universität Basel
81. Bronner U, Brun R, Burri F, Doua F, Ericsson O, Rombo L, Gustafsson LL (1996) Determination of melarsoprol concentrations in plasma, urine and CSF in patients with human African trypanosomiasis using HPLC and bioassay. Trop Med Int Health 1:A37
82. Friedheim EAH, Distefano D (1989) Melarsoprol in the treatment of African sleeping sickness. In: Proceedings 20th meeting ISCTRC, Mombasa. OAU/STRC, Nairobi, p 245
83. Burri C, Blum J, Brun R (1995) Alternative application of melarsoprol for treatment of *T. b. gambiense* sleeping sickness: preliminary results. Ann Soc Belge Méd Trop 75:65-71
84. Anonymous (1986) Epidemiology and Control of African trypanosomiasis. WHO Tech Rep Ser 739:118-121
85. Ceccaldi J (1953) Nouvelle contribution au traitement de la trypanosomiase humaine par l'Arsobal ou 3854 RP. Son efficacité chez les anciens trypanosés atteints de méningo-encéphalite, résistants aux trypanocides généralement utilisés. Bull Soc Path Ex 46:95-111
86. Apted FIC (1957) Four years' experience of Melarsen Oxide/BAL in the treatment of late-stage Rhodesian sleeping sickness. Trans R Soc Trop Med Hyg 5:75-86
87. Ogada T (1974) Clinical Mel B resistance in Rhodesian sleeping sickness. East Afr Med J 51:56-59
88. Bertrand E, Rive J, Serié F, Kone I (1973) Encéphalopathie arsénicale et traitement de la trypanosomiase. Méd Trop 33:385-390
89. Buyst H (1975) The treatment of *T. rhodesiense* sleeping sickness with special reference to its physio-pathological and epidemiological basis. Ann Soc Belge Méd Trop 55:95-104

90. Maathai RG, Masiga RC, Sayer PD, Ngure RM, Ndung'u JM (1991) The role of propylene glycol in melarsoprol toxicity during treatment of *Trypanosoma rhodesiense* infected mice. In: Proceedings 22nd meeting ISCTRC, Yamoussoukro. OAU/STRC, Nairobi, p 152
91. Robertson DHH (1963) The treatment of sleeping sickness (mainly due to *Trypanosoma brucei rhodesiense*) with melarsoprol. I- Reactions observed during treatment. Trans R Soc Trop Med Hyg 57: 122-133
92. Adams JH, Haller L, Boa FY, Doua F, Dago A, Konian K (1986) Human African trypanosomiasis (*T. b. gambiense*): a study of 16 fatal cases of sleeping sickness with some observations on acute reactive encephalopathy. Neuropathol Appl Neurobiol 12:81-94
93. Sina GC, Triolo N, Trova P, Clabaut JM (1977) L'encéphalopathie arsénicale lors du traitement de la trypanosomiase humaine africaine à *T. gambiense* (à propos de 16 cas). Ann Soc Belge Méd Trop 57:67-74
94. Sina GC, Triolo N, Cramet B, Suh Bandu M (1982) L'adrénaline dans la prévention et le traitement des accidents de l'arsobalthérapie. A propos de 776 cas de trypanosomiase humaine africaine à *T. gambiense* traités dans les formations sanitaires de Fontem (R.U. du Cameroun). Méd Trop 42:531-536
95. Schmidt H, Bafort JM (1985) African trypanosomiasis: treatment-induced invasion of brain and encephalitis. Am J Trop Med Hyg 34:64-68
96. Jennings FW, Mc Neil PE, Ndung'u JM, Murray M (1989) Trypanosomiasis and encephalitis: possible aetiology and treatment. Trans R Soc Trop Med Hyg 83:518-519
97. Jennings FW, Hunter CA, Kennedy PGE, Murray M (1993) Chemotherapy of *Trypanosoma brucei* infection of the central nervous system: the use of a rapid therapeutic regimen and the development of post-treatment encephalopathies. Trans R Soc Trop Med Hyg 87:224-226
98. Hunter CA, Jennings FW, Adams JH, Murray M, Kennedy PGE (1992) Subcurative chemotherapy and fatal post-treatment reactive encephalopathies in African trypanosomiasis. Lancet 339:956-958
99. Frommel TO (1988) *Trypanosoma brucei rhodesiense*: effect of immunosuppression on the efficacy of melarsoprol treatment in infected mice. Exp Parasitol 67:364-366
100. Pépin J, Milord F (1991) African trypanosomiasis and drug-induced encephalopathy: risk factors and pathogenesis. Trans R Soc Trop Med Hyg 85:222-224
101. Pépin J, Milord F, Khonde A, Niyonsenga T, Loko L, Mpia B (1995) Risk factors for encephalopathy and mortality during melarsoprol treatment of *Trypanosoma brucei gambiense* sleeping sickness. Trans R Soc Trop Med Hyg 89:92-97
102. Haller L, Adams H, Merouse F, Dago A (1986) Clinical and pathological aspects of human African trypanosomiasis (*T. b. gambiense*) with particular reference to reactive arsenical encephalopathy. Am J Trop Med Hyg 35:94-99
103. Pentreath V (1995) Trypanosomiasis and the nervous system. 1. Pathology and immunology. Trans R Soc Trop Med Hyg 89:9-15
104. Golden MHN (1992) Arsenic, selenium and African trypanosomiasis. Lancet 339:1413
105. Sweins A (1983) Protective effect of selenium against arsenic-induced chromosomal damage in cultured human lymphocytes. Hereditas 98:249-252
106. Ancelle T, Barret B, Flachet L, Moren A (1994) Étude de 2 épidémies d'encéphalopathies arsénicales dans le traitement de la trypanosomiase, Ouganda, 1992-1993. Bull Soc Path Ex 87:341-346
107. Van Nieuwenhove S (1992) Advances in sleeping sickness therapy. Ann Soc Belge Méd Trop 72:39-51
108. Hardenberg J, Claverie N, Tell GP (1991) Eflornithine (Ornidyl) treatment of *Trypanosoma brucei gambiense* sleeping sickness; report on 711 patients treated up to March 1991. In: Proceedings 21rst meeting ISCTRC, Yamoussoukro. OAU/STRC, Nairobi, p 158
109. Pépin J, Guern C, Milord F, Schechter PJ (1987) Difluoromethylornithine for arseno-resistant *T. b. gambiense* sleeping sickness. Lancet 330:1431-1433
110. Milord F, Pépin J, Loko L, Ethier L, Mpia B (1992) Efficacy and toxicity of eflornithine for treatment of *Trypanosoma brucei gambiense* sleeping sickness. Lancet 340:652-655
111. Milord F, Loko L, Ethier L, Mpia B, Pépin J (1993) Eflornithine concentrations in serum and cerebrospinal fluid of 63 patients treated for *Trypanosoma brucei gambiense* sleeping sickness. Trans R Soc Trop Med Hyg 87:473-477
112. Richet P, Lotte M, Foucher G (1959) Résultats des traitements de la trypanosomiase humaine à *Trypanosoma gambiense* par le Mel B. Méd Trop 19:253-265
113. Atouguia J, Costa J, Jennings F (1997) Topical chemotherapy in experimental CNS-trypanosomiasis: drug combinations. Proceedings 24th meeting ISCTRC, Maputo. OAU/STRC, Nairobi (in press)
114. Townson SM, Boreham PLF, Upcroft P, Upcroft JA (1994) Resistance to nitroheterocyclic drugs. Acta Trop 56:125-141
115. Fairlamb AH, Henderson GH, Cerami A (1985) Trypanothione: a novel bis(gluthathionyl)spermidine cofactor for gluthathione reductase in trypanosomatids. Science 227:1485-1487
116. Fairlamb AH (1990a) Future prospects for the chemotherapy of human trypanosomiasis 1. Novel approaches to the chemotherapy of trypanosomiasis. Trans R Soc Trop Med Hyg 84:613-617
117. Fairlamb AH (1990b) Trypanothione metabolism and rational approaches to drug design. Biochem Soc Transactions 18:717-720

118. Docampo R, Moreno SNJ (1986) Free radical metabolism of antiparasitic agents. Federation Proceedings of the Federation of the American Societies for Experimental Biology 45:2471-2476
119. Medenwald H, Brandau K, Schlossman K (1972) Quantitative determination of nifurtimox in body fluids of rat, dog and man. Arzneim Forsch (Drug Res) 22:1613-1617
120. Paulos C, Paredes J, Vasquez I, Thambo S, Arancibia A, Gonzalez-Martin G (1989) Pharmacokinetics of a nitrofuran compound, nifurtimox, in healthy volunteers. Int J Clin Pharmacol Ther Toxicol 9:454-457
121. Duhm B, Maul W, Medenwald H, Patzschke K, Wegner LA (1972) Investigations on the pharmacokinetics of nifurtimox-35S in the rat and dog. Arzneim Forsch (Drug Res) 22:1617-1624
122. Janssens PG, De Muynck A (1977) Clinical trials with nifurtimox in African trypanosomiasis. Ann Soc Belge Méd Trop 57:475-479
123. Van Nieuwenhove, Declercq J (1981a) Nifurtimox (Lampit) treatment in late stage of *gambiense* sleeping sickness. In: Proceedings 17th meeting ISCTRC, Arusha. OAU/STRC, Nairobi, p 206
124. Van Nieuwenhove S, Declercq J (1989) Nifurtimox therapy in late-stage arsenical refractory *gambiense* sleeping sickness. Proceedings 20th meeting ISCTRC, Mombasa. OAU/STRC, Nairobi, p 264
125. Moens F, De Wilde M, Ngato K (1984) Essai de traitement au nifurtimox de la trypanosomiase humaine africaine. Ann Soc Belge Méd Trop 64:37-43
126. Pépin J, Milord F, Mpia B, Meurice F, Ethier L, Degroof D, Bruneel H (1989b) An open clinical trial of nifurtimox for arseno-resistant *Trypanosoma brucei gambiense* sleeping sickness in central Zaire. Trans R Soc Trop Med Hyg 83:514-517
127. Pépin J, Milord F, Meurice F, Ethier L, Loko L, Mpia B (1992b) High-dose nifurtimox for arseno-resistant *Trypanosoma brucei gambiense* sleeping sickness: an open trial in Zaire. Trans R Soc Trop Med Hyg 86:254-256
128. Apted FIC (1960) Nitrofurazone in the treatment of sleeping sickness due to *Trypanosoma rhodesiense*. Trans R Soc Trop Med Hyg 54:225-228
129. Robertson DHH (1961) The haemolytic effect of primaquine and nitrofurazone in cases of sleeping sickness with the haemolytic trait. Ann Trop Med Parasitol 55:278-286
130. Van Nieuwenhove S, Declercq J (1981b) Difluoromethylornithine: a new promising drug against *gambiense* sleeping sickness. Case Report. In: Proceedings 17th meeting ISCTRC, Arusha. OAU/STRC, Nairobi, p 213
131. Bacchi CJ, Nathan HC, Hutner SH, Mc Cann PP, Sjoerdsma A (1980) Polyamine metabolism: a potential therapeutic target in trypanosomes. Science 210:332-334
132. De Gee ALW (1983) Role of antibody in the elimination of trypanosomes after DL-alpha-difluoromethylornithine chemotherapy. J Parasitol 69:818-822
133. Bitonti AJ, Mc Cann PP, Sjoerdsma A (1986b) Necessity of antibody response in the treatment of African trypanosomiasis with alpha-difluoromethylornithine. Biochem Pharmacol 35:331-334
134. Fairlamb AH, Henderson GB, Bacchi CJ, Cerami A (1987) In vivo effects of difluoromethylornithine on trypanothione and polyamine levels in bloodstream forms of *Trypanosoma brucei*. Mol Biochem Parasitol 24:185-191
135. Giffin BF, Mc Cann PP, Bitonti AJ, Bacchi CJ (1986) Polyamine depletion following exposure to DL-alpha-difluoromethylornithine both *in vivo* and *in vitro* initiates morphological alterations and mitochondrial activation in a monomorphic strain of *Trypanosoma brucei brucei*. J Protozool 33:238-243
136. Bass KE, Sommer JM, Cheng QL, Wang CC (1992) Mouse ornithine decarboxylase is stable in *Trypanosoma brucei*. J Biol Chem 267:11034-11037
137. Haegele KD (1981) Kinetics of alpha-difluoromethylornithine: an irreversible inhibitor of ornithine decarboxylase. Clin Pharmac Ther 30:210-217
138. Van Nieuwenhove S, Schechter PJ, Declercq J, Boné G, Burke J, Sjoerdsma A (1985) Treatment of *gambiense* sleeping sickness in the Sudan with oral DFMO (DL-alpha-difluoromethylornithine), an inhibitor of ornithine decarboxylase; first field trial. Trans R Soc Trop Med Hyg 79:692-698
139. Doua F, Boa FY, Schechter PJ, Miézan TW, Diai D, Sanon SR, De Raadt P, Haegele KD, Sjoerdsma A, Konian K (1987) Treatment of human late stage gambiense trypanosomiasis with alpha-difluoromethylornithine (eflornithine). Efficacy and tolerance in 14 cases in Côte d'Ivoire. Am J Trop Med Hyg 37:525-533
140. Taelman H, Schechter PJ, Marcelis L, Sonnet J, Kazyumba G, Dasnoy J, Haegele KD, Sjoerdsma A, Wéry M (1987) Difluoromethylornithine, an effective new treatment for Gambian trypanosomiasis. Am J Med 82:607-614
141. Kazyumba GL, Ruppol JF, Tshefu AK, Nkanga N (1988) Arsénorésistance et difluoro-methylornithine dans le traitement de la trypanosomiase humaine africaine. Bull Soc Path Ex 81:591-594
142. Eozenou P, Jannin J, Ngampo S, Carme B, Tell GP, Schechter PJ (1989) Essai de traitement de la trypanosomiase à *Trypanosoma brucei gambiense* par l'eflornithine en République Populaire du Congo. Méd Trop 49:149-154
143. Pépin J, Ethier L, Kazadi C, Milord F, Ryder R (1992a) The impact of human immunodeficiency virus infection on the epidemiology and treatment of *Trypanosoma brucei gambiense* sleeping sickness in Nioki, Zaire. Am J Trop Med Hyg 47:133-140

144. Bales JD, Harrison SM, Mbwabi DL, Schechter PJ (1989) Treatment of arsenical refractory Rhodesian sleeping sickness in Kenya. Ann Trop Med Parasitol 83:111-114
145. Fozard JR (1987) The contragestational effects of ornithine decarboxylase inhibition. In: McCann AE (ed) Inhibition of Polyamine Metabolism. Biological Significance and Basis for new Therapies. Academic Press, Orlando, p 187
146. Simarro PP, Ndong Asumo P (1996) Gambian trypanosomiasis and synergism between melarsoprol and eflornithine. Trans R Soc Trop Med Hyg 90:315
147. Zweygarth A, Kaminsky R, Sayer PD, Van Nieuwenhove S (1990) Synergistic activity of 5-substituted 2-nitroimidazoles (Ro 15-0216 and benznidazole) and DL-alpha-difluoromethylornithine on *Trypanosoma brucei brucei*. Ann Soc Belge Méd Trop 70:269-279

CHAPTER 16

The nitroimidazoles and human African trypanosomiasis

G Chauvière, J Périé

Introduction

The arsenal of compounds available for the chemotherapy of human African trypanosomiasis (HAT) [1, 2] is essentially limited to suramin and pentamidine, which are active in the hematological stage of the disease, and melarsoprol and eflornithine (DFMO) which have efficacy in the terminal (meningoencephalitic) phase. The first two agents are ineffective at eradicating the parasite if it already has surmounted the blood-brain barrier and the toxicity of melarsoprol too frequently causes a lethal encephalopathy; as for the promising DFMO molecule, it necessitates an expensive and difficult protocol for the less developed countries where HAT is most rampant. There is thus urgent need to develop other drug molecules adapted to the socioeconomic conditions of these countries.

With this perspective, for limiting the added related to the development of novel molecules, research has been focused on drug molecules that already are produced for other treatments and that have been found by *in vitro* testing to be active against the trypanosome.

In the following section, we will assess the nitroimidazole family in which, subsequent to metronidazole (Flagyl®), numerous other molecules have been put on the market and whose activities span antibiotic, antiparasitic or radiosensitizing (treatment of tumors) domains [3, 4].

In vivo activity of nitroimidazoles

A screening of many nitroimidazole compounds, as the pharmaceutical company Hoechst has undertaken [5], has led to emergence of several molecules which are active against *Trypanosoma brucei* and have since been subject to more in-depth *in vivo* studies (Fig. 1).

A single compound from this family has been found to be effective in monotherapy; it is Ro 15-0216, a 2-nitroimidazole made by Hoffman-Laroche. In mouse model trials, it has been found to be effective in isolates resistant to diminazene (Berenyl®) and to melarsoprol [6, 7]. This molecule crosses the blood-brain barrier, which enables it to have activity on the neurological phase of the disease as demonstrated in the sheep model [8]. However, its short half-life *in vivo* would necessitate 3 intravenous injections per day for 6 days to effect cure without relapse.

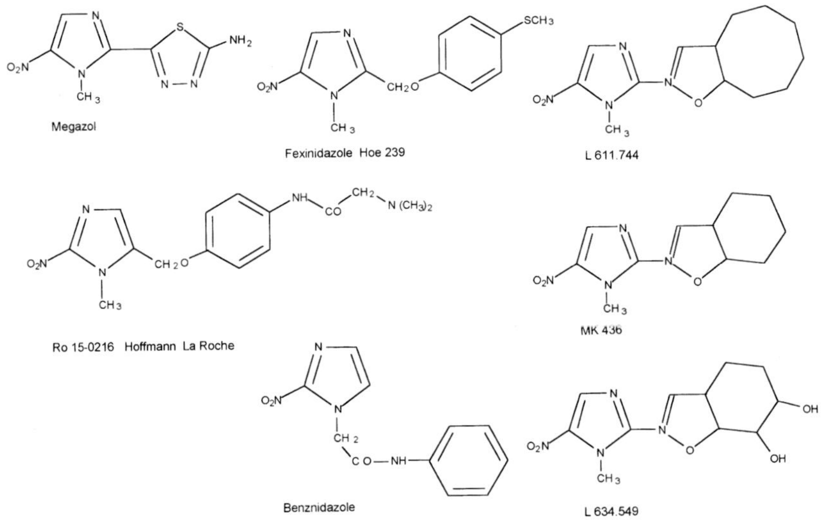

Fig. 1. Chemical structure of substituted 5-nitroimidazoles used in trypanosomiasis

It is thus to increase the efficacy and ease of treatment so that attention has been turned to its combined use. Ro 15-0216 injected once in mice, after two days of treatment with DFMO succeeds in curing mice infected with *T. brucei brucei* [9].

This combination use has been utilized with the various 5-nitroimidazoles listed in Fig. 1, notably fexinidazole Hoe 234, the oxazole ring derivatives made by Merck, Sharp and Dohme, and megazol or CL 64855 of American Cyanamid [10]. These combinations may be explained by the fact that the nitroimidazoles inhibit the peripheral multiplication of the parasite, thus facilitating the ulterior action of the arsenic derivatives, which alone have activity in the meningoencephalitic phase. This enables the arsenics' dose to be reduced, an important point considering their high toxicity. This is, in fact, what has been observed with several trials [11, 12].

What is a priori most surprising is that dual therapy with 5-nitroimidazoles and diminazene or suramin can be effective in the terminal phase of the disease whereas separately, they are not capable of clearing the sequestered cerebral parasites of this phase responsible for ongoing parasitemia. This suggests that their association enables them to broach the blood-brain barrier.

Megazol model and mode of action of nitroimidazoles on the trypanosome

The following discussion will be consecrated to CL 64855 synthesized since 1968 [10] and named megazol by Brazilian researchers [13] who resynthesized it and demonstrated that it was one of the most active compounds against *T. cruzi* responsible for American trypanosomiasis or Chagas' disease.

A collaboration with these researchers has enabled our laboratory to study megazol in much greater depth, particularly its biochemical and physico-chemical behaviour.

One can use 5-nitroimidazole as a model for understanding the behaviour of this family of compounds with respect to trypanosomal action, particularly regarding *T. brucei brucei*. It was recently shown that this compound is active *in vivo* in a mouse model infected with this trypanosome in the acute phase of parasitemia, and active in combination therapy with suramin in the chronic phase [14]. Additionally, in topical application with melarsoprol, it can reduce the melarsoprol dose required [15].

As with the preceding nitroimidazoles, the combination with suramin facilitates the crossing of the blood-brain barrier. A study performed using UV spectrometry [16] and varying the proportion of suramin to megazol appeared to reveal the formation of a complex between these two compounds either by stacking of suramin's napthalene cores and megazol's coplanar rings [17], or by electrostatic effects between the negatively charged sulfonate groups and the positively charged amino groups at physiologic pH. The same results observed with nitroimidazoles made by Merck [18] without the amino group argue in favor of the interaction occurring between the aromatic rings.

Another hypothesis could explain the overcoming of the blood-brain barrier: the surface tension of suramin's structure may disturb cell membrane surfaces and modify the tight junctions of this barrier, thereby facilitating the subsequent penetration of the nitroimidazoles. Results supporting this hypothesis have already been reported [19].

The mechanism of action of megazol is undissociable from the presence of the nitro group whose omission in a homologue eliminates its biological activity. This mechanism supports a more general scheme proposed for the biologically active nitro derivatives [37] which are next discussed (Fig. 2). This scheme first involves a monoelectron bioreduction into free anion radical which is prone to subsequently alter along either of two pathways:

(1) in an aerobic setting, along the pathway of oxygen metabolites, or
(2) in an anaerobic setting, by those leading, via the nitroso derivative, to the final reductions to hydroxylamine and amine.

The first pathway corresponds to the situation of the trypanosome in the blood circulation, whereas the anaerobic conditions match instead its intracellular cycle. It has been effectively demonstrated by electron paramagnetic resonance that the free anion radical of megazol could be generated by the microsomes of *T. cruzi* and by various enzymatic systems having the NADPH cofactor and that in the presence of oxygen, this first radical could give rise to the formation of the superoxide anion [20].

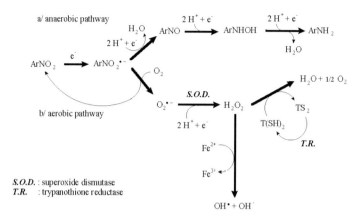

Fig. 2. Metabolism of nitroimidazoles: anaerobic and aerobic pathways

Another study using pulse radiolysis enabled the specific characteristics of the free anion radical of megazol to be revealed [21]. Notably, despite its relatively low redox potential (- 430mV), a priori less favorable to a bioreduction, it was distinguishable from other trypanocidal nitro derivatives such as nifurtimox by a four-fold more rapid reaction with oxygen and a dismutation capacity 70 times greater, a significant property of its aptitude to produce the nitroso derivative which achieves its lethal effect by trapping intracellular thiols [22].

Thus, megazol acts primarily as a catalyst of the oxygen redox cycle, the formation of its free anion radical enables it to amass the superoxide anion and oxygen metabolites; additionally, in anaerobic conditions, it contributes to the formation of nitroso derivatives. One can appreciate megazol's effectiveness via increased production of activated oxygen derivatives if one considers how poorly the trypanosome is equipped to defend itself against these molecules, lacking small anti-oxidant molecules, catalase [23], and glutathione peroxidase [24], having only trypanothione reductase (TR) [25, 26]. TR has limited activity as it only helps to regenerate the reduced form of the cofactor, but the principal reaction involving trypanothione (a glutathione analogue) is uncatalyzed. Noting TR's specificity, a number of efforts have been directed at producing specific inhibitors of this enzyme [27]. Although inhibition on a micromolar scale has been demonstrated in the isolated enzyme, none of these inhibitors have to this point had significant *in vivo* effect [28, 29].

If nifurtimox constitutes a moderate inhibitor of TR (Krauth-Siegel, personal communication), in contrast, megazol has none activity on this enzyme [20, 30].

Among the enzymes having a role in oxygen metabolism, superoxide dismutase (SOD) may itself constitute another target for combating the trypanosome. The trypanosome's version differs from that of the host in first having Fe at the active site [31] compared with Cu/Zn in the host [32], and second, having a different tertiary structure. Preliminary laboratory studies have demonstrated that megazol is a moderate inhibitor of the Fe-SOD of *Escherichia coli*, considered a good model for that of the trypanosome; however, researches of more active agents have been suggested for further study [20].

With respect to megazol, the fact that its trypanocidal activity is greater than that of other nitroimidazoles may be explained besides its redox properties by a preferential internalization by the parasite linked to the presence of the $N=C-NH_2$ group, which is known to be recognized by the purine transporter [33]. As trypanosomes are incapable of synthesizing purine bases, they possess a P2 transporter which serves to internalize the adenosine drawn from the host's blood. Thus, melarsoprol is recognized and internalized by this transporter [33]. However, studies of the internalization process with tritiated megazol [34] have demonstrated that the essential part of its internalization is conducted in a passive manner [35]. This result thus cannot explain megazol's activity as deriving from selective internalization.

From the general scheme outlining the mode of action of nitro derivatives (Fig. 2), it may be deduced that megazol could be a potent oxidant inducer as in the presence of oxygen it generates the superoxide anion and moderately inhibits a Fe-SOD homologue to that of the trypanosome. In contrast, under anaerobic conditions, its activity may better correspond to intracellular thiol depletion given the much greater ease of its anion radical dismutating into its nitroso derivative.

Although it has been discounted as an orphan molecule, as with the other nitro derivatives, because of its mutagenic potential, megazol is no less important as a drug model since permitting the identification of specific trypanosomal targets.

As with many agents with therapeutic effects, it is very likely that this compound acts on several targets. It has been recently shown that it is one of the best inhibitors of the fumarate reductase of *T. cruzi* [36], a Krebs cycle enzyme, where fumarate replaces oxygen as ultimate electron acceptor in the respiratory chain.

Beyond its interest as a model molecule for characterization of various therapeutic targets, megazol represents, because of its great biological activity, a recourse for critical clinical situations. This role will become all the more necessary as nifurtimox will soon be taken off the market because of the level of resistance that has developed to it in *T. cruzi*.

References

1. Wéry M (1994) Drug used in the treatment of sleeping sickness (HAT). Int J Antimicrob Agents 4:227-238

2. Wang CC (1995) Molecular mechanisms and therapeutic approach to treatment of African trypanosomiasis. Annu Rev Pharmacol Toxicol 35:93-127
3. Breccia A, Cavalleri B, Adams GE (1982) Nitroimidazoles Chemistry Pharmacology and Chemical applications. NATO Advanced Study Institute Series-A42, Plenum Press, New York, pp 67-82 and 115-132
4. Nair MD, Nagarajan K (1983) Nitroimidazoles as chemotherapeutic agents. Prog Drug Res 27:163-252
5. Winkelmann E, Raether W, Gebert U, Sinharay A (1977) Chemotherapeutically active nitro compounds-4-5-nitroimidazoles. (Part I). Arzneim-Forsch 27:2251-2263
6. Zweygarth E, Rotcher D (1987) Trypanocidal activity of a 5-substituted 2-nitroimidazole compound (Ro 15-0216) in mice. Trop Med Parasitol 38:171-174
7. Zweygarth E, Rotcher D (1989) Efficacity of experimental trypanocidal compounds against a multidrug-resistant *Trypanosoma brucei brucei* stock in mice. Parasitol Res 75:178-182
8. Bouteille B, Dardé ML, Pestre-Alexandre M, Dumas M, Breton JC, Nicolas JA, Catanzano G, Munoz M (1988) Treatment of experimental trypanosomiasis of sheep caused by *Trypanosoma brucei brucei*: efficacy of Ro15-0216. Bull Soc Path Ex 81:616-622
9. Zweygarth E, Kaminsky R, Sayer PD, van Nieuwenhove S (1990) Synergistic activity of 5-substituted-2-nitroimidazoles (Ro 15-0216 and benznidazole) and DL alpha difluoromethylorinithine on *Trypanosoma brucei brucei*. Ann Soc Belge Med Trop 70:269-279
10. Berkelhammer G, Asato G (1968) 2-amino-5(-1-methyl-5-nitro-2-imidazolyl)-1,3,4,-thiadiazole, a new antimicrobial agent. Science 162:1146
11. Jennings FW (1990) Future prospects for the therapy of human trypanosomiasis. 2. Combination chemotherapy and African trypanosomiasis. Trans R Soc Trop Med Hyg 84:618-621
12. Jennings FW (1993) Combination chemotherapy of CNS trypanosomiasis. Acta Trop 54:205-213
13. Filardi LS, Brener Z (1982) A nitroimidazole-thiadiazole derivative with curative action in experimental *Trypanosoma cruzi* infection. Ann Trop Med Parasitol 76:293-297
14. Bouteille B, Marie-Daragon A, Chauvière G, Albuquerque C, Enanga B, Dardé ML, Vallat M, Périé J, Dumas M (1995) Effect of Megazol on *Trypanosoma brucei brucei* acute and subacute infections in Swiss mice. Acta Trop 60:73-80
15. Jennings FW, Chauvière G, Viodé C, Murray M (1996) Topical chemotherapy for experimental African trypanosomiasis with cerebral involvement: the use of melarsoprol combined with the 5-nitroimidazole, Megazol. Trop Med Int Hlth 1:363-366
16. Albuquerque C (1995) Synthèse et étude physicochimique d'analogues du Mégazol antiparasitaire de type nitroimidazole; étude de leur mode d'action. Thèse de doctorat de l'Université Paul Sabatier N°95, Toulouse
17. Rameau JP, Devillers J, Declercq JP, Chauvière G, Périé J (1996) Molecular structure of Megazol the 2 amino-5-(1-methyl-5-nitro-2-imidazolyl)-1,2,4 thiadiazol. A concerted study by X-Ray cristallography and Molecular Mechanic. Struct Chem 7:187-204
18. Jennings FW, Urquhart GM, Murray PK, Miller BM (1983) Treatment with suramin and 2-substituted 5-nitroimidazoles of chronic murine *Trypanosoma brucei* infections with central nervous system involvement. Trans R Soc Trop Med Hyg 77:693-698
19. Spigelman MK, Zappulla RA, Johnson J, Goldsmith SJ, Malis LI, Holland JS (1986) Etoposide induced blood brain disruption effect of drug compared with that of solvent. Cancer Res 46:1453-1457
20. Viodé C, Bettache N, Cenas N, Krauth-Siegel RL, Chauvière G, Bakalara N, Périé J (1999) Enzyme reduction studies of nitroheterocyles. Biochem Pharmacol 57:549-557
21. Viodé C, Albuquerque C, Chauvière G, Houée-Lévin C, Périé J (1997) Comparative study by pulse radiolysis of the radical anion derived from compounds used in Chagas disease therapy. New J Chem 21:1331-1338
22. Berube LR, Farah S, Mc Clelland RA, Rauth AN (1991) Effect of 1-methyl-2-nitrosoimidazole on intracellular thiol and calcium levels in chinese hamster ovary cells. Biochem Pharmacol 42:2153-2161
23. Morello A (1988) The biochemistry of the mode of action of drugs and the detoxification mechanism in *Trypanosoma cruzi*. Comp Biochem Physiol 90:1-12
24. Boveris A, Sies H, Martino EE, Docampo R, Turrens JF, Stoppani AOM (1980) Deficient metabolic utilization of hydrogen peroxide in *Trypanosoma cruzi*. Biochem J 188:643-648
25. Fairlamb AH, Blackburn P, Ulrich P, Chait BT, Cerami A (1985) Trypanothione a novel bis(glutathionanyl) spermidine cofactor for glutathione reductase in trypanosomatids. Science 27:1485-1487
26. Krauth-Siegel RL, Enders B, Henderson GB, Fairlamb AH, Schrimmer RH (1987) Trypanothione reductase from *Trypanosoma cruzi*. Purification and characterization of the crystalline enzyme. Eur J Biochem 164:123-128
27. Henderson GB, Ulrich P, Fairlamb A, Rosenberg J, Pereira M, Sela M, Cerami A (1988) "Subversive" substrates for the enzyme trypanothione disulfide reductase: Alternative approach to chemotherapy of Chagas disease. Proc Natl Acad Sci 85:5374-5378
28. Girault S, Bailet S, Horvath D, Lucas V, Davioud-Charvet E, Tartar A, Sergheart C (1997) New potent inhibitors of trypanothione reductase from *Trypanosoma cruzi* in the 2 aminodiphenyl-sulfide series. Eur J Med Chem 32:39-52

29. Bonnet B, Soulez D, Davioud-Charlet E, Horvath D, Sergheraert C (1997) New spermine and spermidine derivatives as potent inhibitors of *Trypanosoma cruzi* trypanothione reductase. Biorg Med Chem 5:1249-1256
30. Périé J, Chauvière G, Viodé C, Albuquerque C, Cenas N, Houée-Lévin C, Krauth-Siegel RL (1996) Comparative bioreductives activities of nifurtimox and megazol: enzymatic and pulse radiolysis studies. Trop Med Int Hlth 1:A17-A19
31. Said I (1994) Molecular cloning, characterization and expression in *Escherichia coli* iron superoxide dismutase cDNA from *Leishmania donovani* chagasis. Infect Immun 62:657-664
32. Tainer JA, Getzoff ED, Beems KA, Richardson JJ, Richardson DC (1982) Determination and analysis of the 2A° structure of copper-zinc superoxide dismutase. J Molec Biol 160:181-217
33. Carter N, Fairlamb A (1993) Arsenical-resistant trypanosomes lack an unusual adenosine transporter. Nature 361:173-175
34. Chauvière G, Rousseau B, Pillon F, Périé J (1998) Synthesis of 4^3H Megazol the 2-amino-5-(1-methyl-5-nitro-4^3H-2-imidazolyl)1,3,4-thiadiazol. J Labelled Cpd Radiopharm 41:47-51
35. Barrett MP, Fairlamb AH, Rousseau B, Chauvière G, Périé J (1998) *Trypanosoma brucei*, *Trypanosoma cruzi* Uptake of the 5-nitroimidazole drug Megazol. Exp Parasitol (submitted for publication)
36. Turrens JF, Watts BP, Zhong L, Docampo R (1996) Inhibition *Trypanozoma cruzi* and *brucei* NADPH fumarate reductase of Benzimidazoles and antihelmintic imidazole derivatives. Mol Biochem Parasitol 82:125-129
37. Docampo R, Moreno SNJ (1984) Free radical intermediates in the antiparasitic action of drugs and phagocytic cells. In: Pryor WA (ed) Free radicals in biology. Chap 8, Academic Press, New York, pp 244-288

CHAPTER 17

Experimental models for new chemotherapeutic approaches to human African trypanosomiasis

B Bouteille, M Keita, B Enanga, J Mezui Me Ndong

Introduction

Diminazene aceturate (Berenyl®) has not been commercialized for therapeutic human usage; difluromethylornithine or DFMO (Ornidyl®) has only been made available very recently, and nifurtimox (Lampit®) has had limited usage. Thus presently, there are only three available drug products, all known for over half a century, for treating patients with human African trypanosomiasis (HAT): pentamidine (Lomidine® which is no longer commercially available and has been replaced by the more expensive Pentacarinat®) and suramin (Moranyl®) for cases in early stages which do not yet have CNS involvement, and only melarsoprol (Arsobal®) for the later meningoencephalitic stage. Furthermore, melarsoprol is a very toxic medication which itself causes an often fatal encephalopathy in 5 to 10% of treated patients [1, 2].

The search for newer less toxic medications that are easy to administer and active against the trypanosome at all disease stages is vital, given the current resurgence of HAT and the apparently growing number of patients who are refractory to arsenical therapy [2, 3]. The study of combined therapy using old and newer agents also needs to be amplified [4, 5]. It is vital that further research be done in these areas. It is essential that the drug molecule tested is able to cross the blood-brain barrier, and that its mechanism of action is recognized to be able to inhibit the trypanosome's metabolism (see review by Wang [6] and Chapters 4 and 5). This research is performed via development of experimental models of HAT *in vitro* (cell culture) as well as *in vivo* (rodent, sheep, primate). These models should both permit the gauging of the product's effectiveness for all stages of the disease and also yield an improved understanding of the disease's pathophysiology so that objective criteria for assessing drug effectiveness may be defined. A better understanding of the mechanisms involved in HAT disease progression may likely reveal new therapeutic targets and thus improve the application of existing drug regimens or lead to the rational development of compounds active against specific metabolic pathways of the trypanosome [7].

In vitro models

The first cell cultures of *Trypanosoma brucei* date back to the descriptions of Novy and McNeal in 1904 [8]. It is then necessary to wait until 1977 when Hirumi et al. [9] succeeded in maintaining a permanent cell culture, in bovine embryo fibroblasts, of bloodstream *T. brucei* trypomastigotes which retained their ability to produce host infection. In 1981, Brun et al. [10] cultured *T. b. rhodesiense* and *T. b. gambiense* in human embryo pulmonary fibroblasts. In 1985, two semi-defined culture media were described, making it possible to culture trypanosomes in the absence of nutrient cells. The only poorly defined element in this composition is serum which remains necessary to add to the media. These media are enriched with 2-mercaptoethanol [11] or L-cysteine [12]. It appears that addition of these reducing agents replaces the action of fibroblast cells. The use of semi-synthetic culture media for trypanosome culture has facilitated the *in vitro* search for an agent with trypanocidal activity (see review by Kaminsky and Brun, [13]).

Trypanosoma brucei brucei AnTat 1-9 is a monomorphic clone derived from the isolate EATRO (East African Trypanosomiasis Organization) 1125, isolated in 1966 from the blood of a *Tragelaphus scriptus* in Uganda. For *in vitro* culturing, we have adapted this clone to an acellular semi-defined medium [11]. This medium permits excellent growth of these trypanosomes which conserve their ability to infect mice [14].

Our *in vitro* trials (see Table 1) have confirmed the effectiveness of melarsoprol against *T. b. brucei* at the concentration of 0.1 μg/ml [14].

Table. 1. Effective trypanocidal concentration of products tested *in vitro* on *Trypanosoma brucei brucei* AnTat 1-9 in acellular semi-defined medium

Product tested	Effective concentration in μg/ml
Pentamidine	0.1
Suramin	1
Melarsoprol	0.1
DFMO	50[a]
Nifurtimox	1
Ro 15-0216	1
Megazol	0.1

[a]: trypanostatic concentration

The serum concentrations obtained with the standard melarsoprol doses prescribed in man are 0.2 to 2 μg/ml [15]. One of the current applications of *in vitro* culture is the study of the sensitivity of various isolates and the pharmacokinetics of melarsoprol. *In vitro*, *T. b. brucei* AnTat 1-

9 has been found sensitive to pentamidine at the dose of 0.1 µg/ml, and to suramin at 1 µg/ml [14]. Our *in vitro* trials have shown a trypanostatic effect of DFMO beginning at a concentration of 50 µg/ml [14]. The interference of this product with polyamine metabolism has been well demonstrated [16]. This mechanism may explain in part the trypanostatic effect of DFMO observed in culture and the non-pathogenicity to mice of subinoculation of trypanosomes exposed to DFMO. Nifurtimox is a nitrofurane derivative used in the treatment of Chagas' disease. We have shown its trypanocidal effect *in vitro* at a concentration of 1 µg/ml [17]. The *in vitro* trials of the 2-nitroimidazole Ro-15-0216 (derivative of benznidazole or Radanil® utilized in the treatment of Chagas' disease) show its activity against *T. b. brucei* AnTat 1-9 at the concentration of 1 µg/ml [17]. A comparative study of sensitivity to Ro 15-0216 among trypanosomes of the group *brucei*, using the same semi-synthetic medium, gave the following results: *T. b. rhodesiense* is the most sensitive, *T. b. gambiense* the least sensitive, and *T. b. brucei* is of intermediate sensitivity [18].

Megazol or CL 64,855 is a nitroimidazole-thiadiazole derivative, synthesized for the first time in 1968 by Berkelhammer and Asato [19]. The synthesis of megazol and its derivatives have been achieved by the Biological Organic Chemistry Laboratory of the University Paul Sabatier of Toulouse (Périé and Chauvière, see Chapt. 16). Using *T. b. brucei* AnTat 1-9, we have been able to establish its trypanocidal activity *in vitro* at a concentration of 0.1 µg/ml [20].

On other fronts, a trial involving 45 commercialized drug products, known to cross the blood-brain barrier and belonging to various therapeutic classes (antidepressants, neuroleptics, sedatives, cerebral vasodilators, anti-convulsants, etc.) has proved fruitless [21].

Murine models

The first drug regimen for which the metabolic target in the trypanosome was completely understood was studied using a mouse model infected by *T. b. brucei* or by *T. b. rhodesiense*: it consisted of salicylhydroxamic acid (SHAM) used in conjunction with glycerol [22-25]. Unfortunately, later relapse occurred with this combination, because *in vivo* glycerol was problematic to provide at high and enough a dose, and did not cross the blood-brain barrier [26]. The second identified target, trypanosomal ornithine decarboxylase whose inhibition causes a defect in the biosynthesis of trypanothione, led to the commercial development of DFMO [27, 28]. Novel inhibitors of this trypanosomal polyamine pathway have activity in the infected mouse: most notably MDL 73811, an inhibitor of the S-adenosyl methionine decarboxylase [29, 30].

The mouse model generally yields an acute infection that has no similarity to the human disease. It can only demonstrate that there is *in vivo* activity complementing any activity found in culture.

However, certain isolates of *T. b. brucei*, *T. b. rhodesiense* or *T. b. gambiense* may produce infections with subacute or chronic course in the mouse and rat. It was thus in 1981 that Beckers et al. [31] and Van Marck et al. [32] obtained a chronic infection model: using *T. b. gambiense*, the average survival duration ranged from 24 to 59 days in the mouse and from 82 to 221 days in the rat. Histopathological examination of these models showed that central nervous system (CNS) involvement became increasingly severe with duration of infection, making these chronic infection systems the models of choice for therapeutic experimentation [33]. An additional mouse model with CNS involvement has also been described [34-36].

Beginning in 1977, Jennings et al. [37] developed this type of model for studying new drug therapies for meningoencephalitic sequelae of HAT. The duration of infection prior to treatment was found to play a determinant role in the mouse model: diminazene aceturate, which does not cross the blood-brain barrier, cured mice infected with *T. b. brucei* TREU 667 or *T. b. brucei* GVR 35 if it was administered 3 days after the onset of infection, and did not cure them after 21 days, when there was CNS involvement [38]. The utilization of such models of chronic infection with CNS involvement approaching the human disease permits study of drug effectiveness for the neurological stage of the disease of a product recognized to have trypanosomal activity [4, 5, 38, 39]. The effectiveness of the 5 nitro-imidazoles has been demonstrated using these models. Certain compounds in this chemical family (L 611-744, MK 436, L 634-549, fexinidazole) when used singly, had an effect limited to the lympho-sanguine phase of the disease; when used in conjunction with suramin, they showed considerable effectiveness at lower doses, and on the meningoencephalitic stage [4, 39]. The synergistic effect of combining DFMO and arsenical derivatives has also been shown in this model [38].

For establishing our mouse model system, we selected two clones of *T. b. brucei*, derived from EATRO 1125 cited above. The monomorphic clone *T. b. brucei* AnTat 1-9, causes an acute infection in mice producing death in three to four days. The pleomorphic clone *T. b. brucei* AnTat 1-1E, permits a much longer disease course, 30 days on average, with CNS involvement by around the 21st day [14, 17, 20, 68]. This second clone is not submitted to repeated passage in the Swiss mouse to avoid its evolution toward monomorphism and a more acute pathogenicity: cryostabilats are utilized for each new experiment. The effectiveness of a therapeutic trial is measured in time after treatment without parasitemic relapse: 4 months for the acute model with *T. b. brucei* AnTat 1-9 and 12 months for the chronic model with *T. b. brucei* AnTat 1-1E.

Model of acute infection

Suramin in a single intra-peritoneal dose of 20 mg/kg produced cure in 50% of mice. A much higher dose, 40 mg/kg is necessary to obtain near-complete cure. DFMO is ineffective against these trypanosomes when administered per os or by intra-peritoneal injection at doses 10 times superior to those recommended for the treatment of HAT (4 gm/kg/day 2 times per day for 6 days by intra-peritoneal line or *ad libitum* in water drunk during the same duration). Nifurtimox at the dose 100 mg/kg/d x 15 days provides a 50% cure rate. At a dose of 200 mg/kg/d x 4 days, the product is active but causes significant neurotoxicity, with a mortality rate of 20%. The following protocol: 200 mg/kg/d x 3 days, followed by 100 mg/kg/d x 7 days produces a 90% cure rate with only mild and rapidly reversible signs of

Table. 2. Treatment in the acute experimental mouse model infected with *T. b. brucei* An-Tat 1-9 by suramin or megazol administered by intraperitoneal or per os route

Treatments	Given intraperitoneally		Given per os	
(mg/kg/day)	Number of hours post-infection	Cured mice / treated mice	Number of hours post-infection	Cured mice / treated mice
Control group	–	0 / 32	–	0 / 10
Suramin (20)	12	3 / 5	–	ND
Suramin (40)	12	5 / 5	–	ND
Megazol (20)	12	5 / 5	12	4 / 5
Megazol (80)	24	5 / 5	24	5 / 5
Megazol (80)	48	5 / 5	48	6 / 6

ND: not done

neurotoxicity. Megazol at the dose of only 20 mg/kg produces near-complete cure on the experimental model. Its significant trypanocidal effect is also seen when administered per os at 80 mg/kg, even 48 hours after infection onset when the parasitemia is already quite elevated and the mice difficult to cure.

Model of chronic infection

This model presents CNS involvement by the 21st day of infection manifested by a meningitis with lympho-plasmocytic infiltration on histopathological examination. Mice treated on the 21st day by a single dose of suramin at 20 mg/kg had parasitic relapse and died. After the mice's pre-treatment with suramin, blood subinoculations were not infectious, the mice remained healthy without any detectable parasitemia, while subinoculations of ground brain from these singly treated mice produced infection in the recipient mice, proving that trypanosomes remained in the brain, sheltered from suramin treatment.

Iterative treatments with suramin at a dose of 20 mg/kg for each parasitemic relapse permitted a highly significant survival improvement for parasitized mice, to greater than 300 days, but with progressive worsening of CNS lesions and the appearance of neurological disease manifestations (paralysis, troubles with coordination, somnolent state). The number of mice presenting such symptomatology increased with time; the meningitis became more extensive, evolving toward a meningo-encephalitis with progression into the Virchow-Robin spaces. Trypanosomes could be found in the meninges and the choroid plexus, and then throughout the parenchyma with longer-duration infection. Brain cuts, immunohistochemically stained for GFAP (glial fibrillary acid protein), showed an increasing astrocytosis with time [40]. This model reproduces the clinical and histopathological characteristics observed in the human disease and thus is particularly interesting to employ for identifying the effectiveness of a drug product at the neurological stage of the disease.

Table. 3. Effectiveness of suramin-nifurtimox and suramin-megazol combinations in the chronic experimental mouse model infected with *T. b. brucei* AnTat 1-1E

Doses (mg/kg/d × number of days)	Time of treatment (in days post-infection)	Cured mice / treated mice
Control group	–	0 / 10
Suramin IP[a] (20 × 1)	21	0 / 10
Nifurtimox IP[a] (200 × 3)	25 to 27	0 / 10
Megazol IP[a] (80 × 4)	21 to 24	0 / 10
Suramin IP[a] (20 × 1)	25	
Nifurtimox IP[a] (200 × 3)	26 to 28	5 / 10
Suramin IP[a] (20 × 1)	21	
Nifurtimox IP[a] (200 × 3 then 100 × 14)	26 to 42	10 / 10[c]
Suramin IP[a] (20 × 1)	21	
Megazol Ip[a] (80 × 3)	22 to 24	10 / 10[c]
Suramin IP[a] (20 × 1)	21	
Megazol PO[b] (150 × 3)	22 to 24	10 / 10[c]
Suramin IP[a] (20 × 1)	100	
Megazol IP[a] (80 × 4)	101 to 104	14 / 15[c]
Suramin IP[a] (20 × 1)	100	
Megazol PO[b] (80 × 4)	101 to 104	14 / 15[c]

[a]: given intraperitoneally (IP); [b]: given *per os* (PO); [c]: disappearance of CNS lesions (histopathology)

Indirectly, this model is instructive for determining if a tested product is likely to pass the blood-brain barrier. Treated with suramin, nifurtimox, or megazol beginning in the fourth week of infection, mice had parasitemic relapse and died. Combination treatment using a dose of suramin followed by administration of nifurtimox or megazol led to cure in all the infected animals [20]. Most notably, the suramin-megazol combination, with megazol administered by intra-peritoneal injection or per os, led to cure mice after 100 days of infection – at which time there is significant cellular infiltration of the CNS [71]. These results confirm the interest in testing drugs for their combined effects as previously described [4, 5, 20]. The effects of combination therapy may be synergistic or additive, allowing the possible shortening of treatment duration or reducing its toxicity. Moreover, combination therapy in this case allowed treatment of trypanosomiasis with CNS involvement, whereas it has been shown that treatment with either of the drugs on its own is not effective at this stage of the disease. However, the mechanism of action of suramin and azole derivatives in combination use is still not known.

Sheep model

The sheep is susceptible to infection with *T. brucei* [41]. The raising and handling of this animal are easy. Sampling of biological liquids is reasonably simple, particularly the CSF [42]. In this model, sheep are infected subcutaneously with *T. b. brucei* AnTat 1-9. Survival duration ranges from 34 to 172 days (mean: 75 days). Once infected, the animals display hyperthermia, anemia, weight loss, edema. Beginning in the 3rd week, the animals develop anorexia, little response to stimulation, and unstable gait. Parasitemia appears around the 9th day, and persists at low levels throughout the disease course. Appearance of the trypanosomes and a pleocytosis in the CSF occurs between the 27th and 71st days [43]. Total plasma protein levels are highly elevated, particularly the IgM component, with inversion of usual albumin to globulin ratio [44]. The hemolytic activity of complement in sheep is slight, 23.30 units for 50% hemolysis per ml (CH50). In the infected animal, this hemolytic activity can no longer be detected from the fifth day of infection until the animal's death [45]. On the other hand, circulating immune complexes are detectable from the 13th day. CSF total protein is little increased, never exceeding 1 g/l; IgM percent quickly rises to greater than 10% of CSF total protein. Histopathological analysis reveals involvement of all sampled organs, with exception of the kidneys and lungs. Cerebral involvement, in particular, shows diffuse lymphoplasmocytic and perivascular involvement typical of the human disease.

Using this model, we have studied auto-antibodies directed against CNS components whose presence might be predictive of the neurological involvement characteristically seen in human trypanosomiasis. Using glycolipid chromatography and immunoenzymatic techniques, we have demonstrated significant reactivity of sheep model serum and CSF with galactocerebroside, the major constituent of myelin [46, 47]. These antibodies have since been detected in the serum and CSF of patients with HAT. The proportion of patients with CSF anti-galactocerebrosides goes from 5.5% during the lympho-sanguine stage to 84.6% in the neurological stage with clinical signs [46, 48].

Sheep experimentally infected with *T. b. brucei* AnTat 1-9 adequately reproduce the human disease [44, 45]. This model may be utilized to conduct trials of molecules active against African trypanosomes. We have employed it to demonstrate the activity of Ro 15-0216 beginning at a dose of 25 mg/kg and the effectiveness of melarsoprol starting at 0.9 mg/kg, which is a dose four times lower than that recommended in human treatment [49, 50].

Primate models

Description of primates naturally infected by trypanosomes of the group *brucei* has been infrequently reported in the literature [51]. Nevertheless, monkeys have been utilized for studying trypanosome isolates derived from man or other animals. Generally, acute or subacute infections have been obtained with *T. b. gambiense* and *T. b. rhodesiense* [41] and were reported as far back as 1912 by Laveran and Mesnil [52], then by Peruzzi [53], Van Bogaert and Dewulf [54], and Baker [55]. A recent study on the serum trypanocidal factor against *T. b. gambiense* has demonstrated its presence in *Papio* sp, resistant to trypanosome infection, and its absence in *Cercopithecus* sp, *Erythrocebus patas*, *Colobus* sp and *Macaca* sp, predicting the sensitivity of these primates to experimental infection by *T. b. gambiense* [56].

Cercopithecus aethiops, *Macaca arctoides* and *Pan troglodytes* infected by *T. b. rhodesiense* have led to development of a disease model with CNS involvement and permitted the investigation of reputedly trypanocidal drug products [57-61].

The experimental primate model currently utilized in therapeutic research is *C. aethiops* infected by *T. b. rhodesiense* [60, 62-65]. This model produces CNS lesions and has a mean survival of 65 days in the absence of treatment; trypanosomes are detectable in the CSF 19 to 41 days after infection onset. When CNS involvement is found, treatment by diminazene aceturate or suramin provides a significant lengthening of survival to around a year and thus allows development of a model with more severe meningoencephalitic signs [61, 66]. A number of therapeutic strategies have been described on this model: low dose melarsoprol, nitroimidazoles (MK 436, Ro 15-0216), DFMO, combination therapy [67, 69, 70].

Presently, research on HAT does not have an available experimental primate model with *T. b. gambiense*. The only exisiting model is *C. aethi-*

ops infected with *T. b. rhodesiense* [60, 67]. However, this model has short survival without treatment and significant meningoencephalitic involvement is only realizable by prolonging the animals' survival by drug treatment; *T. b. gambiense* produces a more chronic infection and may prove to be a better trypanosome to create a CNS involvement model [56]. We have attempted development of such a model by infecting *C. aethiops* by a line of *T. b. gambiense* derived from a Congolese (ex-Zaire) patient. Our preliminary results have been encouraging. The primate shows several waves of parasitemia accompanied clinically with febrile spikes. Clinical exam also reveals adenopathy, splenomegaly, weight loss, and anemia. CNS involvement is of late onset, occurring 170 days into the infection, with neurological signs accompanied by a hypercellular CSF containing trypanosomes.

Conclusion

The methodologies used for culturing and studying the chemical sensitivity of trypanosomes have not been standardized, rendering literature results difficult to compare. Animal experiments also need to be conducted using more standardized protocols, which include recognized rules for the ethical treatment of animals, in order to limit the number of animals used and facilitate their coordination of use in laboratories pursuing similar objectives. Additionally, the available experimental models of HAT only imperfectly reproduce the course of the human disease. This deficiency, limiting the reliability of pathophysiological as well as therapeutic research, needs to be remedied by developing an experimental primate model: *C. aethiops* infected by *T. b. gambiense*. Despite the pharmaceutical industry's disinterest in developing new drugs active against HAT, several products merit a more concerted study with view towards human therapeutic use. These are, in our experience, Ro 15-0216 and megazol. These nitro-imidazole derivatives possess a mutagenic capacity *in vitro* that has not been proven *in vivo*. But it seems paradoxical that this could limit further research in face of a disease in recrudescence, fatal without treatment, whose present specific drug treatment engenders an often-fatal encephalopathy.

Acknowledgements

The Authors would like to thank Professors Nestor Van Meirvenne and Dominique Le Ray, and Mr. Yves Claes of the Prince Leopold Institute of Tropical Medicine of Antwerpen (Belgium) for their collaboration. This research was supported by the Conseil Régional du Limousin and the support of the Ministère de la Coopération in the position of "Programme Mobilisateur Trypanosomiase" (projects TR-9601 and TR-9603).

References

1. Arroz JOL (1987) Melarsoprol and reactive encephalopathy in *Trypanosoma brucei gambiense*. Trans R Soc Trop Med Hyg 81:192
2. Kuzoe FAS (1993) Current situation of African trypanosomiasis. Acta Trop 54:153-162
3. Cattand P (1994) Trypanosomiase humaine africaine. Situation épidémiologique actuelle, une recrudescence alarmante de la maladie. Bull Soc Path Ex 87:307-310
4. Jennings FW (1990) Future prospects for the chemotherapy of human trypanosomiasis. 2. Combination therapy and African trypanosomiasis. Trans R Soc Trop Med Hyg 84:618-621
5. Jennings FW (1993) Combination chemotherapy of CNS trypanosomiasis. Acta Trop 54:205-213
6. Wang CC (1995) Molecular mechanisms and therapeutic approaches to the treatment of African trypanosomiasis. Annu Rev Pharmacol Toxicol 35:93-127
7. Périé J, De Albuquerque C, Blonski C, Chauvière G, Gefflaut T, Page P, Trinquier M, Willson M (1994) Conception rationnelle et études de molécules actives contre les différentes trypanosomiases. Bull Soc Path Ex 87:353-361
8. Novy FG, McNeal WJ (1904) On the cultivation of *Trypanosoma brucei*. J Infect Dis 1:1
9. Hirumi H, Doyle JJ, Hirumi K (1977) African trypanosomes. *In vitro* cultivation of animal-infective *Trypanosoma brucei*. Science 196:992-994
10. Brun R, Jenni L, Schonenberger M, Schell KF (1981) *In vitro* cultivation of bloodstream forms of *Trypanosoma brucei*, *T. rhodesiense* and *T. gambiense*. J Protozool 28:470-479
11. Baltz T, Baltz D, Giroud C, Crockett J (1985) Cultivation in a semi-defined medium of animal infective forms of *Trypanosoma brucei*, *T. equiperdum*, *T. evansi*, *T. rhodesiense* and *T. gambiense*. EMBO J 4:1273-1277
12. Duszenko M, Ferguson MAJ, Lamont GS, Rifkin MR, Cross GAM (1985) Cysteine eliminates the feeder cell requirement for cultivation of *Trypanosoma brucei* bloodstream forms *in vitro*. J Exp Med 162:1256-1263
13. Kaminsky R, Brun R (1993) *In vitro* assays to determine drug sensitivities of African trypanosomoses: a review. Acta Trop 54:279-289
14. Bouteille B, Dardé ML, Pestre-Alexandre M (1988a) Action des médicaments testés en milieu acellulaire et chez la souris infectée par *Trypanosoma brucei brucei*. Bull Soc Path Ex 81:533-542
15. Maes L, Doua F, Hamers R (1988) ELISA assay for melarsoprol. Bull Soc Path Ex 81:557-560
16. Fairlamb AM, Henderson GB, Bacchi J, Cerami A (1987) *In vivo* effects of difluoromethylornithine on trypanothione and polyamine levels in bloodstream forms of *Trypanosoma brucei*. Mol Biochem Parasitol 24:185-191
17. Bouteille B, Dardé ML, Pestre-Alexandre M (1988b) Efficacité du nifurtimox sur *Trypanosoma brucei brucei in vitro* et *in vivo* sur souris Swiss. Etude préliminaire. Bull Soc Fr Parasitol 6:15-20
18. Borowy NK, Nelson RT, Hirumi H, Brun R, Waithaka HK, Schwartz D, Polak A (1988) Ro 15-0216: a nitroimidazole compound active *in vitro* against human and animal pathogenic African trypanosomes. Ann Trop Med Parasitol 82:13-19.
19. Berkelhammer G, Asato G (1968) 2-Amino-5-(1-methyl-5-nitro-2-imidazolyl)-1, 3, 4-thiadiazole. A new antimicrobial agent. Science 162:1146
20. Bouteille B, Marie-Daragon A, Chauvière G, De Albuquerque C, Enanga B, Dardé ML, Vallat JM, Périé J, Dumas M (1995) Effect of megazol on *Trypanosoma brucei brucei* acute and subacute infections in Swiss mice. Acta Trop 60:73-80
21. Marie-Daragon A, Rouillard MC, Bouteille B, Bisser S, De Albuquerque C, Chauvière G, Périé J, Dumas M (1994) Essais d'efficacité sur *Trypanosoma brucei brucei* de molécules franchissant la barrière hématoméningée et du mégazol. Bull Soc Path Ex 87:347-352
22. Clarkson AB, Brow FH (1976) Trypanosomiasis: an approach to chemotherapy by the inhibition of carbohydrate catabolism. Science 194:204-206
23. Opperdoes FR, Aarsen PN, Van Der Meer C, Borst P (1976) *Trypanosoma brucei*: an evaluation of salicylhydroxamic acid as a trypanocidal drug. Exp Parasitol 40:198-205
24. Evans DA, Brightman AJ, Holland MF (1977) Salicylhydroxamic acid/glycerol in experimental trypanosomiasis. Lancet ii 8041:769
25. Fairlamb AH, Opperdoes FR, Borst P (1977) New approach to screening drugs for activity against African trypanosomes. Nature 265:270-271
26. Evans DA, Brightman AJ (1980) Pleomorphism and the problem of recrudescent parasitaemia following treatment with salicylhydroxamic acid (SHAM) in African trypanosomiasis. Trans R Soc Trop Med Hyg 74:601-604
27. Schechter PJ, Sjoerdsma A (1986) Difluoromethylornithine in the treatment of African trypanosomiasis. Parasitol Today 23:223-224
28. Milord F, Pépin J, Loko L, Ethier L, Mpia B (1992) Efficacy and toxicity of eflornithine for the treatment of *Trypanosoma brucei gambiense* sleeping sickness. Lancet 340:652-655
29. Bitonti AJ, Byers TL, Bush TL, Casara PJ, Bacchi CJ (1990) Cure of *Trypanosoma brucei brucei* and *Trypanosoma brucei rhodesiense* infections with an irreversible inhibitor of S-adenosylmethionine decarboxylase. Antimicrob Agents Chemother 34:1485-1490

30. Bacchi CJ, Nathan HC, Yarlett N, Goldberg B, McCann PP (1992) Cure of murine *Trypanosoma brucei rhodesiense* infections with an S-adenosylmethionine decarboxylase inhibitor. Antimicrob Agents Chemother 36:2736-2740
31. Beckers A, Wéry M, Van Marck E, Gigase P (1981) Experimental infections of laboratory rodents with recently isolated stocks of *Trypanosoma brucei gambiense*. 1. Parasitological investigations. Z Parasitenkd 64:285-296
32. Van Marck EAE, Gigase PLJ, Beckers A, Wéry M (1981) Experimental infections of laboratory rodents with recently isolated stocks of *Trypanosoma brucei gambiense*. 2. Histopathological investigations. Z Parasitenkd 64:187-193
33. Wéry M (1981) Improvements in laboratory models for drug-testing using stocks of *Trypanosoma brucei gambiense* of low virulence. In: Canning EU (ed) Parasitological Topics. Special Publication N°1 of the Society of Protozoologists, pp 275-283
34. Poltera AA (1980) Immunopathological and chemotherapeutic studies in experimental trypanosomiasis with special reference to the heart and brain. Trans R Soc Trop Med Hyg 74:706-715
35. Poltera AA, Hochmann A, Rudin W, Lambert PH (1980) *Trypanosoma brucei brucei*: a model for cerebral trypanosomiasis in mice – an immunological, histological and electronmicroscopic study. Clin Exp Immunol 40:496-507
36. Poltera AA, Hochmann A, Lambert PH (1981) *Trypanosoma brucei brucei*: the response to Melarsoprol in mice with cerebral trypanosomiasis. An immunopathological study. Clin Exp Immunol 46:363-374
37. Jennings FW, Whitelaw DD, Urquhart GM (1977) The relationship between duration of infection with *Trypanosoma brucei* in mice and the efficacy of chemotherapy. Parasitology 75:143-153
38. Jennings FW (1988) The potentiation of arsenicals with difluoromethylornithine (DFMO): experimental studies in murine trypanosomiasis. Bull Soc Path Ex 81:595-607
39. Jennings FW, Urquhart GM, Murray PK, Miller BM (1983) Treatment with suramin and 2-substituted 5-nitroimidazoles of chronic *Trypanosoma brucei* infections with central nervous system involvement. Trans R Soc Trop Med Hyg 77:693-698
40. Keita M, Bouteille B, Enanga B, Vallat JM, Dumas M (1997) *Trypanosoma brucei brucei*: a long-term model of human African trypanosomiasis in mice, meningo-encephalitis, astrocytosis, and neurological disorders. Exp Parasitol 85:183-192
41. Losos GJ, Ikede BO (1972) Review of pathology of diseases in domestic and laboratory animals caused by *Trypanosoma congolense*, *T. vivax*, *T. brucei*, *T. rhodesiense* and *T. gambiense*. Vet Pathol 9 (Suppl):1-71
42. Ndo D, Nicolas A, Caix M, Bouteille B, Dumas M, Pestre-Alexandre M (1991) Techniques for collecting cerebrospinal fluid from sheep. Progr Vet Neurol 2:77-79
43. Bouteille B, Dardé ML, Dumas M, Catanzano G, Pestre-Alexandre M, Breton JC, Nicolas JA, N'Do DC (1988c) The sheep (*Ovis aries*) as an experimental model for African Trypanosomiasis. I. Clinical study. Ann Trop Med Parasitol 82:141-148
44. Bouteille B, Dardé ML, Dumas M, Catanzano G, Pestre-Alexandre M, Breton JC, Nicolas JA, Munoz M (1988d) The sheep (*Ovis aries*) as an experimental model for African Trypanosomiasis. II. Biological study. Ann Trop Med Parasitol 82:149-158
45. Bouteille B, Dardé ML, Monteil J, Pestre-Alexandre M (1988e) Le complément: témoin d'infestation par *Trypanosoma brucei brucei* chez le mouton, modèle expérimental; son évolution après traitement. Bull. Soc Path Ex 81:522-529
46. Jauberteau MO, Ben Younes-Chennoufi A, Amevigbe M, Bouteille B, Dumas M, Breton JC, Baumann N (1991) Galactocerebrosides are antigens for immunoglobulins in sera of an experimental model of trypanosomiasis in sheep. J Neurol Sci 101:82-86
47. Amevigbe M, Jauberteau-Marchan MO, Bouteille B, Doua F, Breton JC, Nicolas JA, Dumas M (1992) Human African trypanosomiasis: presence of antibodies to galacto-cerebrosides. Am J Trop Med Hyg 47:652-662
48. Jauberteau MO, Bisser S, Ayed Z, Brindel I, Bouteille B, Stanghellini A, Gampo S, Doua F, Breton JC, Dumas M (1994) Détection d'autoanticorps anti-galactocérébrosides au cours de la trypanosomose humaine africaine. Bull Soc Path Ex 87:333-336
49. Bouteille B, Dardé ML, Pestre-Alexandre M, Dumas M, Breton JC, Nicolas JA, Catanzano G, Munoz M (1988f) Traitement de la trypanosomiase expérimentale du mouton à *Trypanosoma brucei brucei*: recherche d'une dose minima active de Mélarsoprol. Bull Soc Path Ex 81:548-554
50. Bouteille B, Dardé ML, Pestre-Alexandre M, Dumas M, Breton JC, Nicolas JA, Catanzano G, Munoz M (1988g) Traitement de la trypanosomiase expérimentale du mouton à *Trypanosoma brucei brucei*: efficacité du Ro150216 (dérivé 2-nitroimidazolé). Bull Soc Path Ex 81:616-622
51. Bruce D, Hamerton AE, Bateman HR, Mackie FP, Lady Bruce (1911) Experiments to ascertain whether the mammals, birds or reptiles in Uganda, living on or near the lake-shore, harbour *Trypanosoma gambiense*. Reports of the Sleeping Sickness Commission of the Royal Society 11:100-104
52. Laveran A, Mesnil F (1912) Trypanosomes et trypanosomiases. Masson, Paris, pp 702-716
53. Peruzzi MRI (1928) Pathologico-anatomical and serological observations on trypanosomiasis. Final Report. League of Nations International Committee on Human Trypanosomiasis 3:245-328

54. Van Bogaert L, Dewulf A (1938) Etudes sur le mode d'extension et l'histopathologie des trypanosomiases expérimentales. 1. La méningo-encéphalite à *Trypanosoma gambiense* chez *Papio jubilaeus*. J Belge Neuro Psychiatr 38:559-582
55. Baker JR (1962) Infection of the chimpanzee (*Pan troglodytes verus*) with *Trypanosoma rhodesiense* and *T. brucei*. Ann Trop Med Parasitol 56:216-217
56. Seed JR, Sechelski B, Loomis MR (1990) A survey for a trypanocidal factor in primate sera. J Protozool 37:393-400
57. Godfrey DG, Killick-Kendrick R (1967) Cyclically transmitted infections of *Trypanosoma brucei*, *T. rhodesiense* and *T. gambiense* in the chimpanzees. Trans R Soc Trop Med Hyg 61:781-791
58. Sadun EH, Johnson AJ, Nagle R, Duxbury R (1973) Experimental infections with African trypanosomes. V. Preliminary parasitological, clinical, hematological, serological and pathological observations in rhesus monkeys infected with *Trypanosoma rhodesiense*. Am J Trop Med Hyg 22:323-330
59. Raether W, Seidenath H (1976) Trypanocidal effect of diamidine 98: 202 in experimental *Trypanosoma rhodesiense* infection of the stumptailed macaque (*Macaca arctoides*). Tropenmed Parasit 27:238-244
60. Schmidt H, Sayer P (1982a) *Trypanosoma brucei rhodesiense* infection in vervet monkeys. I. Parasitology, hematology, immunology and histologic results. Tropenmed Parasit 33:249-254
61. Schmidt H, Sayer P (1982b) *Trypanosoma brucei rhodesiense* infection in vervet monkeys. II. Provocation of the encephalitic late phase by treatment of infected monkeys. Tropenmed Parasit 33:255-259
62. Poltera AA, Sayer PD, Schmidt H, Njogu AR (1981) Une étude comparative des trypanosomiases cardiaques africaines chez l'homme, les souris et les singes. Conseil Scientifique International de Recherche sur les Trypanosomiases et leur Contrôle. OUA/STRC publication n° 112, Arusha, Tanzania, pp 254-258
63. Poltera AA, Sayer PD (1983) Cardiac lymph drainage in experimental African trypanosomiasis in vervet monkeys. Bull Soc Path Ex 76: 614-621
64. Rudin W, Poltera AA, Jenni L (1983) An EM study on cerebral trypanosomiasis in rodents and primates. Contr Microbiol Immunol 7:165-172
65. Poltera AA, Sayer PD, Rudin W, Bovell D (1985a) Trypanosomal cardiac valvulitis in vervet monkeys. Trop Med Parasitol 36:77-80
66. Poltera AA, Sayer PD, Grichouse G, Bovell D, Rudin W (1985b) Immunopathological aspects of trypanosomal meningoencephalitis in vervet monkeys after relapse following Berenil treatment. Trans R Soc Trop Med Hyg 79:527-531
67. Sayer PD, Onyango JD, Gould SS, Waitumbi JN, Raseroka BH, Akol GWO, Ndung'u JM, Njogu AR (1987) Treatment of African trypanosomiasis with combinations of drugs with special reference to suramin and nitroimidazoles. International Scientific Council for trypanosomiasis Research and Control. 19th Meeting, Lomé, Togo, pp 205-210
68. Bouteille B, Dardé ML, Pestre-Alexandre M (1990) Trypanosomose à *Trypanosoma brucei brucei*: traitement par des associations. Bull Soc Fr Parasitol 6:15-20
69. Burudi EME, Karanja SM, Njue AI, Githiori JB, Ndung'u JM (1995) Establishment of a partly DFMO-sensitive primate model of *Trypanosoma rhodesiense* sleeping sickness. Acta Trop 59:71-73
70. Sayer PD, Gould SS, Waitumbi JN, Murray PK, Njogu AR (1985) Guérison des singes infectés par *Trypanosoma rhodesiense* au moyen de faibles doses de mélarsoprol ou de suramine et de nitroimidazole. Conseil Scientifique International de Recherche sur les Trypanosomiases et leur Contrôle. OUA/STRC publication n° 113, Harare, Zimbabwe, p 131
71. Enanga B, Keita M, Chauvière G, Dumas M, Bouteille B (1998) Megazol combined with suramin: a chemotherapy regimen which reversed the CNS pathology in a model of human African trypanosomiasis in mice. Trop Med Int Health 3:736-741

CHAPTER 18

Prophylactic strategies in human African trypanosomiasis

A STANGHELLINI

Introduction

Prophylaxis concerns all measures that can be undertaken to prevent the onset or transmission of a disease. In trypanosomiasis, these prophylactic measures have arisen from epidemiological data concerning man and his vector, and involve action taken against the vector population and health care for the human population. Strategies directed against the tsetse fly can reduce the contact between man and his vector, and thus, the risk of being bitten. Two management approaches are possible in the human population: first, routine screening for the disease and treatment of infected persons, and secondly, drug prophylaxis of exposed populations.

We will successively describe the measures available to control the tsetse fly, screen for disease, treat patients, and then discuss the different strategies to be undertaken depending on the type of epidemiological situation.

Measures available

Reduction of the contact between man and vector (measures against the tsetse fly)

Techniques aimed at reducing the contact between the genus *Glossina* and man have progressively developed, concomitant with our knowledge of the biology of the vector and the development of insecticides. The goal in controlling the vector is not its eradication, but rather to maximally reduce tsetse fly densities in areas populated by man. Historically, three periods can be distinguished in this effort: the pre-insecticide era, the insecticide era, and the ecological era.

During the pre-insecticide era many techniques were perfected, occasionally with very good results. These involved either measures directed against the flies' resting sites (agronomical prophylaxis) or the use of mechanical means to which the flies were attracted and retained (black screens impregnated with glue).

Agronomical prophylaxis, widely used for several decades and codified in July 1920 by the Commission de la Société de Pathologie Exotique [1], consisted of cutting down gallery forest and other forest tracts (for example, the sacred woods of Tenkodogo in the Upper Volta) of all bushes, grass, stumps, and fallen trunks which were excellent fly resting sites. Furthermore, by modifying the exposure of soil to sunlight, these cleared tracts of land diminished the egg-laying potential of tsetse flies whose larvae require cool, moist, shady soil to develop in. Lastly, a salubrious effect was achieved indirectly because the land cleared of forest was used for market gardening, thus improving the diet of the various populations.

The use of black tissue impregnated with glue, worn on the back of field workers, was not as widespread as agronomical prophylaxis, however, this made it possible in certain islands in the Gulf of Guinea to totally eradicate local populations of tsetses. It was an ingenious method developed before its time. These islands were subsequently reinfested from the continent by boats.

The insecticide era was introduced during the Second World War which lead to considerable development in chemical research, especially insecticides. During this period, the chlorobenzene derivatives were developed, in particular DDT, a miraculous insecticide widely used in the countryside to control vectors of insect-transmitted diseases during the fifties and sixties. The initial results were promising; however, increasing resistance of the tsetse fly to DDT rapidly developed.

The chemical industry then improved new halobenzene insecticides such as dieldrin, whose effects were longer lasting, and endosulfan, of shorter duration but more effective and more soluble, and then the synthetic pyrethrins, with the regular appearance of newer derivatives and presentations. These insecticides were extensively used by aerosol application from airplanes over the savanna, and from helicopters over forests in areas needing treatment. These applications from aircraft were later gradually abandoned because, as stated by Wéry in 1990 [2], "their application in the forest amounted to a sporting event, and on the savanna to a financial prowess".

Because of the problems which arose from the massive use of insecticides (the appearance of resistance, toxicity of other exposed fauna, in particular mammals), medical entomologists, whose mission was to rid affected zones of the vector, recommended the more rational use of insecticides by spraying selectively only those areas that the tsetse fly was susceptible to use as resting sites. Thus, in endemic zones this method was developed, and produced very encouraging results using considerably less quantities of insecticides, with a concomitant reduction in the risk of resistance.

To more effectively target control of the tsetse fly, techniques using traps were developed. These traps derived from a biconical trap [3] are of two types: traps in the strict sense of the term, and attractive screens.

The former can be divided in two types: those used to eliminate flies and those used for capturing them for study purposes. These traps are composed of diverse sheets of blue or black tissue which attract the tsetse fly, with an upper portion made of mosquito netting to direct the insect toward the top of the trap. They are impregnated with a long-acting insecticide (generally deltamethrin), requiring reimpregnation approximately every four to six months depending on the rainfall, or they are furnished with an internal capturing system (a hoop net made of tissue or plastic) that does not require the use of an insecticide. There are two advantages from the latter system. First, the use of an insecticide is not necessary, and secondly, the persons responsible for the trap can verify if it is functioning properly by visualizing tsetse flies in the hoop net. If these traps are kept in good conditions (regular repair of tears, for example), they can normally last for 12 to 18 months before needing to be replaced.

Screens represent a marked simplification of the trap, being comprised of blue and black tissue placed in vertical sheets (to attract flies) that are impregnated with a long-lasting insecticide. After landing, the contact time between the fly and the tissue is sufficient to deliver a lethal dose. Certain screens have, in addition, lateral sheets of black netting; others emit odors (CO_2, acetone) that attract tsetse flies within a radius of several kilometers. Their cost is relatively low; however, as with traps, they must be reimpregnated every 4 to 6 months [4].

Very interesting results have been obtained in Western Africa against *Trypanosoma brucei gambiense* [5], as well as in Eastern Africa against *T. b. rhodesiense* [6]. It should be emphasized that the good results obtained by these authors were due to permanent intervention by specialized technical personnel, within the framework of a strict program with help from rural communities in carrying out simple, carefully monitored tasks.

Action against the parasite reservoir

The fight against trypanosomiasis is directed principally at the parasite reservoir to stop its transmission. This can be accomplished in two ways: by screening for and treating those persons with the disease, and by providing prophylactic drugs to exposed populations at risk.

Screening for patients in endemic areas for trypanosomiasis must be systematic. Inhabitants of all the villages in the area should be examined, illustrating what Jamot has stated, "you must seek out the disease". Indeed, passive screening of all patients consulting in a dispensary will not solve the problem. The use of specialized mobile teams, on the other hand, has two advantages: first, it allows for an early diagnosis when the majority of the persons screened would be in the first phase of the disease and could receive less aggressive treatment, and secondly, it blocks very effectively transmission of the disease by clearly reducing the size of the parasite reservoir.

This approach lead to the creation of the Autonomous General Service for Sleeping Sickness, and later the General Service of Mobile Hygiene and Prophylaxis. With the alarming resurgence of the disease in very numerous areas (for example in Angola and former Zaïre), these methods have by no means lost their interest.

In the past, patients were suspected to have the disease when certain clinical signs were present, among which lymphadenopathy was foremost, especially the nodes of the posterior cervical triangle, referred to as Winterbottom's sign. Lymph node aspiration was performed with a search for the parasite in wet mounts and, after staining, in thin blood films. The problem of varying numbers of false negative tests was offset by the regular passage of teams in the villages.

Later, progress in biology and immunology made it possible to have more accurate methods of detection, thus improving the rate of discovery of infected individuals. Evidence of infection can be accomplished by a search for increased IgM serum levels [7] by gel precipitation using Ouchterlony's technique or by titration of the serum level. A search for serum antibodies can be made using immunological reactions: indirect fluorescent antibody tests [8] using antigen from human trypanosomes obtained from inoculation of laboratory animals or culture media (Institute of Tropical Medicine, Antwerp); indirect hemagglutination tests (Cellognost®) using red blood cells sensitized to surface glycoproteins of a selected trypanosome serotype; a direct agglutination test (Test Tryp CATT®) [9] using lyophilized trypanosomes of a determined serotype stained blue. The latter technique can be performed using whole blood, serum, or an eluate from filter paper. After determining reference thresholds, serum titration allows one to establish the diagnosis in certain patients suspected of disease, even without visualizing parasites.

The widespread use of these screening techniques during systematic screening visits has made it possible to identify individuals most likely to be infected. Confirmatory diagnostic tests to visualize the parasite can be performed in these persons only.

Diagnostic methods [10]

Generally speaking, the diagnosis of trypanosomiasis (see also Chapt. 14) can not be affirmed unless the parasite is identified. To accomplish this, a drop of lymph node aspirate can be placed between a slide and cover slip and examined directly, demonstrating actively motile refringent parasites which cause the blood cells to vibrate. A search can be performed after spreading the aspirate on a slide and staining with Giemsa's stain; however, because of the abundance of lymphocytes the reading is difficult and is seldomly used.

The peripheral blood can be examined using a thin film of fresh blood stained with Giemsa's stain, but is rarely used because of the low concentrations of parasites in the blood. Methods to concentrate specimens are then employed. One method consists of performing a differential (triple) centrifugation, requiring a venipuncture of 10 ml of blood; this method is no longer in use. Capillary tube centrifugation or CTC [11] is performed by fingerstick using 75 µl of blood in a heparinized capillary tube that is centrifuged for 3 min at 12,000 rpm/min. A search for parasites is made at the cell-plasma interface. Another method is by filtration and centrifugation. After filtering 90 µl of blood over a column of DEAE cellulose, the eluate is collected in a pipette with a long, narrowed extremity which is then centrifuged for 10 min at 2,000 rpm/min. A search for parasites is carried out at the extremity with the centrifugate.

In the cerebrospinal fluid (CSF), a search for the parasite is performed by centrifuging 2 ml of freshly obtained CSF at 2,000 rpm/min for 5 min; a drop of the pellet is then examined. A double centrifugation method uses centrifugation as above after which the liquid is pipetted into a microhematocrit tube and one extremity is sealed. The tube is then centrifuged at 12,000 rpm/min for 2 min, and a search is made for mobile parasites.

Drug prophylaxis

In dealing with sleeping sickness, Health authorities quickly realized that in addition to the screening and treatment of patients, it was important to treat the healthy population. For this reason, prophylactic measures were used during the fifties and sixties in all households. After having screened and treated those in the population who had contracted the disease, the remainder of the population was given a single dose of pentamidine intramuscularly. This injection was repeated every six months during the routine visit of the mobile health team. This prophylactic activity, thoroughly performed, was undertaken concomitantly with regular screening measures and yielded the excellent results known today, since, in the sixties, the illness was controlled in all endemic areas.

Strategies in the fight against sleeping sickness

Classically, in each area of trypanosomiasis, at least one team of health workers, either assigned specifically or as a part of a larger multidisciplinary team, would regularly canvass all villages located in transmission zones and begin treatment of all new patients. This team also administered systematically pentamidine to the healthy population ("lomidinization").

At the end of the sixties, when the disease in all endemic zones had been under control, this method of management came under criticism. Certain people felt that canvassing from village to village was somewhat coercive and that obligatory measures were no longer in line with current tastes. Thus, the presence of visits by health workers regressed, and each area was left to do as it pleased. It was forgotten that the success of these measures depended on involving the entire population to maximally reduce the parasite reservoir, and that, in general, the fight against a transmissible disease systematically concerns all persons exposed to the disease. The eradication of smallpox was made possible, among other measures, by the obligatory vaccination of each individual; likewise, the success in the fight against tuberculosis in Europe was, to a great extent, obtained by obligatory systematic screening in schools and workplaces.

Chemoprophylaxis with pentamidine also drew criticism. With the populations tired of receiving regular injections, there was the risk of masking an early, previously undetected case of trypanosomiasis, and the action of pentamidine was relatively limited over time.

It is a remarkable constatation that despite more effective techniques than those used in the past, we have returned almost everywhere to the situation that existed in 1925-1930 [12]. We have no one to blame but ourselves for this astonishing and serious situation leading to the death of thousands of individuals. Such a resurgence in these areas is due to many factors [13], nevertheless, the principal causes must be sought for in the Health system itself, in the will of Health authorities to consider policies objectively and not as the latest trend, to apply methods and strategies related to the desired goals and epidemiological situation, and their ability to convince the different internal and external parties concerned, despite government reticence.

At the present time the problem is disquieting and clear, according to the 1996 World Health Organization (WHO) report on trypanosomiasis: 55 million people are at risk of infection, only 3 to 4 million are under surveillance (the type of surveillance is not explicit), at least 300,000 individuals at the present time are infected and thus the majority is at risk of dying since only 25,000 new cases are reported annually.

What strategy must be adopted? The priority must be given to areas with increased disease activity. For example, in Angola and Zaïre (recently the Democratic Republic of Congo), prevalences above 20% have been recorded in numerous provinces, and in many villages the prevalence exceeds 50%. In these situations, active systematic screening by specialized mobile teams is essential. It is evident from the outset that such a screening method is inconceivable unless effective treatment structures are available with adequate supplies to care for newly diagnosed patients.

Active mobile screening

A mobile team can be comprised of five to six well-trained, efficacious technicians. Before undertaking any mission in the field, it is essential to establish a realistic work schedule, in which 300 to 350 persons can be examined daily. Care must be taken to visit each village and hamlet, avoiding as often as possible "gathering places" in which a large number of the inhabitants will be missed because they will not make the trip. After carefully planning a work calender, it must be transmitted to the local administrative and traditional authorities, neither too early for fear of it being lost or forgotten, nor too late because the messages will not be forwarded in due time.

When all the populations have been notified, the canvassing can start, taking care to scrupulously respect the established time–table previously circulated. When appointments are not kept, the credibility of the team in the eyes of the population comes immediately into question.

After emphasizing to village leaders and heads of families that everyone must be present, the population to be examined is assembled. The first post is secretarial: for registration. Presentation can proceed by family or by groups of dwellings thus facilitating the task. The name of each village and the family identification number is indicated on each registration sheet; each sheet can contain as many as ten names. The surname, first name, age and sex of each individual is then noted in the appropriate column. In this manner, these sheets can be used during subsequent visits and kept up to date at each visit. This responsibility can be assumed by a nonmedical worker of the team.

The second post is that of obtaining blood samples. For this, blood is withdrawn in a capillary tube for the CATT. It is preferable for the population to come in groups of ten individuals, in the order established during registration. This order of sequence will be maintained while awaiting the test results. The filled capillary tubes are placed in racks in the same order as specified on the sheet, and then forwarded to the next post. At this point, subjects can be examined for lymphadenopathy and the results noted on the sheet under the corresponding heading "presence of lymph nodes".

The third post is the post where the CATT is performed. The results are written on the sheet in the corresponding space. If the positivity of a blood test is questionable, a test using undiluted serum, or a range of dilutions is performed.

The fourth post is the "parasitology" post, an area where all subjects with adenopathy and/or a positive CATT are assembled. It is kept by two laboratory technicians.

Lastly, the fifth post is where the lumbar punctures and the cerebrospinal fluid analyses are carried out in persons found to have trypanosomiasis. Thus, at the end of the chain, subjects with disease will have been detected and the stage of the illness determined. The sheet of ten individuals is thus forwarded from post to post. It can be designed as in Fig. 1.

Such an endeavour requires strict organization with each post having sufficient space to function properly, in the presence of only the groups of ten individuals awaiting for their results.

N°	Surname	First name	Age	Sexe	CATT blood	CATT serum	Lymph node	Trypanosome	CTC	CSF
1										
2										
3										
4										
5										
6										
7										
8										
9										
10										
TOTAL										

Date: Village:
Family: n° of family (or n° of sheet)

Fig. 1. At the end of the chain, subjects with disease will have been detected and the stage of the illness determined. The sheet of ten individuals is thus forwarded from post to post

The health care team can be deployed as shown in Fig. 2. The work areas must be well-equipped and isolated from the population waiting to be tested.

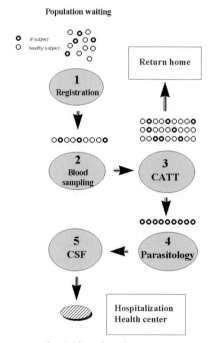

Fig. 2. The health care team can be deployed as shown

Once subjects with the disease have been detected (they will be quite numerous in areas with high rates of transmission that have not been screened for long periods of time), they must be treated under the best possible conditions. Trained, efficient, and well-supervised health personnel must be available, including the necessary medication and supplies to care for these patients. Accommodations for the ill with neurological involvement can be simple: for example, the temporary construction of large tents in proximity to a preexisting health care facility, which can be rehabilitated in case of need. It is more important to invest time and money in the formation of personnel, the necessary supplies and medication, and the supervision of activities than into expensive buildings which will be of only temporary use.

This strategy to detect disease using a mobile team capable of treating trypanosomiasis is crucial in areas of high prevalence; it is the only way to halt transmission of the disease.

One difficulty is to determine at which prevalence threshold an epidemic occurs, allowing one to answer, among others, Peter de Raadt's pertinent question of several years ago: "At which prevalence can screening become passive, with the use of only permanent health care facilities?" At the present time, no one has been able to answer this question, making it difficult to justify controlling the disease with the use of only permanent health care structures.

Passive screening in permanent health care facilities

In providing primary health care, screening at a permanent center has the disadvantage of detecting, at best, only those motivated to consult. This approach is inadequate since it disregards an essential element, i.e. the long, asymptomatic, first phase of sleeping sickness. It is crucial that screening be performed as early as possible during the natural course of the disease to avoid progression to the neurologic phase which is always serious and more difficult to treat. Furthermore, from an epidemiological point of view, this approach more quickly reduces the parasite reservoir. This aspect of the problem receives increasingly less attention because public health issues are perceived more and more on an individual rather than a collective level.

In the fight against trypanosomiasis, treatment provided by permanent health care facilities is far from being negligible, but it can not be the sole element. To be efficacious, care must be active, not passive, meaning that surveillance must be strictly programmed and executed, in conformity with clear and realistic directives. Previous areas of high transmission where the disease is under control or those of low transmission must be carefully followed. It is important to specify which type of examination must be performed, the population on which it should be carried out, under which conditions and frequency, and the information to be sought. This mode of surveillance is recommended by WHO authorities.

Surveillance of an area can be accomplished by performing serologic surveys using the CATT with either whole blood or serum, or even capillary blood on filter paper. The population sampling must be representative of the overall population, comprising in particular those persons susceptible of being infected during their daily activities (farmers, fishermen, those using water from watering holes, etc.).

It is unreasonable to expect more from permanent health stations personnel. Serologic screening of all subjects would require considerable quantities of reactive agents and would constitute a work load incompatible with the normal activities of the health station which are already substantial: daily office visits, vaccinations, follow-up of pregnant patients, etc. The amount of work could possibly be diminished by focusing the screening on patients with signs of the first phase of sleeping sickness (headache, fever), nonspecific signs which constitute the principal reasons for seeking medical advice; the work load would be only modestly lightened.

The local team accomplishes this surveillance by periodically performing serologic tests (once per trimester or yearly) according to the schedule previously established, and reports results to the Health authorities in charge of the program. Examination of the samples is performed either locally by the personnel of the center, or by a specialized laboratory if capillary blood dried on filter paper is used. If the prevalence, determined serologically, exceeds a certain predetermined threshold, then systematic screening of the entire population or the involved villages is undertaken by the mobile team that includes these populations into their work calender. In this instance, we return to the classic procedure of active mobile screening (Fig. 3).

All these activities of verification and surveillance require careful organization and the rapid transfer of information, making it possible to respond efficaciously to epidemiologic changes. The trypanosomiasis unit of the WHO is endeavouring to progressively incorporate this plan of action in all concerned countries. Three types of information (general information concerning the country and the program established, epidemiological data and geographical data) are linked to a common structure: the village, and a geographical information system (GIS), enabling health officials to analyze chronologically and geographically the epidemiological situation, the course of action to take and its cost.

This method of processing information also makes it possible to know how funds are managed in the fight against sleeping sickness by the government and other Health organizations. This analysis is essential to convince sponsors of the proper usage of their funds and the necessity of continuing their aid, not for a provisional 2 or 3-year period, but for much longer periods of 10 or 15 years. Indeed, the fight against sleeping sickness requires a sustained effort involving health workers and dependable

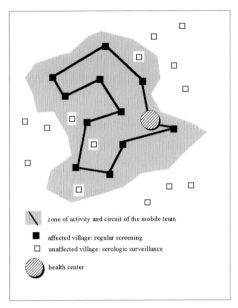

Fig. 3. The classic procedure of active mobile screening

technical and financial means over long periods of time while adapting regularly to changing needs according to the epidemiological data.

Control of the insect vector

A recent publication by Gouteux and Artzroumi [14] has reviewed the problem of vector control which has often divided the medical professionals and entomologists. The authors conclude that screening and treatment is more efficacious than controlling the vector in endemic areas, whereas the opposite holds true during epidemics. Furthermore, control of the vector is all the more justified when the degree of man-insect exposure, measured by studying the number of blood meals, is elevated. This interesting study and its results, however, raise questions similar to those already referred to concerning the surveillance of areas of transmission, i.e. "how does one define an endemic from an epidemic area?". Furthermore, the main objective of any method to eliminate a parasitic disease is to detect and treat the maximum number of affected individuals, which, as the authors recognize, made it possible to control the epidemics at the beginning of the century because of the efficacy of the mobile screening teams. The crucial question, however, is to know whether the introduction of a program to eliminate the vector would control the disease more rapidly and especially for a longer duration than without it.

Whatever the case, a program directed against the vector requires a considerable financial effort and qualified personnel. It is essential that the fight be considered an affair of qualified technicians; nevertheless, the population of the involved area can possibly participate in certain tasks. The idea that the population can manage "through community participation" must be abandoned, just as it was thought that the CATT could be performed by anyone (community health workers, for example) "under the mango tree".

Conclusion

According to the epidemiologic situation, the approach to follow is shown in Fig. 4. In an epidemic area, several mobile screening teams must be activated to provide regular and systematic examination of the entire population of the villages in the affected area. Subjects found to have the disease are treated, possibly with assistance from the permanent health facilities. The latter also provide passive screening for persons that seek medical advice. The fight against the vector can be lead by specialized personnel with help from the population.

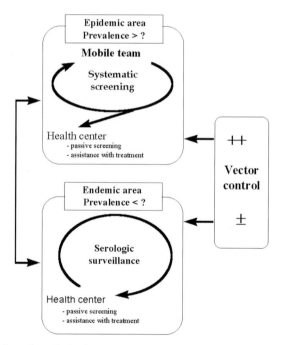

Fig. 4. Approaches of prophylactic measures according to epidemiologic situations

In endemic areas, the approach is essentially that of screening the population of the concerned villages with serologic tests, performed by personnel from the permanent health facilities or a specialized team. The health center also participates in passive screening of potential patients and provides treatment. Control of the vector can be considered. If the epidemiologic situation changes, the approach is modified according to the proposed diagram.

References

1. Martin G (1920) Notes sur la prophylaxie de la maladie du sommeil. Bull Soc Path Ex 13:477-485
2. Wéry M (1990) Les lents progrès du contrôle de la maladie du sommeil. Ann Parasitol Hum Comp 65 (Suppl 1):89-93
3. Challier A, Laveissière C (1973) Un nouveau piège pour la capture des glossines (*Glossina*: Diptera, Muscidae): description et essais sur le terrain. Cahiers ORSTOM Série Entomol Méd Parasitol 11:251-262
4. Laveissière C (1988) Les glossines vectrices de la trypanosomiase africaine. Biologie et contrôle. WHO/UBC, doc n° 958
5. Laveissière C, Grebaut P, Couret D (1987) Les communautés rurales et la lutte contre la maladie du sommeil en zone forestière de Côte d'Ivoire. Institut Pierre Richet (Bouaké) OCCGE, doc n° 17/IPR/RAP
6. Lancien J, Obayi H (1993) La lutte contre les vecteurs de la maladie du sommeil. Bull Soc Fr Parasitol 11:107-117
7. Mattern P (1968) Etat actuel et résultats des techniques immunologiques utilisées à l'Institut Pasteur de Dakar pour le diagnostic et l'étude de la trypanosomiase humaine. Bull OMS 38:1-15
8. Wéry M, Wéry-Paskoff S, Van Mattere P (1970) The diagnosis of human African trypanosomiasis (*T. gambiense*) by the use of fluorescent antibody test. Standardisation of an easy technique to be used in survey. Ann Soc Belge Méd Trop 50:613-620
9. Magnus E, Vervoot T, Van Meirvenne N (1978) A card agglutination test with stained trypanosomes (CATT) for the serological diagnosis of *Trypanosoma brucei gambiense* trypanosomiasis. Ann Soc Belge Méd Trop 58:169-176
10. Stanghellini A, Roux JF (1984) Techniques de dépistage et de diagnostic de la trypanosomiase humaine africaine. Méd Trop 44:361-367
11. Duvallet G, Stranghellini A, Saccharin C, Vivant JF (1979) Trypanosomiase africaine: centrifugation en tube capillaire. Presse Méd 8:214-215
12. OMS/Division de la lutte contre les maladies transmissibles/THA (1996), Rapport Annuel
13. Stanghellini A, Gampo S, Sicard JM (1994) Rôle des facteurs environnementaux dans la recrudescence de la trypanosomiase humaine africaine. Bull Soc Path Ex 87:303-306
14. Gouteux JP, Artzroumi M (1996) Faut-il ou non un contrôle des vecteurs dans la lutte contre la maladie du sommeil. Une approche bio-mathématique du problème. Bull Soc Path Ex 89:299-305

CHAPTER 19

International co-operation: past and present

P de Raadt, J Jannin

Introduction

Sleeping sickness is a disease of the African continent South of the Sahara. It was known amongst several tribes in Africa before Arabs and Europeans found it. The earliest written reference dates from the 15th century. It was reported by two Egyptian historians describing Tombouctou and Mali, well–known outside it from the trans-Saharan trade route. Some three hundred years later, the first reports by Europeans appeared from doctors who had noticed the disease amongst slaves on the West coast of Africa. Until the 20th century the disease was seen in Europe merely as one of the curiosities of the "Dark Continent" until Europeans started to explore the interior of Africa. The "opening up" of Africa became possible early during the 19th century, first because the extraction of quinine had been standardised and malaria, the major obstacle for European to go inland, could be prevented more or less successfully. Then, in 1885 in Berlin, the European Powers had agreed on their respective "zones of influence", thus avoiding territorial disputes and confrontations amongst themselves. The acceleration of activities that followed the establishment of colonial administrations entailed an unprecedented displacement of African people and disturbance of social relationships. As a result, a major epidemic of sleeping sickness broke out across Central Africa which caused over a million deaths in less than 10 years. The colonial administrations, facing a dramatically fatal disease without knowing its cause or how it was spreading, had serious doubts about the feasibility of exploiting their African colonies, so sleeping sickness became a high priority on the international agenda. The various colonial offices concerned sent scientists to Africa who were remarkably successful. In terms of trypanosomiasis research, the first years of the century were the most ever productive: the trypanosome was identified, as the causative agent, the vector identified, and in 1905 the trypanocidal properties of atoxyl had been discovered and the drug was ready to be applied in Africa.

Now that the essentials about the disease were known, the basic principles for control were defined, initially isolation of patients, restrictions of peoples' movements and, where necessary, evacuation of the whole population. In 1917, the Frenchman Jamot in Oubangui-Chari (Central

African Republic) introduced the principle of large scale case detection and treatment ("atoxylisation") by means of mobile teams. Its convincing success made the French administration decide to apply this system for the whole of French Equatorial Africa, later also in West Africa. Whereas systematic case detection and treatment became the standard for all the *T. b. gambiense* zones, in the *T. b. rhodesiense* endemic areas in East Africa control was, and still is, focused on reducing exposure to tsetse in combination with passive case detection.

By the mid thirties, each of the colonial administrations had "their" strategy. It seemed as if every overseas medical service became locked up in his own system and the need to share experience was no longer as strongly felt as at the beginning of the century. This was the first occasion that an international organization, the League of Nations, took an initiative in trypanosomiasis by convening an international Expert Committee in 1924.

The second World War had made new tools available for trypanosomiasis control such as insecticides and melarsoprol, a new drug for late stage sleeping sickness. Their use required new methodologies and strategies and there was again a strong urge to compare results and to resume international exchanges. Besides the newly established technical agencies of the United Nations, the World Health Organization (WHO) and the Food and Agricultural Organization (FAO), the regional "Permanent Inter-African Bureau for Tsetse and Trypanosomiasis" was founded, which later evolved in the International Scientific Council for Trypanosomiasis Research and Control (ISCTRC).

With the independence of most African colonies during the fifties and sixties, the new governments faced budgetary constraints, shortages of equipment and lack of professionals. It was necessary to reconsider the priority of sleeping sickness control amongst overburdening other health problems. Ironically, the overall low prevalence rates maintained by the control systems in the past became an argument against giving high priority to sleeping sickness control. As a consequence, serious outbreaks occurred again during the late sixties and seventies notably in Democratic Republic of Congo (DRC, ex-Zaire), Sudan, Angola and Uganda. International support was provided by UN agencies, bilateral aid departments and NGOs by means of technical assistance as well as financial support.

The present chapter reviews the wide spectrum of international linkages (see Fig. 1) that has developed over the years and their historical background (see Table 1). For details on the history of the disease the reader is referred to publications elsewhere [1-4].

International co-operation: past and present 317

Table 1. International collaboration, chronology of major events

Manson convenes international meeting in London to define treatment standards	1907
International meeting on common policy in London (organised by the British Foreign Office)	1907-1908
Sleeping Sickness Bureau (later: Tropical Diseases Bureau)	1909-1912
Agreement between Britain and Germany on common policy in trypanosomiasis control in West Africa	1911
International Expert Committee on Tuberculosis and Sleeping Sickness in London (League of Nations)	1924
International Research Team stationed in Entebbe (League of Nations)	1926-1927
International conference on tsetse and trypanosomiasis, Brazzaville (French and Belgian governments)	1948
First meeting of ISCTR, London	1949
Permanent Inter-African Bureau for Tsetse and Trypanosomiasis established, Léopoldville (Kinshasa)	1949-1962
OCCGE established by eight states (agreement of Abidjan)	1960
OCEAC established by four Central African states	1963
ISCTR brought under OAU and becomes ISCTRC	1964
Expert Committees on Trypanosomiasis convened jointly by FAO and WHO, Geneva and Rome	1968, 1979
UNDP/WORLD BANK/WHO's Special Programme TDR established, Geneva	1975
WHO Primary Health Care Approach for the prevention and control of trypanosomiasis initiated	1984
WHO Expert Committees on Trypanosomiasis	1986, 1995
WHO/OCCGE/OCEAC programme for prevention and control of *T. b. gambiense* established	1995
FAO/IAEA/OAU/WHO programme on integrated control PAAT established, Vienna	1994

First international communications

The first note known on sleeping sickness was by Ibn Khaldun and Alqalaqshandiy. It concerned King Diata II, sultan of Mali, who probably died of it in 1373. More important of this publication was that it indicates that the disease was locally recognised in West Africa: "Lethargy, a disease that frequently befalls the inhabitants" [1]. There was a long interval until 1734 when the British naval surgeon Atkins published a remarkable paper on "Sleeping Distemper", prophetic in the sense that without access to autopsy results, Atkins alludes to cerebral deem ("serum extravasations in the brain"). His compatriot Winterbottom drew, in 1803, attention to enlarged cervical lymph nodes as being typical for sleeping sickness and known to the slave merchants. It is not surprising that amongst some 10 million slaves transported to the Americas there were people who already had been infected before their shipment. Some 150 of these have been described in detail by French clinicians in the Antilles [5].

At the end of the 19th century, Professors Stephen Mackenzie and Patrick Manson had three Congolese sleeping sickness patients expatriated to London for investigations. Although the cause of the disease was not found, the histopathology of the brain was studied in detail and reported in Mott's classic paper [6].

The European scientists sent to Africa from 1902 faced more than a scientific challenge, the long journeys on foot, the restricted means of communication and a high mortality risk due to malaria and other diseases [7]. The hardships must have tempered the enthusiasm of many a scientist and the borders of the colonial territories were wide open, irrespective of nationalities, for those who volunteered. For example, Emile Brumpt, the later prominent French parasitologist, embarked as a young man in 1901 on a two-year journey across equatorial Africa from Djibouti via Ethiopia, Kenya and Sudan to the French Congo. Excellent observer and naturalist as he was, he had noticed that the geographic distribution of tsetse flies and sleeping sickness were the same, and upon his return, he suggested the possible connection between the two [22]. Castellani, Italian, became member of the first Commission of the British Royal Society to Uganda; Dutton, British and Todd, Canadian; were assigned by King Leopold II from Belgium to a mission to the Congo Free State (now Democratic Republic of Congo). Also Robert Koch on his sleeping sickness expedition to German East Africa in 1906 did the greater part of his studies on epidemiology and treatment on British territory, the Ugandan Sese Islands.

International conferences and across-the-border agreements

The remarkable progress made in research in Africa during the first years of the century resulted in a strong demand for sharing results at international level. Special sessions on sleeping sickness, now recognised as trypanosomiasis, were organised at international conferences. In 1907, a group of high calibre experts including Ehrlich, Laveran and the Portuguese Kopke were invited to London by Manson, to define appropriate treatment strategies in particular for atoxyl, that two years earlier had been discovered as a trypanocidal drug.

The British Foreign Office in London organised in 1907 and 1908 two international conferences entirely devoted to trypanosomiasis with as primary objective to agree upon a common policy amongst the various colonial departments and devise administrative measures such as regulating across-the-borders traffic. One of the recommendations made at these meetings was to create a central international Bureau "to extract and circulate all new literature on Sleeping Sickness". This was done by the British Foreign Office and since 1909 the "Sleeping Sickness Bureau" in London produced comprehensive and high quality extracts of publications and meetings which, even today, offer excellent reading. So positive was the international response to it that in 1912 the task of the Sleeping Sickness Bureau was extended to all tropical diseases and it became the "Tropical Diseases Bureau".

Apart from collaboration amongst scientists, there was also a will for international co-operation at administrative level. An example was the formal agreement concluded between the British and German Governments in 1911 on common policies for restricting movements of people and on data exchange between the respective territories in West Africa, the British Gold Coast colony, Ashanti protectorate and German Togoland [1]. Another example is the agreement between the colonial administrations of the Anglo-Egyptian Sudan, the Uganda protectorate and French equatorial Africa and Belgium on co-operation in the control of sleeping sickness across the North eastern border of Belgian Congo [7].

The League of Nations

The First World War brought international scientific contacts virtually to a standstill and also meant the end of German research in Africa. The Health Committee of the League of Nations took an initiative in 1922 to form an International Expert Committee on Tuberculosis and Sleeping Sickness. It was convened in London, in 1924, under A Balfour (Table 1). Their report (League of Nations, 1925) provided a detailed account of the epidemiological data, preventive measures, diagnosis and treatment from the French, Belgian, British, Portuguese and Spanish territories and the Abyssinian Empire. It recommended to assign a special mission to Africa to study the role of animal reservoir hosts, the relationship between *T. b. rhodesiense* and *T. b. gambiense* and the host preferences of tsetse flies.

The League's international research team to Africa consisted of seven experts from different European countries. They were stationed in Entebbe from 1926 to 1927. Their international status permitted free access to any of the Central African territories. Amongst the remarkable results obtained by this group were the development of an immune precipitin test for bloodmeal analysis, a retrospective study of the epidemics in the Semliki valley in Belgian Congo, the comparison of human pathogenic trypanosomes from Tanganyika and Congo and the classic studies on experimental histopathology by Peruzzi (League of Nations, [8]). Their final report remained for many years the standard work on trypanosomiasis [8]. The pre-war political tensions during the thirties apparently did overrule further initiatives in trypanosomiasis.

World Health Organization and other United Nations agencies

The United Nations Organization (UNO) was established in 1945, and in 1948 WHO was founded as one of its specialised agencies. WHO started specific activities in African trypanosomiasis during the sixties. A series of expert reviews appeared in the Bulletin of WHO and in 1968, 40 years after the League of Nations' expert meeting in Paris, WHO, jointly with FAO, convened an Expert Committee again [9]. Noteworthy is this Committee's address to the newly independent states recommending "to reassess the relative priority to be accorded to the control of trypanosomiasis", a difficult issue that continues preoccupying the Health authorities up till the present day. After the Expert Committee, WHO established a special unit for Trypanosomiasis in Geneva from where, through its office in Brazzaville, direct support is given to the newly independent member states for the planning and implementation of national programmes.

WHO organised, jointly with FAO, field research in the Lambwe valley on appropriate control strategies for animal and human trypanosomiasis in East Africa. The project was financed by the United Nations Development Programme (UNDP) and became operational in 1968. During its three years existence new serological tests have been developed, the importance of animal reservoirs and vectors were evaluated and a variety of insecticides were tested under field conditions [10]. A similar applied research project on control methods against *T. b. gambiense* disease was launched by WHO in West Africa in 1975 [11]. Nineteen countries in the moist savannah took part in it and financial support of over two million US dollars contributed by UNDP. The research was primarily focused at vector ecology and control and carried out by an international team including African professionals. FAO and the International Atomic Energy Agency (IAEA) were associated agencies in monitoring sterile tsetse release. Voluntary contributions from Germany and the Netherlands allowed for essays with helicopter spraying and ecological monitoring. Two of the most pertinent findings were that, contrary to the common opinion, tsetse flies are capable of covering, on their own wings, distances as far as 25 kilometres and that domestic pigs could act as a reservoir host of *T. b. gambiense*.

The programme in West Africa had also a clinical research component for which hospital and laboratory facilities were established in Daloa, Côte d'Ivoire. The latter was as from 1979 continued under the UNDP/WORLD

BANK/WHO Special Programme for Research and Training in Tropical Diseases (TDR) [12]. Since 1985, the year that TDR support had come to an end, it was continued under the national budget of Côte d'Ivoire. The laboratory facilities and the clinical and epidemiological baseline data available at this centre attract many research groups from elsewhere, for clinical trials (Ornidyl), pharmacokinetic studies, and post-mortem neuropathological investigations [13].

During the early seventies there was great concern about the global resources for research being largely spent on the major health problems of the industrialised world. A joint initiative was taken by WHO, the World Bank and UNDP to create a special programme for research and training of tropical diseases (TDR). With the formal approval of the World Health Assembly, it was established in 1975. The TDR programme would support research on six tropical diseases included trypanosomiasis. The first remarkable impact of the TDR programme has been made by its choice of six priority diseases which had a catalysing effect beyond expectations all over the world. Numerous research centres anticipating new emphasis on these issues responded promptly by adjusting their programmes. The second stimulating aspect of the programme was its regular convening of groups of scientists to advise on the priorities and practical planning of its scientific programme. This allowed for numerous occasions for research professionals from the industrialised as well as the endemic countries to meet and establish collaborating project.

In addition, there was financial support to research. A few years after its establishment, TDR obtained a highly favoured place amongst the international donor community and by the seventies, its budget for trypanosomiasis reached over a million US dollars per year. Over the recent years, the programme adopted horizontal approaches rather than specific disease oriented activities so that direct contributions to trypanosomiasis were considerably reduced. To mention a few examples of TDR supported basic research over the last twenty years, it supported the development of new trypanocidal compounds such as difluoromethylornithine (DFMO, Ornidyl) and invested in cell biochemistry, antigenic variation and experimental pathology. On the applied side, TDR seconded Mehlitz and colleagues to demonstrate in West Africa that a variety of animals may act as reservoir host of *T. b. gambiense* [14], and it coordinated the evaluation of serological tests and tsetse traps.

In the early eighties the sleeping sickness situation was deteriorating with a 100% increase of reported cases over a period of less than 10 years and the World Health Assembly of 1983 adopted a resolution whereby the endemic countries were urged to ensure effective use of control tools and co-operate at intercountry level. A year later, WHO initiated the programme "Primary health care approach for the prevention and control of trypanosomiasis". This programme aimed at assisting in the planning of national projects, stimulating inter country collaboration and providing training. It collected over one million USD from voluntary contributions, additional budgetary allocations from WHO's regular budget and donations in kind from industry. The programme was operating mainly at country level by means of in-service training of medical surveillance teams, *ad hoc* evaluation of national control programmes and by mobilising bilateral resources. It also was one of the first to establish an international supply line for drugs and diagnostic materials through a revolving fund.

Despite WHO efforts and the successful reinforcement of some of the national control programmes, it worsened in others owing to lack of resources, political upheavals or ecological disturbances that interfered. In several countries the situation got out of hand and "fire brigades" of Non Governmental Organizations (NGOs) or international agencies took care of some of the most seriously affected foci.

At present, the WHO programme aims at sustaining a core of national expertise in each endemic country plus propagating international coherence of control efforts. A typical task of WHO is also co-ordinating the contributions from outside, the NGOs, bilateral teams and international organizations in order to ensure their appropriate adaptation to local standards and national strategies.

WHO's plan of action concerns standardisation of programme planning and evaluation criteria. It has defined case detection protocols and reporting, and collects and updates epidemiological surveillance data, administrative and strategic information which is made widely accessible through an intercountry information system. The WHO programme has recently been supported by a recommendation of the World Health Assembly of 1997 emphasizing the need for co-ordinated control and its pertinence to rural development.

A concrete example forms the programme launched in 1995 with OCCGE (Organisation de coordination et de coopération pour la lutte contre les grandes endémies) and OCEAC (Organisation de coordination pour la lutte contre les endémies en Afrique centrale) for prevention and control of practically the whole *T. b. gambiense* belt from Guinea to the West Nile province of Uganda. Altogether, 16 countries are participating (Angola, Democratic Republic of Congo, Ghana, and Uganda plus the member states of OCEAC and OCCGE [see section Regional Organizations and Services]). This programme receives financial support from Belgium and France and AGFUND (The Arab Gulf Programme for United Nations Development Organizations).

PAAT or Programme against African Trypanosomiasis is a joint undertaking of FAO, IAEA and WHO, three UN agencies with OAU. It was initiated by FAO and IAEA in 1994. Its overall objective is to approach animal and human trypanosomiasis together in the broad context of rural development and sustainable agriculture. PAAT intends to stimulate participation of a highly diverse range of groups concerned from farmers and village responsibles to research scientists and government officials. Its members consist of representatives of donor agencies and technical advisers from Africa. Its programme consists of international coordination, data collection and dissemination providing strategic guidance and maintaining a database of resources. The secretariat is shared amongst the four main participating organizations.

Regional organizations and services

Regional initiatives were taken already during the thirties by the French and British authorities in order to rationalise research and control within their territories. After independence, these structures formed a base for three sub-regional organisations: OCCGE for francophone West Africa, OCEAC for francophone Central Africa, and the East African Trypanosomiasis Research Organization (EATRO) for East Africa. Whereas EATRO lost its regional status in 1976 with the collapse of the East African Community, the others continue to fulfil their useful role in coordinating control, research and training. A fourth one, the "Permanent Inter-African Bureau for Tsetse and Trypanosomiasis" (BPITT, Bureau Permanent Inter-africain de la Tsé-tsé et de la Trypanosomiase) had been created shortly after World War II for the whole of Africa. This became the predecessor of the present International Scientific Council for Trypanosomiasis Research and Control (ISCTRC).

ISCTR

In 1948, after an interval of almost 10 years due to World War II, the French colonial services organised the first post-war international conference on tsetse and trypanosomiasis in Brazzaville. Priority items on the agenda were trypanocidal treatment, bushclearing and policies for human settlements. One of the recommendations of this conference addressed to the French and Belgian governments was to set up a central bureau for distribution of information. The following year the "Permanent Inter-African Bureau for Tsetse and Trypanosomiasis" (BPITT) was set up, located in Léopoldville (Kinshasa). During 12 years this office redistributed published materials all over Africa and Europe translated in English, French and Portuguese. These included unpublished papers such as national reports that otherwise were not accessible. The other recommendation made by the Brazzaville meeting was to create an international standing committee of experts to provide regular exchanges for research. A year after the conference, this committee, the "International Scientific Committee for Trypanosomiasis Research" (ISCTR) called for its first meeting in London. In 1950, the ISCTR and the BPITT became part of the "Commission for Technical Co-operation in Africa South of the Sahara" (CCTA), a group of independent experts active in a wide range of sciences pertaining to Africa: geology, mapping, climatology, oceanography, etc. The financing and overall responsibility for it was in the hands of the governments of Belgium, France, Portugal, Rhodesia and Nyassaland, Union of South Africa, United Kingdom. It had initially two secretariat offices one in London and the other in Bukavu, Belgian Congo. During the sixties, these were transferred to Lagos and Nairobi.

As from 1958 with the formation of independent African nations, an increasing number of new members joined the CCTA and the colonial administrations either stayed on as "Founder Member" without further financial commitments or terminated their membership. Owing to the diminishing financial resources, the BPITT was closed down in 1962. Now that the members of CCTA and ISCTR were African governments it was a logic step to let it become part of the Organization of African Unity. This took place in 1964. At the same time, ISCTR's mandate was extended to manage control as well as research and its name was changed to "International Scientific Council for Trypanosomiasis Research and Control" or ISCTRC [15]. Up till the present, the biannual ISCTRC meetings provide a unique forum for scientists from Africa, Europe and USA, the only international occasion where workers on animal trypanosomiasis and sleeping sickness can meet each other.

OCCGE

The West African OCCGE was established by the agreement of Abidjan in 1960 whereby eight countries (Bénin, Burkina Faso, Côte d'Ivoire, France, Mali, Mauritania, Niger, Sénégal, and later Togo) agreed upon transforming the sleeping sickness service of the general health and preventive service (SGHMP, "Service Général d'Hygiène Mobile et de Prophylaxie") of the colonial administration into an intercountry facility for technical support, research, training and coordination of control. The former had been established by Muraz since 1939, exclusively as a sleeping sickness control service (SGAMS, "Service général autonome de prophylaxie de la maladie du sommeil en Afrique occidentale et au Togo"), during the fifties it was extended by Muraz and Richet to include leprosy, malaria, meningitis, onchocerciasis and vaccinations. OCCGE headquarters are the offices and laboratories at the Centre Muraz in Bobo Dioulasso. Its programme for trypanosomiasis control is now located in the Pierre Richet Institute in Bouaké.

OCEAC

Following the establishment of OCCGE in West Africa, OCEAC was created in 1963, on the same model, based on the former technical services under the colonial administration and now an intercountry organization serving five and later six countries (Cameroon, Chad, Central African Republic, Congo, and later Equatorial Guinea). Apart from a trypanosomiasis department, it has one for malaria and another for public health in general. It is very active in public health training through CIESPAC (Centre Inter-Etats d'Enseignement de la Santé publique en Afrique centrale), an international centre for training in public health and under the aegis of OCEAC and seconded by WHO. All departments of OCEAC are in Yaoundé, Cameroon.

EATRO

In East Africa the "East African Trypanosomiasis Research Organization" (EATRO) was established in Tororo, Uganda, in 1955 before independence by amalgamation of the common research facilities in Shinyanga (Tanzania) and Tinde (Tanzania). The research centre in Tororo established a high international reputation within a few years' time and became the first WHO reference centre for its collection of cryopreserved trypanosomes, serum bank and its unique clinical research hospital. After independence during the sixties, it became part of the East African Common Services of Kenya, Tanzania and Uganda. Sadly, by 1976, the East African Community collapsed and the staff and equipment were split up between the three participating countries.

International research centres

ORSTOM

The French institute for scientific research for co-operative development (Institut français de Recherche scientifique pour le Développement en Coopération) in Paris, was founded in 1943 as the Organization for scientific and technical research overseas (ORSTOM, "Organisation de la Recherche Scientifique et Technique Outre-Mer"). It is an independent centre of technical expertise with research centres all over the world, in a wide variety of scientific disciplines. From its headquarters in Paris, ORSTOM produces documentation, films and computerised training programmes and sends a large number of specialists on secondment to Africa, the Americas and Asia. Several ORSTOM professionals work in regional organisations such as OCCGE, OCEAC and ICIPE. ORSTOM's noteworthy accomplishments in African trypanosomiasis have been the propagation of serology during the seventies, the device of a simple tsetse trap for control purposes [16] and its assistance to Uganda in the successful large scale application of such traps since the late eighties. ORSTOM had a special department for trypanosomiasis research at its laboratories in Brazzaville, but this disappeared in 1993 when the centre was abandoned.

ICIPE

In 1967, Thomas Odhiambo published a paper in *Science* listing the shortcomings in research policies or better, the lack thereof in East Africa. His paper [23] was published the same year when Nyerere in his Arusha declaration proposed a novel African approach to socialism as opposed to copying the western model. Odhiambo questioned how African culture and philosophy should compare with the objective scientific standards in accordance with the European tradition [Odhiambo argues in his paper that African culture and philosophy did not favour a distinction between objective and subjective and that the foundation of science in Africa should begin from the beginning, by rewriting and remodeling the primary school teaching materials]. He also drew attention to the low proportion of African scientists holding research posts, a problem aggravated by grants favouring employment of expatriates. He suggested as a short-term solution to concentrate research efforts in a few large African centres of excellence, open for African scientists and visiting scientists from abroad. Odhiambo, who had charisma and scientific credibility, succeeded to assemble a group of donors and founded in 1970 the International Centre of Insect Physiology and Ecology (ICIPE) in Nairobi. This institute developed multidisciplinary programmes on agricultural pests and vectors of animal and human diseases. Amongst others, ICIPE focused on tsetse physiology and trap development and carried out a field project of several years on their application and maintenance by rural communities themselves. Odhiambo left ICIPE after almost 25 years and was replaced by Herren, a Swiss specialist in biological control. If those 25 years did not provide the answers to the former questions about science in Africa, ICIPE certainly gave East Africa a unique international centre for insect science.

ILRAD

In the early seventies, Sadun's group at Walter Reed did a series of vaccination experiments leaving no doubt that antigenic variation would brake through any vaccine no matter how effective against homologous challenge [17]. A vaccine for trypanosomiasis would make a radical change in cattle raising in Africa and deserved a special investment in research. In 1973, the Kenyan government and the Rockefeller Foundation signed an agreement for setting up the International Laboratory for Research on Animal Diseases (ILRAD) in Nairobi with a main mandate research on the diseases, which seriously limits world food production. It would concentrate on trypanosomiasis and East Coast Fever (theileriosis). World Bank, UNDP, IAEA and 11 donor countries participated in the funding, and by 1977 the laboratory was fully operational. Trypanosomiasis research at ILRAD was focused on clinical pathology, the host's immune response, and the immunological changes occurring at each stage of the trypanosome cycle. It was, in many respects, relevant to animal as well as human trypanosomiasis, and soon the ILRAD staff was in close collaboration with WHO and TDR especially in training courses and as members of Expert Committees and various TDR Committees. Amongst the various contributions made by ILRAD, should be mentioned the technique for *in vitro* cultivation of bloodstream forms of trypanosomes and the genetic control of differentiations from the proliferating slender bloodstream forms to the non-proliferating stumpy forms and their respective antigenic properties. The laboratory has recently been merged with the International Livestock Centre for Africa in Addis Ababa under the name of International Livestock Research Institute (ILRI).

WAITR

In British West Africa, an inter-territorial research centre for Gambia, Ghana, Sierra Leone and Nigeria was established in 1948 under the name of "West African Institute for Trypanosomiasis Research" (WAITR) located at Kaduna, Nigeria. The history of this centre is very similar to the one in East Africa. After having built up a high reputation, particularly in the fields of entomology and protozoology, the inter-territorial function came to an end in 1962 with the independence's of the British West African territories and became the Nigerian Institute for Trypanosomiasis Research.

The tropical schools

Since the very beginning, the tropical schools of the four main countries in Europe who had interest in Africa had taken the lead in trypanosomiasis research and training. The Liverpool School with Dutton and Todd, Thomas and Breinl; the Pharo in Marseille with illustrious students such as Jamot and Muraz; the London School associated with Manson; the Prince Léopold Institute in Antwerp with Broden and Rhodain; the Lisbon Tropical Institute with Kopke, Bettencourt and da Costa, and of course the Pasteur Institutes in Dakar, Brazzaville and Yaoundé. The famous German investigators Kleine, Koch and Ehrlich came all from different institutes. From outside Europe, a major contribution was made by the Rockefeller University of New York with the discovery of tryparsamide in 1919.

By the end of the colonial period, other tropical schools became involved: the Bernard Nocht Institute in Hamburg, for example, with the group of Mehlitz; the series of field studies by Geigy and Swiss Tropical Institute in East Africa, and the Amsterdam Royal Institute with epidemiological work of Wijers. Since then, a large number of institutes became interested notably in the ultra-structure and the cell biology of trypanosomes pathogenicity and chemotherapeutics. The present range of laboratories involved is world-wide.

Medical schools in Africa which have contributed are the schools of Dakar with the work by Mattern and colleagues and the one of Makerere, through its association with the brothers Cook at Mengo hospital and with the historical studies by Foster. Both have become universities.

Non-governmental organizations

Amongst the numerous NGOs there are at present about 15 active in trypanosomiasis control. Angola, Sudan and Uganda are since long on the high priority list of NGOs.

The following is a short list of NGOs presently involved in trypanosomiasis.

AMREF

The AMREF (African Medical and Research Foundation) in Nairobi, better known for its flying doctor service, is also active in training and consultant services. It participated in the preparation and distribution of WHO's trypanosomiasis control manual [18] and provided managerial assistance to Uganda for the control of the epidemic during the seventies and eighties.

CARITAS INTERNATIONAL

This organization provides substantial assistance to Angola (Caritas Angola) trypanosomiasis programme for case detection and treatment.

CIDOB

It is the international centre for information and documentation (Barcelona) which supported since many years Equatorial Guinea. Besides substantial contributions in diagnostics and drugs it has an expert medical officer who is permanently based there.

EPICENTRE

This is a European group for expertise in applied epidemiology that was created in 1986 by MSF (Médecins Sans Frontières). At the time MSF started its programme in the North West of Uganda. Epicentre has a permanent office in Kampala from where training and applied research are organised.

IMC

The IMC (International Medical Corps) maintains since 1995 a treatment centre in Sudan, the district of Tambura and Eizo. It is also occupied with case detection.

MEMISA

The Belgian branch has several support activities in the Democratic Republic of Congo.

MSF

The MSF (Médecins Sans Frontières, France) has stationed since 1986 an important team of professionals in the West Nile province in the North West of Uganda to combat the epidemic which exists due to uncontrolled movements and resettlements of refugees. Since 1996, MSF (France) is also present in the Democratic Republic of Congo and MSF (Belgium) in Angola. It is anticipated that MSF (Netherlands) will establish in 1998 a trypanosomiasis control programme.

NPA

Norwegian People's Aid assists Angola since several years in the province of Cuanza Norte.

FOMETRO

FOMETRO (Fonds Médical Tropical) is a non-profit private fund for medical support to the ex-Belgian colonies in Africa: Congo (now Democratic Republic of Congo), Burundi and Rwanda. It was established in 1961 on the initiative of Prince Stéphane d'Arenberg and Count d'Aspremont Lynden under the aegis of various governmental and scientific institutions. For a long time, it was the sole Belgian agency for health support to Africa. Its programme consisted of training, health education, vaccinations and supplementary provisions for the health services. Since 1962, FOMETRO has been the national executive agency responsible for trypanosomiasis and tuberculosis control. Apart from the Belgian co-operation, it collaborates with WHO, the European Union, MSF and OCEAC.

Bilateral support

Since independence bilateral agreements and grants for trypanosomiasis, researches have been numerous. It would be beyond the scope of this review, if at all possible, to list them here. Traditionally, European donor agencies were more inclined to support trypanosomiasis programmes than those from elsewhere. In many cases, close ties have been maintained between the previous colonies and the corresponding ex-colonising states in the form of financial support as well as through providing professional experience. France has several hundreds of professionals permanently employed in Africa. UK provides expertise at shorter term in several tropical sciences. Most of the time, this is done through bilateral agreements but in some cases staff has been made at the disposal of international programmes, for instance in the case of the UNDP/WHO applied research project in West Africa (see section World Health Organization and other United Nations agencies). Several new bilateral partnerships have been established since the sixties. To mention one example, Sweden gave during several years considerable financial support to the trypanosomiasis control programme in Angola with Brazil as unexpected second partner.

Multilateral support

Important regular contributions for control are made by Belgium and France to the WHO/OCEAC/OCCGE programme for the control of *T. b. gambiense* (see section Regional Organizations and Services). In the past, the majority of multilateral support for trypanosomiasis used to be for research through the Special Programme TDR. Since recent years, the proportion spent on trypanosomiasis through this programme has been considerably reduced and the expenditure on trypanosomiasis is no longer comparable to what it has been during the eighties.

Industry

Collaboration between industry and other partners in trypanosomiasis research and control exists since 1905 when Robert Koch in Uganda and Ayres Kopke in Angola did the first trials with atoxyl against sleeping sickness. At present, from the commercial point of view, investments of drugs against the human disease are difficult to justify unless such compounds can also be used against animal trypanosomiasis or other diseases in man. Industry has the experience in drug development and marketing, but for clinical trials for tropical diseases it relies on national authorities, most of the time, together with WHO and TDR. These were involved in trials and evaluation of several commercialised diagnostics such as the indirect agglutination test, the card agglutination and recently the antigen detection test, in addition to a long series of clinical trials with DFMO. TDR also acts as mediator to negotiate cheaper production plants or basic materials for orphan drugs like DFMO which, as it stands, would be too expensive to take in regular production in USA.

An exceptional arrangement has been offered to WHO by Rhône Poulenc Rorer who provides pentamidine free of charge on the condition that WHO assures its exclusive use for sleeping sickness.

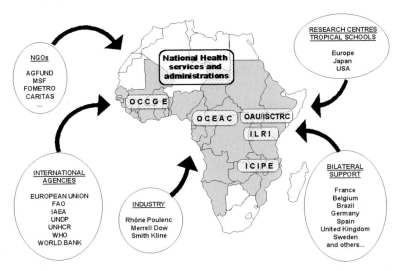

Fig. 2. International support to trypanosomiasis Control and Research (African Development Bank, AGFUND, Australia, Belgium, Canada, Denmark, France, Germany, Italy, Japan, Luxembourg, Netherlands, Norway, Switzerland, United Kingdom.)

Abbreviations used – AGFUND: Arab Gulf Programme for United Nations Development Organizations; FAO: Food and Agriculture Organization; FOMETRO: Fonds Médical Tropical; IAEA: International Atomic Energy Agency; ICIPE: International Centre of Insect Physiology and Ecology; ILRI: International Livestock Research Institute; MSF: Médecins Sans Frontières; NGOs: Non Governmental Organizations; OAU/ISCTRC: Organization of African Unity/International Scientific Council for Trypanosomiasis Research and Control; OCCGE: Organisation de Coordination et de Coopération pour la lutte contre les Grandes Endémies; OCEAC: Organisation de Coordination pour la lutte contre les Endémies en Afrique Centrale; UNDP: United Nations Development Programme; UNHCR: United Nations High Commissioner for Refugees.

Conclusion

Political developments, socio-economic changes and the waxing and waning of sleeping sickness transmission has influenced international co-operation in African trypanosomiasis and exchange of knowledge.

Thereby have we seen over the last hundred years a spectacular evolution of transport and communication facilities. Whilst in 1904 it took two months for Dutton to pass a message by telegram from Kasongo (Congo) to Liverpool, today it should be a matter of seconds. At the turn of the century, intercontinental travel took three to four weeks, whereas at present it is a matter of one day or less. Nevertheless, it is quite remarkable that such high quality reports as those published early this century by the Royal Society [19], the French geographical Society [20], and the Portuguese Ministry of the Navy and Colonies [21], detailed bookworks of 250 to 800 pages long could appear printed within a year after the end of the missions reported.

The present international community does benefit a great deal from unprecedented facilities for sharing information. The reverse side of the practically unlimited flow of information is the complexities caused by so much information and opinions that have be taken into account. If one compares the work of the League of Nations during the twenties with the current programmes of the UN agencies, it is clear that at the time the League organised the first Expert Committee, the principal issue for international co-operation was to facilitate exchange of information and bring isolated groups and different nationalities together. Today, there are practically no borders any longer for the information flow or for individuals to communicate each other. More and more the issue for international co-operation has become to co-ordinate and integrate action and propagate acceptable international standards.

References

1. Scott HH (1939) A History of Tropical Medicine. Edward Arnold, London, pp 454-532
2. Duggan AJ (1970) An historical perspective. In: Mulligan HW (ed) The African Trypanosomiases. George Allen and Unwin Ltd, London, pp XLI-LXXXVIII
3. Kean BH, Mott KE, Russell AJ (1978) Tropical Medecine and Parasitology Classic Investigations. Cornell University Press, Ithaca and London, II, pp 181-210
4. de Raadt P (1998) The history of sleeping sickness. In: Gilles HM (ed) Handbook of Protozoal Infections. Chapman & Hall, London (in press)
5. Guérin PA (1869) Maladie du sommeil (résumé d'une thèse). Archives Générale de Médecine Série VI 14:605-607
6. Mott FW (1899) The changes in the central nervous system of two cases of Negro lethargy: sequel to Dr Manson's clininical report. Brit Med J II: 1666
7. Lyons M (1985) From "death camps" to cordon sanitaire: the development of sleeping sickness policy in the Uele district of the Belgian Congo, 1903-1914. J Afr History 26:69-91
8. League of Nations (1928) Final Report of the League of Nations International Commission on Human Trypanosomiasis. League of Nations, Geneva, CH 629: 1-404
9. World Health Organization (1979) African trypanosomiasis. Report of a joint FAO/WHO expert committee. Technical Report Series N° 635, World Health Organization, Geneva
10. World Health Organization (1972) Trypanosomiasis. Bull Wld Hlth Org 47:785-813
11. World Health Organization (1976) African trypanosomiasis. WHO Wkly Epidemiol Rec 4: 21-24
12. World Health Organization (1979) Epidemiology and control of African trypanosomiasis. Report of a WHO expert committee. Technical Report Series N° 739, World Health Organization, Geneva
13. Adams JH, Haller L, Boa FY, Doua F, Dago A, Konian K (1986) Human African trypanosomiasis (*T. b. gambiense*): a study of 16 fatal cases of sleeping sickness with some observations of acute reactive arsenical encephalopathy. Neuropathol Appl Neurobiol 12:81-94
14. Mehlitz D, Zillmann U, Scott CM, Godfrey DG (1982) Epidemiological studies on the animal reservoir of gambiense sleeping sickness. Part III. Characterization of *Trypanozoon* stocks by isoenzymes and sensitivity to human serum. Tropenmed Parasit 33:113-118
15. Odelola (1965) Address by the Acting Secretary-General of CCTA, to the ISCTR meeting. In: International Scientific Committee for Trypanosomiasis Research, Tenth Meeting, Kampala 1964. Stephen Austin and Sons, Hertford, England

16. Challier A, Laveissière C (1973) Un nouveau piège pour la capture des glossines (Diptera Muscidae: *Glossina*): description et essais sur le terrain. Cah ORSTOM sér Ent méd Pararsitol XI: 251-262
17. Wellde BT, Schoenbechler MJ, Diggs MJ, Langbehn HR, Sadun EH (1975) *Trypanosoma rhodesiense*: variant specificity of immunity endured by irradiated parasites. Exp Parasitol 37: 125-129
18. World Health Organization (1983) Trypanosomiasis control manual. World Health Organization, Geneva
19. Royal Society (1903-1913) Reports of the Sleeping Sickness Commission nrs I-XIII. Harrison and Sons London
20. Martin G, Leboeuf A, Roubaud E (1909) Rapport de la mission d'études de la maladie du sommeil au Congo français, 1906-1908. Société de Géographie, Masson & Cie, Paris
21. Bettencourt A, Kopke A, de Rezende G, Mendes C (1903) La Maladie du Sommeil, Rapport par la Mission envoyée en Afrique Occidentale Portugaise. Libanio da Silva, Lisboa
22. Bouet J (1950) Allocution au Professeur Brumpt. Jubilé du Professeur Emile Brumpt. A. Lahure, Paris, pp 33-36
23. Odhiambo TR (1967) Science for Development. Science 158:876-881

Subject Index

αβ T cell receptors, 142
γδ T cells, 142
γGCS (see g-glutamylcysteine synthetase)
γ-glutamylcysteine synthetase (gGCS), 92
α-methylornithine, 84
α-oxoglutarate dehydrogenase, 60
1,3-diaminopropane, 87, 93
1,5-diaminopentane
  (see Cadaverine)
1,8-diaminooctane, 93
2-phosphoglycolate, 68
5-hydroxytryptamine, 105
6PGDH (see 6-phosphogluconate dehydrogenase)
6-phosphogluconate dehydrogenase (6PGDH), 63, 71
Abnormal movements, 223, 224
Acetate, 58, 59, 60, 61
Acetate production, 61
Acetate : succinate CoA-transferase, 60
Aconitase, 60
Active mobile screening, 306, 310
Adenopathy, 4, 216, 219, 227, 297, 307
Adenosine-5'-thioethers derivatives, 91
Adenosylspermidine, 91
Adenylate
  – cyclase, 35, 43
  – kinase, 55
AdoMac, 89
AdoMao, 90
AdoMet (see S-adenosylmethionine)
AdoMetDC (see S-adenosylmethionine decarboxylase)
Adverse effects, 256, 258, 259, 260, 262, 268, 271, 272, 274
AEP (see Auditory evoked potentials)
Agaricic acid, 66, 69
AGFUND, 322
Agronomical prophylaxis, 301, 302
Alanine, 60
Aldolase, 55, 67, 73, 74
Alkoxyphospholipid, 63
Alternative oxidase, 56, 57, 62, 74

Amitryptyline, 97
AMREF, 328
Angiotensin, 186
Animal inoculation, 242
Anion exchange column, 240, 246
Antibody, 169, 230
  – response, 33, 47, 140, 243, 244
Antigen detection, 244
Antigenic
  – structure, 243
  – variation, 9, 33, 34, 36, 37, 40, 42, 43, 47, 119, 123, 137, 158, 218, 236, 321, 326
Antizyme, 82, 84
Arsenical encephalopathy, 111, 164, 209
Arsobal® (see Melarsoprol)
Ascofuranone, 57
Association indices, 15, 16
Astrocytes, 108, 110, 113, 114, 125, 159, 169, 172, 220
Astrocytosis, 162, 294
Atkins, 318
Atoxyl®, 253, 315, 319, 330
Auditory evoked potentials (AEP), 209, 210, 211
Autoantibodies, 137, 140, 143, 144, 172, 226, 236, 295
Autocoids, 105
Availability, 256, 259, 265, 268, 271, 273, 274

B cells, 121, 123, 129, 130, 141, 142, 143, 144, 145, 148, 150
Bacteraemia, 112
Bacterial meningitis, 110, 114
Bantu, 2, 139
Bayer, 205 (see Suramin)
BBB (see Blood-brain barrier)
BCNU, 96
Berenyl® (see Diminazene acetaturate)
BIIT, 9
Bioreduction, 283
Bis amino diphenylsulfur, 97

Bis methylenecycloheptane derivative, 92
Blastomycosis, 162
Blood-brain barrier (BBB), 87, 105, 106, 107, 108, 110, 111, 112, 113, 114, 169, 197, 198, 215, 218, 220, 221, 227, 229, 250, 253, 261, 270, 272, 281, 282, 283, 289, 291, 292, 294
Bloodstream form, 8, 31, 33, 34, 35, 36, 37, 38, 43, 44, 45, 46, 47, 53, 54, 55, 59, 61, 62, 64, 66, 67, 68, 70, 72, 73
BPITT, 323
Bradykinin, 105, 174
Brain (o)edema, 165
Bruce, 8
Brumpt, 318
BSO (see Buthionine sulfoximine)
Buffy coat, 239, 240
Bushbuck, 23
Buthionine sulfoximine (BSO), 92, 93

Cachectin (see Tumor necrosis factor-α)
Cadaverine (1,5-diaminopentane), 87, 93
Cameroon, 8
Capillary permeability, 105
Carbohydrate metabolism, 53, 62
Card agglutination trypanosomiasis test (CATT), 23, 139, 193, 216, 229, 246, 247, 304, 307, 310, 312
Cardiovascular manifestation, 220
Caritas International, 328
Castellani, 318
CATT (see Card agglutination trypanosomiasis test)
Cellular differentiation, 43, 44, 46, 47
Central nervous system (CNS), 3, 23, 93, 97, 105, 111, 113, 114, 123, 125, 126, 128, 130, 144, 148, 149, 164, 172, 185, 215, 217, 218, 220, 228, 235, 253, 254, 256, 261, 262, 263, 264, 265, 268, 274, 292, 294, 295, 296
Cercopithecus aethiops, 296
Cerebral edema, 110
Cerebrospinal fluid (CSF), 1, 8, 105, 107, 108, 109, 110, 112, 113, 128, 144, 150, 158, 168, 172, 177, 191, 193, 205, 226, 229, 230, 236, 241, 243, 245, 247, 250, 253, 254, 255, 256, 257, 258, 261, 262, 264, 265, 267, 269, 270, 272, 274, 294, 295, 296, 297, 305, 307
Cf16 (see Tryparedoxine)
Cf21, 98
c-fos, 175, 176, 177
CGP 40215A, 88
Chagas' disease, 1, 68, 245, 265, 267, 273, 283, 291

Chancre, 123, 139, 140, 151, 157, 158, 215, 235
Chemoprophylaxis, 255, 258
Chemotherapy, 56, 74, 112, 113, 126, 128, 151, 241, 245, 269, 281
Chinifur, 98
Choroid plexus, 106, 107, 159, 161, 167, 168, 169, 170, 172, 221, 294
CIDOB, 328
CIESPAC, 325
Circadian
– disruptions, 194
– rhythms, 126, 173, 175, 177, 183, 184, 187, 191, 192, 195, 196, 197, 198, 206, 223
Circumventricular organs, 107, 169, 170, 177
Citrate synthase, 60
Citric-acid cycle, 54, 58, 59, 60, 61, 62
Cladistic methods, 19
Clinical forms, 226
Clonality, 16
CNS (see Central nervous system)
CNS involvement, 88, 108, 109, 111, 113, 221, 226, 289, 292, 293, 294, 296, 297
Combination, 5, 230, 255, 258, 260, 271, 272, 273, 274, 282, 283, 291, 294, 296
Compartmentation, 57, 60, 73, 74
Complement, 106, 111, 140, 146, 151, 171, 246, 247, 295
Control of glycolysis, 73
Corticosteroids, 262, 263, 265, 274
Cortisol, 183, 184, 185, 186, 187, 188, 192, 196, 223
Cost, 256, 259, 265, 268, 271
Cryopreservation, 9
CSF (see Cerebrospinal fluid)
Culture, 290, 291
Cyclohexylamine, 90
Cymelarsan® (see MelCy)
Cystamine, 93
Cytochrome, 56, 57, 62, 218
Cytokines, 32, 105, 106, 107, 108, 110, 112, 113, 114, 119, 120, 121, 123, 124, 125, 128, 129, 130, 131, 137, 139, 141, 142, 143, 144, 145, 146, 147, 148, 149, 150, 151, 169, 172, 219, 220, 221, 264

dcAdoMet (see Decarboxylated S-adenosylmethionine)
DDT, 302
Decarboxylated AdoMet (see Decarboxylated S-adenosylmethionine)

Index

Decarboxylated S-adenosylmethionine (dcAdoMet), 85, 86, 90
Deltamethrin, 303
Dementia, 225, 226
Demyelinating encephalitis, 226
Demyelination, 123, 160, 161, 162, 164, 211, 222, 224, 226
Dendrogram, 17, 19
DFMO (difluoromethylornithine, eflornithine, Ornidyl®), 20, 82, 84, 85, 88, 89, 110, 254, 256, 264, 265, 268, 269, 270, 271, 272, 273, 274, 281, 282, 289, 291, 293, 296, 321, 330
DHAP (see Dihydroxyacetone phosphate)
Diagnosis, 159, 193, 205, 216, 217, 219, 220, 227, 227, 228, 229, 235, 236, 243, 245, 255, 303, 304, 319
Diamidino-4,4'-diphenoxypentane (see Pentamidine)
Dicyclohexylamine, 90
Dieldrin, 302
Difluoromethylornithine (see DFMO)
Dihydroxyacetone phosphate (DHAP), 54, 55, 56, 67, 69
Diminazene (see Diminazene aceturate)
Diminazene aceturate (Diminazene, Berenyl®), 254, 259, 260, 264, 272, 273, 281, 282, 289, 292, 296
Direct agglutination, 243, 246
Distance methods, 18
DNA
 – rearrangements, 36, 40, 41, 47
 – sequence, 12
Dogs, 166
Double centrifugation, 241, 305
Drug development, 54, 73, 92
Dutton, 318, 327, 331

EATRO, 290, 323, 325
Edema (Oedema), 110, 157, 161, 220, 235, 294
EEG (see Electroencephalogram)
Effectiveness, 255, 258, 261, 266, 270, 274, 289, 290, 292, 294, 296
Eflornithine (see DFMO)
Ehrlich, 318
Electroencephalogram (EEG), 191, 192, 194, 204, 205, 206, 209
Electroencephalographic, 203, 204, 205, 227
ELISA (see Enzyme-linked immunosorbent assay)
Embden-Meyerhof-Parnass pathway (EMP pathway), 53, 54, 62, 66, 73

EMP pathway (see Embden-Meyerhof-Parnass pathway)
Endosulfan, 302
Endotox(a)emia, 111, 112, 130
Endotoxin antibodies, 113
Endotoxins, 108, 111, 112, 113, 114, 124, 149
Enolase, 55, 59, 70
Enzyme-linked immunosorbent assay (ELISA), 249, 250
Epicentre, 328
Epidemiology, 19
Epimastigotes, 93
ESAG (see Expression site-associated genes)
Ethiopia, 24
Evoked potentials, 203, 205, 209, 210, 212
Evolution, 19
Experimental models, 138, 157, 221, 289, 297
Expression site-associated genes (ESAG), 35, 36, 43
Extrapyramidal, 221, 228

FAO, 316, 320, 322
Fever, 111, 114, 125, 126, 147, 193, 219, 227, 235, 256, 259, 260, 262, 271, 310
Fexinidazole, 292
Field's stain, 238, 239, 242
Flagellar pocket, 32, 151, 171, 172
FOMETRO, 329
FR (see Fumarate reductase)
Friedheim, 253, 261
Fru-2,6-P2 (see Fructose 2,6-bisphosphate)
Fructose, 54, 58, 64, 65
Fructose 2,6-bisphosphate (Fru-2,6-P2), 57, 58, 59, 67, 71, 74
Fumarate reductase (FR), 60, 285

G3P (see Glycerol-3-phosphate)
G3PDH (see Glycerol-3-phosphate dehydrogenase)
G6PDH (see Glucose 6-phosphate dehydrogenase)
GA3P (see Glyceraldehyde-3-phosphate)
Galactocerebrosides, 144, 168, 226, 230, 295
GAPDH (see Glyceraldehyde-3-phosphate dehydrogenase)
Gene conversion, 37, 38, 39, 42
Genetic control, 138
Genetic exchange, 7, 15, 20 24
Geographical information system (GIS), 310

Germanin® (see suramin)
GH (see Growth hormone)
GH 8693, 98
GHRH, 185, 186
Giemsa stain, 238, 239, 241, 304
GIS (see Geographical information system)
GK (see Glycerol kinase)
Glial, 211
– cells, 113, 159, 226
Gluconeogenesis, 59
Glucose, 53, 54, 56, 58, 59, 64, 65, 73, 217, 218, 238
Glucose 6-phosphate dehydrogenase (G6PDH), 63
Glucose transport, 64, 73
Glutamate receptors, 175, 176, 177
Glutathione, 81, 92, 93, 94, 265, 269, 284
Glutathione reductase (GR), 94, 95, 96, 97
Glutathione synthetase, 93
Glutathionylspermidine, 88, 93, 94
Glutathionylspermidine synthetase, 93, 94
Glyceraldehyde-3-phosphate (GA3P), 63, 67, 69
Glyceraldehyde-3-phosphate dehydrogenase (GAPDH), 55, 59, 68, 73, 74
Glycerol, 54, 56, 57, 58, 59, 218, 291
Glycerol kinase (GK), 55, 56, 57, 70, 218
Glycerol-3-phosphate (G3P), 55, 56, 69, 70
Glycerol-3-phosphate dehydrogenase (G3PDH), 55, 62, 69, 73, 74
Glycerol-3-phosphate oxidase, 55, 56
Glycolysis, 54, 55, 56, 57, 58, 60, 65, 70, 73, 74, 218, 255, 260
Glycosome, 54, 56, 57, 59, 65, 66, 67, 68, 69, 70, 72, 73, 74, 255
Glycosyl-phosphatidylinositol (GPI), 31, 33, 171
GM-CSF, 141
Goats, 168
Gossypol, 69
GPI (see Glycosyl-phosphatidylinositol)
GR (see Glutathione reductase)
Growth hormone (GH), 183, 184, 185, 187, 192, 196
Gut, 112

Hardy-Weinberg equilibrium, 14, 16
Headache, 111, 126, 193, 219, 228, 235, 259, 262, 268, 271, 310
Healthy carriers, 3
Heart, 220

Hemolymphatic
– phase, 123, 191, 203, 216, 217, 225
– stage, 227
Hemorrhagic leucoencephalopathy, 165
HETA, 91
Hexokinase, 55, 57, 58, 65, 66
Histamine, 105, 173
Homologous recombination, 38, 47, 48
Homotrypanothione, 93
Hormones, 183, 185, 186, 188, 189, 192
Hydrogenosome, 61
Hypergammaglobulin(a)emia, 111, 137, 148, 150, 218
Hypersomnia, 191, 192, 195, 223
Hypothalamopituitary axis, 186
Hypothalamus, 110, 160, 161, 169, 175, 176

IAEA, 320, 322, 326
ICIPE, 325, 326
IFN-γ (see Interferon-γ)
IFN-γ receptor (see Interferon-gamma receptor)
IgG, 3, 143, 227, 244, 247, 248
IgM, 3, 123, 143, 144, 146, 161, 172, 177, 226, 227, 229, 236, 244, 250, 295, 304
IL-1β, 124, 129, 141, 149
IL-1, 113, 114, 130, 147, 148, 172, 174
IL-2, 121, 124, 129, 130, 141, 142, 143, 150
IL-3, 141
IL-4, 121, 130, 139, 142, 144, 149, 151, 172
IL-5, 121, 142, 144
IL-6, 121, 124, 129, 142, 144, 149, 172
IL-7, 141
IL-8, 141
IL-10, 119, 121, 129, 130, 131, 141, 142, 148, 149
IL-12, 129, 141, 151
IL-13, 142
IL-15, 141
ILRAD, 326, 327
ILRI, 327
IMC, 328
Immune complexes, 106, 140, 167, 172, 219, 236, 248, 264, 295
Immune trypanolysis, 246
Immunodiffusion, 248
Immunofluorescence, 167, 193, 247, 248
Immunoglobulins (see also IgG, IgM), 172, 221
Immunopathology, 166

Immunosuppression, 8, 111, 119, 123, 124, 130, 131, 137, 143, 149, 150, 173, 219, 245, 263
In situ activation, 36, 37, 42, 43
In vitro cultures, 8, 69, 242
In vitro models, 290
Indications, 255, 258, 261, 266, 270
Indirect agglutination, 249
Inducible nitric oxide synthase (INOS), 126, 130, 149
Innate immunity, 2, 137, 139
INOS (see Inducible nitric oxide synthase)
Insect stage, 58, 59, 60, 61, 62, 70
Insecticide, 301, 302, 303, 316, 320
Interferon-γ (IFN-γ), 106, 107, 121, 124, 126, 129, 130, 131, 141, 142, 143, 144, 148, 149, 150, 172, 174, 175, 220
Interferon-gamma receptor (IFN-γ receptor)
Interleukins (see also IL-), 113, 220
Internalization, 285
Invariant antigen, 236, 243, 244, 245, 246, 248, 249
ISCTRC, 316, 323, 324
Isocitrate dehydrogenase, 22, 60
Isoenzyme, 10, 11, 12, 15, 16, 19, 20, 22, 23, 24, 68, 70, 72
Isolation, 7, 8, 9

Jamot, 4, 303, 315

K complexes, 204, 206, 207
Kenya, 16, 22, 23, 24
Kerandel, 222
Kinetoplastida, 57, 58
Kinins, 111, 173, 174, 236
Kit for in vitro isolation (KIVI), 8, 9, 242, 243
KIVI (see Kit for in vitro isolation)
Knock-out mice, 128, 130, 151, 175
Koch, 318, 327, 330
Kopke, 318, 327,330

L 611744, 292
L 634549, 292
Lampit® (see Nifurtimox)
Laveran, 318
League of Nations, 316, 319, 320, 331
Leishmania, 61, 68, 71, 86, 92, 121, 140, 149, 245
Leishmaniasis, 56, 138, 142
Leucoencephalitic, 211
Leucoencephalitis, 159, 160, 162, 212
Linkage, 15, 16
Lomidine®, 257, 289

Long slender (bloodstream) form, 53, 56, 84
Lymph examination, 237
Lymph nodes, 1, 141, 142, 143, 150, 158, 165, 193, 216, 217, 218, 220, 229, 237, 304, 307, 318
Lymphocytes, 157, 160, 161, 163, 165, 167, 169, 174, 175, 184, 193, 220, 221, 224

Mac Kenzie, 318
Macroglobulinaemia, 227
Macrophages, 124, 126, 137, 144, 145, 146, 147, 148, 149, 150, 151, 157, 161, 165, 167, 172, 173, 174, 197, 216, 224
Major histocompatibility complex (MHC), 124, 146, 169, 170
Malaria, 3, 138, 142, 148, 324
Malate, 59
Malate dehydrogenase (MDH), 59, 60, 72
Malic enzyme, 59, 60, 72
Mannose, 54, 58, 64, 65
Mannose-6-phosphate isomerase, 65
Manson, 318, 327
M-CSF, 141
MDH (see Malate dehydrogenase)
MDL 27695, 91
MDL 73811, 89, 90, 291
Megazol, 231, 282, 283, 284, 285, 291, 293, 294, 297
Mel B (see Melarsoprol)
Mel Cy (Cymelarsan®), 69, 95
Mel T, 95, 96, 260
Mel W, 253
Melarsen oxide, 92, 95, 96, 99, 260
Melarsoprol (Arsobal®, Mel B), 96, 99, 110, 126, 129, 193, 194, 195, 196, 216, 229, 230, 231, 253, 254, 255, 256, 258, 260, 261, 262, 263, 264, 265, 266, 272, 273, 281, 283, 285, 289, 290, 296, 316
Melarsoprol-refractory, 254, 266, 267, 268, 269, 270, 273, 274
Melatonin, 192, 196, 223
MEMISA, 329
Menaquinone, 61
Meninges, 106, 184, 221, 222
Meningitis, 106, 108, 158, 167, 228, 293, 324
Meningoencephalitic phase, 203, 204, 205, 206, 209, 221, 225
Meningoencephalitis, 108, 123, 167, 168, 183, 191, 192, 193, 203, 206, 230, 264, 270, 294
MEP (see Motor evoked potentials)
Mepacrine, 97

Metacyclic, 9, 31, 245
Metacyclic VSG, 35, 38
Methionine, 81, 85
Methylglyoxal bis guanylhydrazone (MG-BG), 87
Methylthioadenosine (see MTA)
Metronidazole, 230, 264, 272, 273, 281
MGBG (see Methylglyoxal bis guanylhydrazone)
MHC (see Major histocompatibility complex)
MHZEA, 90
Microfilariae, 237, 239
Microglia, 114, 128, 161, 173, 220
Microglial cells, 125, 159, 162
Microhaematocrit technique, 239
MK 436, 292, 296
Mobile team, 4, 303, 306, 309, 310, 312, 316
Model
 – Sheep, 281, 294, 295
 – Mouse, 292
Mode of action, 254, 257, 260, 265, 269, 282, 285
Molecular characterization, 7, 9, 20
Monkeys, 166, 168, 296
Monocyte-macrophage, 121, 124, 126, 129, 130, 131
Monocytes, 124, 129, 140, 147, 149
Moranyl® (see Suramin)
Morular cells (see Mott's cells)
Motor evoked potentials (MEP), 209, 210, 211
Mott, 158, 159, 160, 318
Mott's cells (Morular cells), 150, 159, 161, 162, 163, 164, 166, 167, 168, 172, 177, 184, 228
MSF, 328, 329
MTA, 90, 91
Muraz, 324, 327
Murine models, 291
Mutagenic, 285, 297
Mutagenicity, 230, 231, 259
Myelin, 144, 159, 160, 162, 164, 167,169, 170, 222, 226, 295
Myeloma, 227

NADH dehydrogenase, 62
Naphtoquinones, 97
Natural killer cells (NK cells), 129, 141, 151
Neujean, 255
Neurofilament, 144, 226, 230
Neuronal, 128, 165, 196, 226
Neurons, 113, 144, 162, 169, 170, 171, 174, 175, 176, 197, 221

NGO (see Non governmental organization)
Nifurtimox (Lampit®), 254, 256, 264, 265, 266, 267, 268, 272, 273, 274, 284, 285, 289, 291, 293, 294
Nitosoureas, 96
Nitric oxide (NO), 124, 126, 128, 129, 131, 137, 144, 148, 149, 150, 151, 173, 175, 192, 197, 198, 199, 220, 221
Nitric oxide synthase (NOS), 148, 149, 174, 175, 197, 198, 199, 221
Nitrofurazone 268
Nitroimidazole, 230, 256, 272, 273, 281, 282, 283, 285, 296
NK cells (see Natural killer cells)
NMDA, 177, 199
NO (see Nitric oxide)
NO synthase (see Nitric oxide synthase)
Non governmental organization (NGO), 316, 322, 328
Non-REM sleep (see NREM sleep)
Non-telomeric genes, 35, 41
NOS (see Nitric oxide synthase)
NPA, 329
NREM sleep (Non-REM sleep), 186, 191, 194, 195, 206
Numerical methods, 14, 17, 19
Nycthemeron, 192, 195

OAU, 322
OCCGE, 322, 323, 324, 325, 330
OCEAC, 322, 323, 324, 325, 329, 330
ODC (see Ornithine decarboxylase)
Oedema (see Edema)
Onchocerciasis, 256, 324
Ornidyl® (see DFMO)
Ornithine, 82
Ornithine decarboxylase (ODC), 81, 82, 83, 84, 86, 89, 91, 269, 291
Ornithine decarboxylase inhibitors, 84
ORSTOM, 325
Oxaloacetate, 59
Oxygen metabolism, 284
Oxygen redox cycle, 284

P1 transporter, 99
P2 transporter, 99, 285
PAAT, 322
Pain, 174, 177, 222, 228, 235, 259, 262, 268, 271
Panmixia, 15, 16
Parasite reservoir, 303, 306, 309
PARP (see Procyclin)
Parsimony, 18, 19
Passive screening, 4, 309, 312
PCR, 12, 250

Index

Pentacarinat® (see Pentamidine)
Pentalenolactone, 69
Pentamidine (Diamidino-4,4'-diphenoxypentane, Lomidine®, Pentacarinat®), 88, 99, 220, 229, 254, 255, 257, 258, 259, 260, 272, 273, 274, 281, 289, 291, 305, 306, 330
Pentose Phosphate Pathway (PPP), 62, 63, 66, 71, 72
PEP (see Phosphoenolpyruvate)
PEPCK (see Phosphoenolpyruvate carboxykinase)
Peripheral nerves, 160, 164
Perivascular, 159, 160, 161, 166, 191, 197, 220, 221, 222, 224, 295
Perivascular inflammation, 162
Perivascular spaces, 107
Peroxisome, 54, 63, 66, 74
Peroxyredoxine, 98
Peruzzi, 320
PEST, 82, 84, 86
PFK (see Phosphofructokinase)
PGE2, 113, 147
PGF2, 113
PGI (see Phosphoglucose isomerase)
PGK (see Phosphoglycerate kinase)
Pharmacokinetics, 255, 257, 258, 260, 266, 269, 274, 290
Phenetic methods, 17, 18
Phenothiazine, 97
Phosphoenolpyruvate (PEP), 58, 59, 71, 72
Phosphoenolpyruvate carboxykinase (PEPCK), 59, 60, 72
Phosphofructokinase (PFK), 55, 57, 66, 67, 73, 74
Phosphoglucose isomerase (PGI), 55, 66, 71
Phosphoglycerate kinase (PGK), 54, 55, 56, 69, 70, 73, 74
Phosphoglycerate mutase, 55, 70, 71
Phylogenetic, 14, 17, 18
Phylogeny, 19
Phytomonas, 56
Pigs, 23, 166, 168
Plasma renin activity (PRA), 183, 186, 188, 192
Plasmodium, 121, 237
Polyamines, 81, 83, 84, 88, 91, 95, 99, 257, 269, 291
Polyclonal, 111, 123, 143, 144, 150, 218
Polygraphic, 191
Polyphosphates, 54, 57
Polysomnographic, 192, 205
Polysomnographic recording, 184, 186, 193, 194, 206

Polysomnography, 191
Population genetics, 7, 14, 17
Post-transcriptional regulation, 45
Post-treatment follow-up, 236, 241, 250, 254
PPi-dependent PFK, 67
PPP (see Pentose Phosphate Pathway)
PRA (see Plasma renin activity)
Pregnancy, 259
Primate models, 296, 297
Procyclic, 8, 10, 31, 33, 44, 45, 46, 47, 58, 59, 60, 61, 66, 72, 140, 243, 244, 246, 247, 248, 249, 250, 269
Procyclic stage, 53, 59, 61, 72
Procyclin (PARP), 31, 33, 44, 45, 46, 47, 48
Prolactin, 183, 184, 185, 186, 187, 188, 192, 196, 223
Proline, 59, 60
Prophylaxis, 301, 304, 305
Prostaglandins, 106, 112, 113, 124, 144, 147, 150
Pruritus, 193, 219, 227, 228, 256, 259
Psychiatric, 193, 203, 225, 228
Pulsatile, 186, 189
Pulse, 187, 192, 194, 194, 220
 – hormone, 188
 – radiolysis, 284
Purine transporter, 89, 91, 285
Putrescine, 81, 84, 85, 87, 88, 89, 90, 93
Pygmy(ies), 2, 139
PYK (see Pyruvate kinase)
Pyrethrin, 302
Pyruvate, 54, 56, 58, 59, 61, 65, 72, 74, 86
Pyruvate dehydrogenase, 54, 58, 60, 61, 62
Pyruvate kinase (PYK), 54, 55, 57, 58, 59, 71, 72, 73, 74
Pyruvate transport, 65
Pyruvoyl-dependent enzymes, 86

QBC (see Quantitative buffy coat)
Quantitative buffy coat (QBC), 240

Rabbits, 168, 169
Randomization methods, 15, 16
RAPD, 12, 13, 16, 19, 20
Raphe nuclei, 196, 197
Reactive arsenical encephalopathy, 110, 164, 165, 262, 263, 264, 265
Reciprocal recombination, 37, 38, 40
Recombination
 – DNA, 37
 – methods, 15
 – tests, 15

Recommended regimen, 256, 259, 265, 268, 271, 272
Regulation of glycolysis, 57
Relapse, 204, 221, 254, 255, 258, 260, 262, 263, 267, 270, 271, 281, 291, 292, 293, 294
REM sleep, 186, 187, 188, 191, 194, 195, 197, 206, 207, 208
Renin, 186, 223
Reservoir host, 7, 19, 22, 23, 24, 235, 319, 320, 321
Resistance, 2, 7, 22, 99, 119, 138, 139, 140, 150, 204, 253, 257, 258, 260, 262, 285, 302
Respiratory chain, 56, 62, 285
RFLP, 11, 12, 19, 20, 22, 23, 24
Rhodoquinone, 61
Ribulose-5-phosphate epimerase, 63
Richet, 324
RNA elongation, 45, 47, 48
RNA polymerase, 38, 44
Ro 15-0216, 230, 281, 282, 291, 296, 297
Rodents, 166

S-adenosyl-1,12-diamino-3-thio-9-aza-dodecane, 91
S-adenosylmethionine (AdoMet), 85, 86, 87, 89
S-adenosylmethionine decarboxylase (AdoMetDC), 86, 87, 88, 89, 91, 257, 291
S-adenosylmethionine decarboxylase inhibitors, 87
S-adenosylmethionine synthetase, 85
S-adenosylmethionine synthetase inhibitors, 85
Salicyl hydroxamic acid (SHAM), 56, 57, 62, 218, 291
SCN (see Suprachiasmatic nuclei)
Screen, 301, 303
Screening, 4, 247, 301, 304, 305, 306, 309, 310, 311
SDH (see Succinate dehydrogenase)
Segmental gene conversion, 39, 41
SEP (see Somatosensory evoked potentials)
Serotonin, 144, 197
Serotoninergic, 197
SGAMS, 324
SGHMP, 324
SHAM (see Salicyl hydroxamic acid)
Sheep, 168
Short stumpy (bloodstream) form, 58, 84
Sinefungin, 85, 86
Skeletal muscles, 165

Sleep, 130, 157, 171, 175, 183, 184, 185, 186, 187, 188, 189, 191, 192, 193, 194, 195, 196, 203, 204, 206, 207, 208, 208, 211, 212, 221, 222, 223, 227
Sleep onset REM (SOREM), 195, 196
Sleep spindle, 204, 207
Sleeping Sickness Bureau, 319
Sleep-wake cycle, 123, 175, 183, 184, 187, 191, 192, 194, 195, 196, 197, 198, 212
Slow-wave, 204, 206, 207
Slow-wave sleep (SWS), 185, 186, 187, 188, 191, 194, 195, 204, 223
Somatosensory evoked potentials (SEP), 209, 210, 211
Somatosensory system, 173
SOREM (see Sleep onset REM)
Spermidine, 81, 87, 88, 89, 90, 91, 92, 93, 95, 265, 269
Spermidine synthase, 90, 91
Spermidine/spermine-N1-acetyltransferase, 91
Spermine, 93
Spinal cord, 160, 164
Spleen, 141, 145, 149, 150, 165
Splenomegaly, 137, 219, 297
Succinate, 59, 60, 61
Succinate dehydrogenase (SDH), 60, 61
Succinate production, 60
Succinate thiokinase, 60, 61
Succinyl CoA synthase, 60, 61
Sulphorhodamine B, 110
Superoxide, 97, 98, 283, 284, 285
Superoxide dismutase (SOD), 284, 285
Suprachiasmatic nuclei (SCN), 170, 175, 176, 177, 187, 196, 197, 198
Suramin (Bayer 205, Germanin®, Moranyl®), 66, 68, 69, 70, 71, 72, 220, 230, 231, 254, 255, 256, 258, 264, 266, 271, 272, 273, 274, 281, 282, 283, 289, 291, 292, 293, 294, 296
Surface proteins, 31, 33, 35
SWS (see Slow-wave sleep)
Syphilis, 158, 162, 229

T-cells, 107, 120, 123, 124, 129, 130, 131, 138, 140, 141, 142, 143, 144, 148, 150, 151
T helper cells, 120, 142
Tanzania, 22, 24
Taxonomy, 17, 19
TDR, 321, 327, 330
Telomeres, 34, 36, 38, 42
Telomeric genes, 34, 35, 40, 41
Teratogenicity, 262
TGF-β (see Transforming growth factor-b)

Th1, 121, 125, 129, 131, 137, 142, 149, 150, 151
Th2, 121, 129, 130, 131, 137, 142, 149, 150
Thalamus, 110, 160, 161, 169, 207, 221
Thick blood smear, 238
Thin blood smear, 238
Thioredoxine, 98
Three dimensional structure, 68, 69, 74, 95
TIM (see Triose-phosphate isomerase)
TLF (see Trypanosome lytic factors)
TLTF (see Trypanosome lymphocyte triggering factor)
TNF-α (see Tumor necrosis factor-α)
Todd, 318, 327
Tolerance, 2
TPX (see Tryparedoxine)
TR (see Trypanothione reductase)
trans-4-methylcyclohexylamine, 91
Transaldolase, 63
Transcription promoter, 39, 44
Transcriptional regulation, 44
Transcriptional silencing, 38, 42
Transferrin receptor, 32, 40, 43
Transforming growth factor-β (TGF-β), 121, 129, 130, 131, 141, 144, 148
Transient activation phase (TAP), 206, 207, 208
Transketolase, 63
Transmethylation
Trap, 2, 302, 303, 321, 325, 326
Treatment, 1, 4, 74, 81, 110, 119, 126, 128, 141, 142, 148, 161, 164, 194, 195, 196, 204, 205, 209, 215, 217, 218, 222, 223, 224, 226, 229, 230, 236, 250, 253, 254, 255, 256, 258, 259, 260, 261, 262, 264, 265, 266, 268, 269, 270, 272, 273, 274, 281, 282, 292, 293, 294, 296, 297, 297, 301, 303, 305, 306, 309, 311, 312, 316, 318, 319, 323
Treatment-induced encephalopathy, 264, 265
Trichomonas, 61
Tricyclic antidepressant, 97
Trifluoperazine, 97
Triose-phosphate isomerase (TIM), 55, 67, 68
Tropical Diseases Bureau, 319
Trypanolytic activity, 3
Trypanolytic factors, 9, 121, 216
Trypanosome lymphocyte triggering factor (TLTF), 124, 130, 148, 172, 174
Trypanosome lytic factors (TLF), 139
Trypanothione, 81, 84, 88, 92, 93, 95, 98, 260, 265, 269, 273, 284, 291

Trypanothione reductase (TR), 92, 94, 95, 96, 97, 98, 265
Trypanothione synthetase, 93
Trypanotolerance, 138, 139
Trypanotoxin, 171
Tryparedoxine (TPX, Cf16), 98
Tryparsamide, 20, 253, 327
Tryptophan, 144, 171, 197, 224, 226
Tryptophol, 111, 171
Tuberculosis, 162, 229, 230, 263, 319, 329
Tumor necrosis factor-α (TNF-α, Cachectin), 106, 113, 114, 119, 121, 124, 125, 126, 128, 129, 130, 131, 141, 147, 148, 149, 151, 168, 172, 220
Turncoat inhibition, 97

Ubiquinol, 55, 56
Ubiquinone, 55, 61, 62
Uganda, 16, 22, 23, 24, 258, 259, 290
UK5099, 65
UNDP, 320, 321, 326, 330
UNO, 320
Untranslated region, 46
UPGMA, 17, 22

Variable antigen, 243
Variable antigen type (VAT), 140, 143, 236, 243, 244, 245, 246, 247, 248, 249
Variant surface glycoprotein (VSG), 31, 32, 33, 34, 37, 40, 42, 43, 44, 46, 47, 48, 123, 137, 140, 142, 143, 147, 148, 151, 171, 218, 236, 243, 244, 245, 247, 248, 249, 250
Vasogenic (o)edema, 110, 111
VAT (see Variable antigen type)
Vector control, 311, 312
VEP (see Visual evoked potentials)
Virchow-Robin spaces, 107, 113, 168, 221, 222, 294
Virulence, 8, 20
Visual evoked potentials (VEP), 209, 210, 211
Voltammetry, 198
VSG (see Variant surface glycoprotein)
VSG expression sites, 35, 36, 38, 39, 40, 43, 45, 47
VSG gene, 11, 12, 13, 20, 24, 37

WAITR, 327
Wakefulness, 191, 192, 194, 195, 204, 206, 209, 223
WHO, 306, 309, 310, 316, 320, 321, 322, 325, 327, 328, 329, 330

Winterbottom, 217, 304, 318
World Bank, 321, 326

Zambia, 16, 22, 24

Cet ouvrage a été achevé
d'imprimer en juin 1999
sur les presses de l'imprimerie de l'Indépendant
53200 Château-Gontier - France
Dépôt légal : 2ᵉ trimestre 1999